BASIC MECHANICAL ENGINEERING

A Textbook for First Year Students of Engineering

(All Branches)

By

Prof. D.K. Chavan

Professor,
Mechanical Engineering Department,
Marathwada Mitra Mandal's
College of Engineering
(M.M.C.O.E.) Pune – 52

Ex. Assistant Professor
Mechanical Engineering Department,
Maharashtra Institute of Technology
M.I.T., Pune – 38

Prof. G.K. Pathak

Sr. Faculty Member,
Mechanical Engineering Department,
Maharashtra Institute of Technology
M.I.T., Pune – 38

STANDARD BOOK HOUSE

Unit of: **RAJSONS PUBLICATIONS PVT. LTD.**

1705-A, Nai Sarak, PB. No. 1074, Delhi-110006 Ph.: +91-(011)-23265506
Show Room: 4262/3, First Lane, G-Floor, Gali Punjabian, Ansari Road, Darya Ganj, New Delhi-110002 Ph.: +91-(011)43551085 Tele Fax : +91-(011)43551185, Fax: +91-(011)-23250212

E-mail: sbh10@hotmail.com www.standardbookhouse.com

Published by:
RAJINDER KUMAR JAIN
Standard Book House
Unit of: Rajsons Publications Pvt. Ltd.
1705-A, Nai Sarak, Delhi - 110006
Post Box: 1074
Ph.: +91-(011)-23265506 Fax: +91-(011)-23250212

Showroom:
4262/3. First Lane, G-Floor, Gali Punjabian,
Ansari Road, Darya Ganj,
New Delhi-110002
Ph.: +91-(011)-43551085, +91-(011)-43551185
E-mail: sbhl0@ hotmail.com
Web: www.standardbookhouse.com

First Edition : 2009
Second Edition : 2014
Third Edition : 2016
Fourth Edition : 2020

Price : **₹ 220.00**
US $: 20

ISBN: 978-81-89401-31-3

Typeset by:
C.S.M.S. Computers, Delhi.

Printed by:
R.K. Print Media Company New Delhi-110039

PREFACE

It gives us great pleasure, to present a Text book on *'Basic Mechanical Engineering'*. This introductory text is intended to first year students of Engineering and has been strictly written according to the New—Revised syllabus of Various University.

In this book we will be mainly studying three main topics

(*i*) *Thermodynamic Principles* — Thermodynamics is a basic science related to heat and energy as such regarded as the core of Mechanical Engineering curriculum.

(*ii*) *Design Consideration* — Need of design of any component, related stresses, strains. Material selection for the component to be manufactured, also its Aesthetic and ergonomic considerations are dealt with.

(*iii*) *Manufacturing Processes* — This part is intended to make all the students to become conversant with all the machines which are used in industries like Lathe, Drilling, Grinding, Welding, Brazing, Soldering, Metal cutting, Power saws, NC/CNC machines etc.

The knowledge and clear understanding of all these basics is essential to all branches of Engineering.

Every care has been taken to present the matter in precise and simple language. Self explanatory figures are included to enable the students to draw the same in the exams very easily.

From a long teaching experience, it has been observed that,

- The students should know how much and what should be written in the examinations.
- Secondly, mistakes are committed in units, as such, for every step in the solution of problems, units have been made clear.

Hence, in order to give clear idea regarding examination approach and to develop more confidence in the students, appropriate questions from university papers are given at the end of each chapter.

In the entire book S. I. system of units is used. Sufficient number of University problems have been solved, including recent papers.

All necessary care has been taken to avoid mistakes and misprints in this book. However, it is quite likely that some mistakes and misprints might have passed unnoticed. Such mistakes and misprints of the book, if brought to the notice, will be gratefully acknowledged. Any suggestions to improve the usefulness of the book will also be gratefully accepted:

We express our sincere thanks to Prof. A. D. Desai, Prof. N. K Joshi, Prof. S. A. Dayane, Prof. S. S. Shinde, Prof. K. D. Deshmukh, Prof. S. S. Udgirkar, Mr. Srinivas for their help in bringing out this book.

We also express on sincere thanks to entire team of **Standard Book House Unit of Rajsons Publication Pvt. Ltd.** for publishing the book in this shortest possible time.

Authors

ABBREVIATIONS USED

A — Area
a — Acceleration
C — Specific heat, velocity
°C — Degree centigrade
Cp — Specific heat at constant pressure
Cv — Specific heat at constant volume
E — Young's Modulus or Modulus of Elasticity
F — Force
G — Modulus of rigidity
g — Acceleration due to gravity
g_c — Constant of proportionality in Newton's law
h — Specific enthalpy, heat transfer coefficient
H — Enthalpy
J — Mechanical equivalent of heat or Joules equivalent
K — Degree kelvin, Thermal Conductivity
kg — Kilograms
KE — Kinetic Energy
L — Length
m — Meter, mass
M — Molecular weight
n — Number of moles, Polytropic index
N — Revolutions Per Minute (RPM)
P — Pressure
PE — Potential energy
q, Q — Heat transfer
R — Characteristic gas constant
R — Universal gas constant
s S — Entropy
t — Temperature Celsius
T — Absolute temperature K
u, U — Internal energy
v, V — Volume

w, W — Work

x — Dryness fraction

Z — Elevation above datum

y — Adibatic index

Δ — Change in a property

δ — Small change in a property

η — Efficiency

θ — Temperature

ρ — Density

σ — Stress, Stefan Boltzmann constant

ϵ — Strain, emissivity

τ — Shear stress

f — Shear strain

Contents

1. Fundamental Concept and Definition 1 – 65

1.1	Introduction	1
1.2	Thermodynamic Systems	1
1.3	Thermodynamic Properties, Processes and Cycles	6
1.4	Working Substance	9
1.5	Units and Dimensions	9
1.6	Mechanical and Thermodynamic Work	11
1.7	*Pdv* - Work	13
1.8	Work is a Path Function and Properties are Point Functions	14
1.9	Equations for Work Done in Various Processes	15
1.10	Heat	19
1.11	Pressure	20
1.12	Pressure Measurement	20
1.13	Pressure Exerted due to a Column of Fluid	21
1.14	Barometer	22
1.15	Bourdon Pressure Gauge	23
1.16	Manometers	24
1.17	Zeroth Law of Thermodynamics (Thermal Equilibrium)	24
1.18	Thermometric Property and Thermometers	26
1.19	Principle of Temperature Measurement	26
1.20	Scale of Temperature	27
1.21	Microscopic and Macroscopic Point of View	29
1.22	Quasi-Static Process	30
1.23	Energy in Transit	31

2. First Law of Thermodynamics 65 – 127

2.1 Introduction 65
2.2 Joule's Experiment 65
2.3 First Law for a Closed System Undergoing a Process 67
2.4 Energy — a Property of The System 67
2.5 Different Forms of Stored Energies 68
2.6 Internal Energy (U) 69
2.7 Enthalpy (H) 70
2.8 Specific Heat at Constant Pressure and Specific Heat at Constant Volume 70
2.9 Adiabatic Index 72
2.10 Perpetual Motion Machine of First Kind [PMM– 1] 73
2.11 Flow Process; Control Volume; Control Surface 73
2.12 Flow Work or Flow Energy 74
2.13 Conditions of Steady Flow System 75
2.14 Steady Flow Energy Equation (S.F.E.E.) 75
2.15 Steady Flow Energy Equation on Mass Basis 77
2.16 In a Steady Flow System W.D. = $-\int v.dP$ when Changes in KE and PE are Neglected 78
2.17 Significance of $\int Pdv$ in Case of Steady Flow Process and Non-Flow Process 79
2.18 Proofs 80
2.19 Applications of Energy Equations 81
2.20 Continuity Equation or Law of Conservation of Mass 86
2.21 Limitations of First Law of Thermodynamics 87

3. Power Producing Devices 128 – 159

3.1 Introduction 128
3.2 Forms of Matter 128
3.3 Steam Boilers 131
3.4 Classification of Boilers 131
3.5 Types of Boilers 134
3.6 Advantages of Superheating The Steam 143
3.7 Essentials of a Good Boiler 143
3.8 Comparison Between Water Tube and Fire Tube Boilers 144
3.9 Steam Turbines 144
3.10 Internal Combustion Engines 149

3.11 Two Stroke SI Engine 152
3.12 Gas Turbine Power Plant 156
3.13 Hydraulic Turbines 156
3.14 Compressed Air Motors 157

4. Power Absorbing Devices **160 – 177**

4.1 Introduction 160
4.2 Hydraulic Pumps 160
4.3 Reciprocating Pumps 160
4.4 Reciprocating Pump 161
4.5 Centrifugal Pump 162
4.6 Air Compressors 165
4.7 Uses of Compressed Air 165
4.8 Reciprocating Air Compressor 166
4.9 Rotary Positive Displacement Compressors 168
4.10 Centrifugal Compressor 170
4.11 Axial Flow Compressor 171
4.12 Household Refrigerator 172
4.13 Air Conditioning 175
4.14 Room Air Conditioner (Window Air Conditioner) 175

5. Conventional and Non-conventional Energy Resources **178 – 197**

5.1 Introduction 178
5.2 Sources of Energy 178
5.3 Importance of Electric Energy 178
5.4 Classification of Sources of Energy 180
5.5 Advantages/Disadvantages of Non-Conventional and Conventional Energy Sources 180
5.6 Solar Energy 180
5.7 Wind Energy 182
5.8 Ocean Thermal Energy 183
5.9 Geothermal Energy 184
5.10 Tidal and Wave Energy 184
5.11 Biomass 186
5.12 Bio-Gas Plant 188
5.13 Simple Steam Power Plant (or Thermal Power Plant) 189
5.14 Nuclear Power Plant 190
5.15 Hydro Electric Power Plant 193
5.16 Fuel Cell 194

6. Introduction to Heat Transfer 198 – 245

6.1 Introduction 198
6.2 Modes of Heat Transfer 198
6.3 Film Coefficient of Heat Transfer 203
6.4 Composite Walls 205
6.5 Heat Exchanger 207
6.6 Basic Types of Heat-Exchangers 209
6.7 Insulation 211
6.8 Types of Insulations 212
6.9 Extended Surfaces (Fins) 213

7. Introduction to Metal Cutting Process 246 – 265

7.1 Machine Tool 246
7.2 Centre Lathe 246
7.3 Types of Lathes 250
7.4 Lathe Operations 250
7.5 Drilling Machines 251
7.6 Drilling Machines 252
7.7 Different Drilling Operations 256
7.8 Grinding Machine 257
7.9 Power Saw 261
7.10 Introduction to NC and CNC Machine 262
7.11 Components of NC Machine 263
7.12 Advantages of NC Machines 264
7.13 Disadvantages of NC Machines 264
7.14 Uses of Nc Machines 264
7.15 Illustrative Example of NC Programming 264
7.16 CNC Machines 265

8. Introduction to Metal Joining Process 266 – 280

8.1 Welding 266
8.2 Welding Applications 266
8.3 Requirements of Good Welded Joints 266
8.4 Welding Classification 267
8.5 Brazing 277
8.6 Soldering 277
8.7 Gas Cutting Process and Equipments or (Oxyacetylene Gas Cutting Process) 278

9. Introduction to Sheet Metal Working 281 – 292

9.1 Sheet Metal 281
9.2 Uses of Sheet Metals 281
9.3 Advantages of Sheet Metal Parts 282
9.4 Metal Working 282
9.5 Press Work 282
9.6 Sheet Metal Working 284
9.7 Drawing Operation 284
9.8 Bending 285
9.9 Blanking 286
9.10 Piercing 287
9.11 Notching 287
9.12 Interchangeability 288
9.13 Elements of Interchangeability 288
9.14 Limits 289
9.15 Tolerance 289
9.16 Allowance 290
9.17 Fits 290
9.18 Hole and Shaft Based Fit System 291

10. Principles of Design 293 – 309

10.1 Need of Design 293
10.2 Stress and Strain 294
10.3 Stress-Strain Curve 298
10.4 Modes of Failures 301
10.5 Factor of Safety 301
10.6 Selection of Factor of Safety 302
10.7 Material Properties and Selection 303
10.8 Material Selection 304
10.9 Common Engineering Materials 304
10.10 Aesthetics 305
10.11 Ergonomics 307

11. Machine Devices 310 – 326

11.1 Individual and Group Drive System 310
11.2 Belt Drive 312
11.3 Ropes 314
11.4 Chain Drive 315

11.5 Gear Drives 317
11.6 Clutches 320
11.7 Brakes 323

12. Machine Elements 327 – 343

12.1 Power Transmission Shafts 327
12.2 Types of Shafts 327
12.3 Shaft Material 327
12.4 Application of Shafts 327
12.5 Axle 328
12.6 Application 328
12.7 Keys 328
12.8 Coupling and Their Types 330
12.9 Flanged Coupling 331
12.10 Oldham's Coupling 333
12.11 Universal Coupling or Hooke's Joint 333
12.12 Bearings and Their Types 334
12.13 Flywheel 336
12.14 Flywheel Construction and Types 337
12.15 Governor 339

Practicals Experiments (TW 25 Marks) 344 – 349

Model Question Paper 1 350 – 352

Model Question Paper 2 353 – 354

Index 355 – 359

Unit 1

Fundamentals of Thermodynamics

Syllabus

Fundamental Concepts and Definition — Thermodynamic system, Surroundings and boundary, thermodynamic properties, temperature and temperature scale. Macro and microscopic approach.

Law of Thermodynamics — Principle of conservation of mass and energy, continuity equation, First law of Thermodynamics, Joule's experiment, application of first law to flow and non flow processes and cycle.. Concept of internal energy, flow energy and enthalpy. Application of steady flow energy equation to nozzles, turbines and pumps.

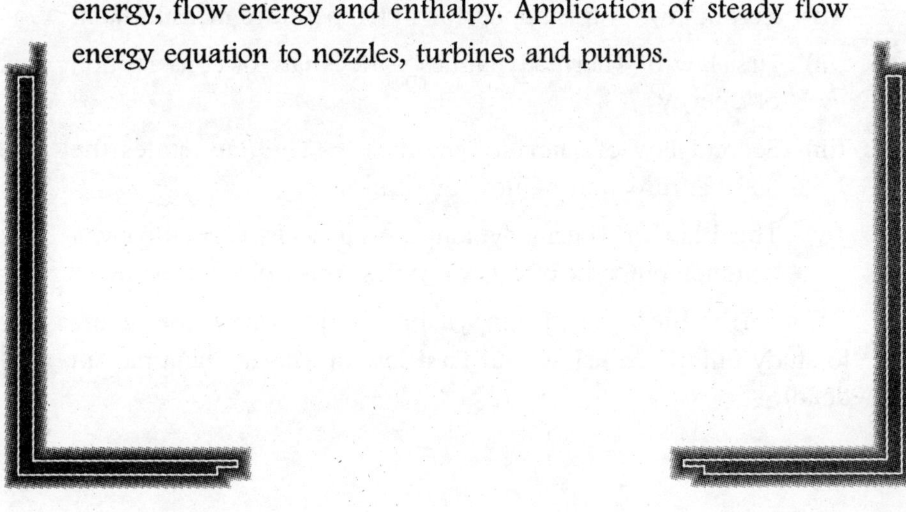

Introduction to Unit 1

This unit gives the introduction to the science known as Thermodynamics. This science has an immense importance is Mechanical Engineering field as it provides the fundamental guidelines and criteria for the further development

Thermodynamics is a branch of a fundamental science 'physics' that deals with the transformations of one form of energy into other form.

For example :

- The internal combustion engine used to drive the vehicle converts heat energy to mechanical energy (i.e. shaft power)
- A hydraulic turbine converts the potential energy of water to mechanical energy.
- A compressor, on other hand, converts the mechanical energy into the pressure energy of fluid.

As stated above, the operation of all these devices are developed using various laws provided by thermodynamics.

Major contribution to the foundation of thermodynamics is by the scientists Joule, Kelvin, Plank and Clausius. The basic laws established by these scientists on the basis of large experimentations are : —

(i) Zeroth law — It is applicable in temperature measurement

(ii) First law of Thermodynamics — It is law of conservation of energy

(iii) Second law of Thermodynamics — This law states the feasible direction of flow of heat energy.

(iv) Third law of Thermodynamics — It deals with entropy, a common outcome of every process taking place in practice.

According to the scope of our syllabus, in this unit we are going to study only Zeroth law and First law of Thermodynamics in detail.

CHAPTER 1

FUNDAMENTAL CONCEPTS AND DEFINITIONS

1.1 INTRODUCTION

In order to understand any science, everyone must get familiar with its unique terminology. This chapter is dedicated for introduction to the basic terms and concepts involved in thermodynamics. Along with the introduction to various concepts and definitions of different terms we shall also discuss the basic system of units which will be followed in the numerical part of this chapter.

1.2 THERMODYNAMIC SYSTEMS

Sometime we have to visit a general physician for certain health related problem. At that time, the physician examines our body for the accurate diagnosis of the illness. But, when we consult an ophthalmologist, he concentrates only a part of our body i.e. Eye. Thus, every doctor decides the portion of body upon which concentration is made, according to his scope of diagnosis and treatment.

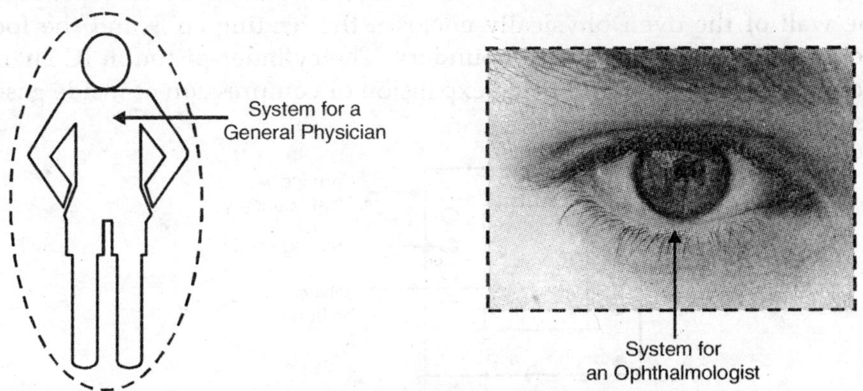

Fig. 1.1 (a) Development of the concept—'System'.

Above mentioned fact is elaborated in view of thermodynamic studies and is specified with the terms like System, Surrounding and Boundary.

A Thermodynamic System is defined as *quantity of matter or region of space upon which attention is concentrated in the analysis of the problem.*

Thermodynamic system may be defined as any part of the *Universe* which we want to study.

In order to fix the limits of a system, there must be a partition around the system. This partition is known as the *boundary of the system.*

Everything external to the boundary is called the *Surrounding.* The system along with surrounding forms an *Universe.* [See Fig. 1.1(b)].

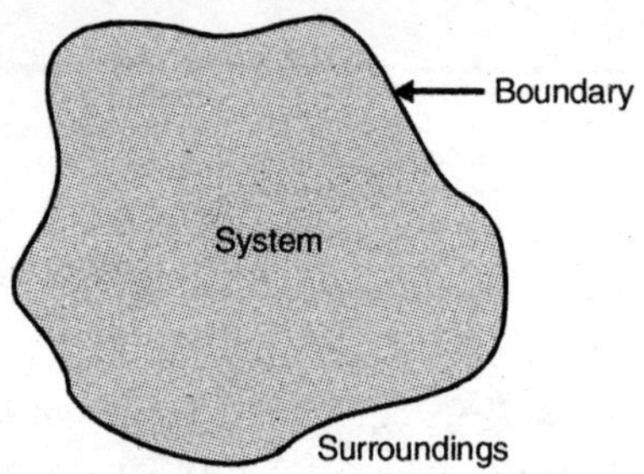

Fig. 1.1 (b) Thermodynamics System.

1.2.1 Types of Boundary

The working of a thermodynamic system depends on the energy interactions between the system and surrounding, which takes place across the boundary. Hence we must start our studies with the type of boundaries and its properties.

The types of boundaries on the basis of its existence are—

(i) Real boundary

(ii) Imaginary boundary

(i) Real Boundary. Consider an oven used for baking the food products. The wall of the oven physically encloses the heating coils and the food products. Thus it forms a real boundary. The cylinder-piston in IC engine provides a real boundary during expansion or compression of inside gases. [See Fig. 1.2 (a)].

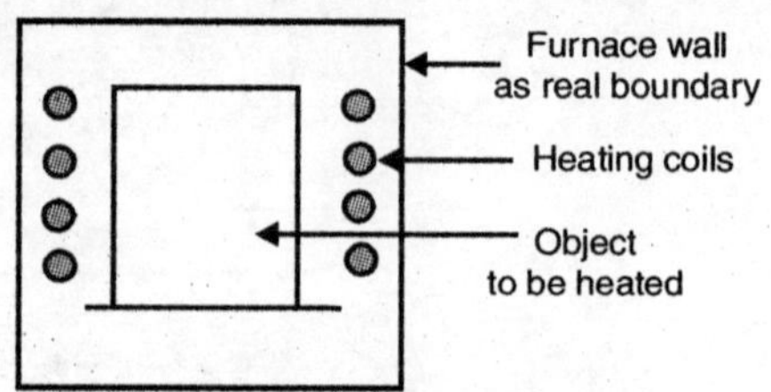

Fig.1.2 (a) A real boundary.

(ii) Imaginary Boundary. When we wish to study the working of a tap of water, we separate the valve system from rest of the thing using an imaginary boundary. In thermodynamics the systems like nozzles, turbines, carburettor of vehicle can be defined with a concept of such an imaginary boundary. [See Fig. 1.2 (b)].

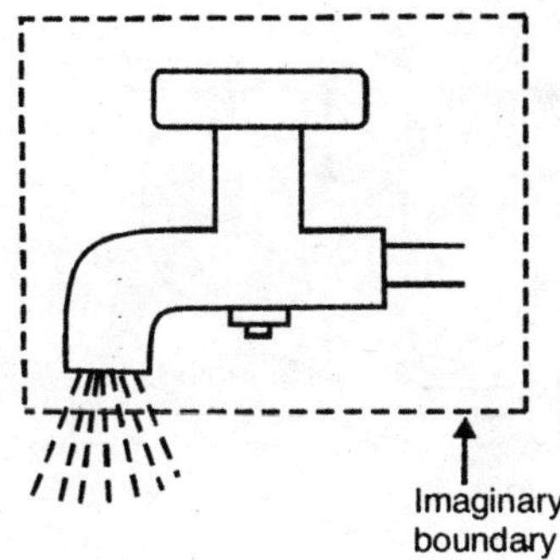

Fig.1.2 (b) An imaginary boundary.

The types of boundaries on the basis of its position are as discussed below:

(i) Fixed Boundary. This type of boundary has a definite shape and encloses a fixed region of universe. A wall of a furnace is example of fixed real boundary while the system of turbine has fixed imaginary boundary.

(ii) Moving Boundary. This type of boundary may change its shape and the region of universe enclosed by it. A wall of balloon which changes its size and shape with the pressure of inside air is an example of moving real boundary.

1.2.2 Types of System

The types of systems may be viewed in a tree form as shown in Fig. 1.3:

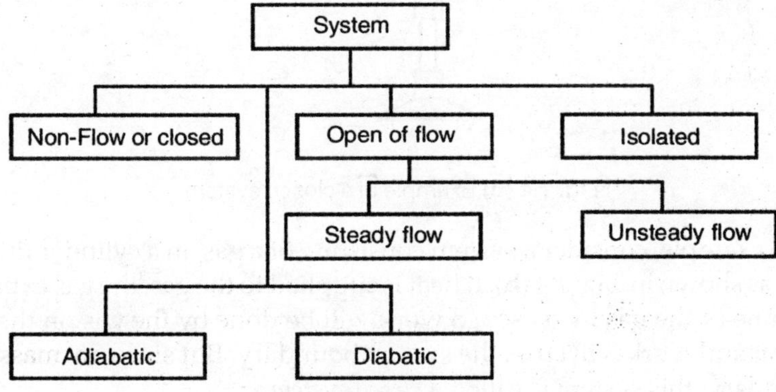

Fig. 1.3

(a) Closed or Non-flow System. It is fixed mass i.e. there is no mass transfer across the boundary. But there may be energy transfer into and out of the system as shown in Fig. 1.4(a).

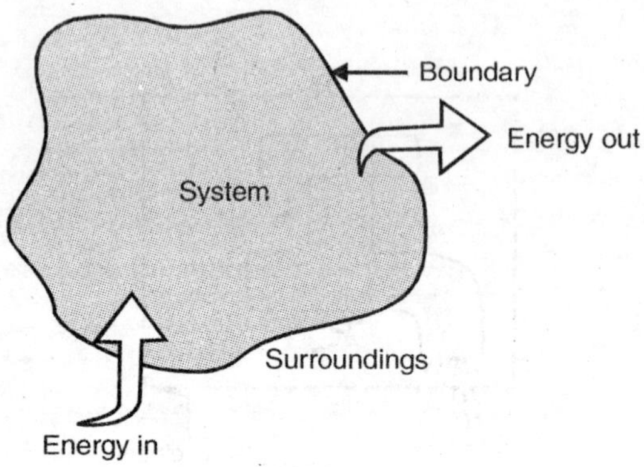

Fig. 1.4 (a) A closed system.

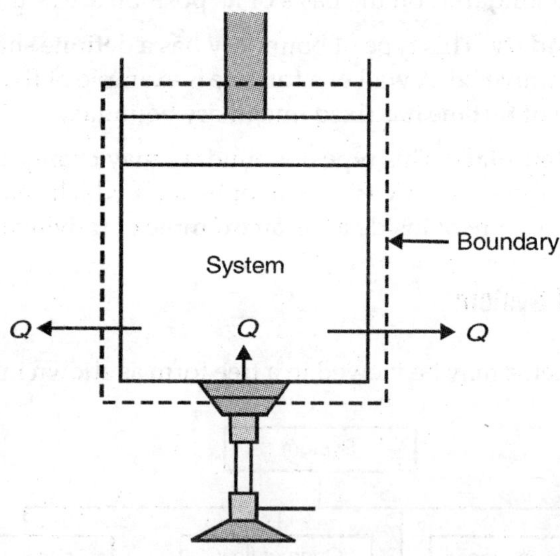

Fig. 1.4 (b) Example of a closed system.

As an example, consider a system consisting of a gas, in a cylinder fitted with a piston as shown in Fig. 1.4 (b). If heat is supplied to the gas, the gas expands i.e. the volume of the gas increases so work will be done by the gas on the piston. Thus heat and work will cross the system boundary. But since no mass crosses the boundary, this system is called a *closed system.*

(b) Open System or Flow System. It is one in which both mass and energy can enter and leave the system as shown in Fig. 1.5. Most of the engineering devices are open systems.

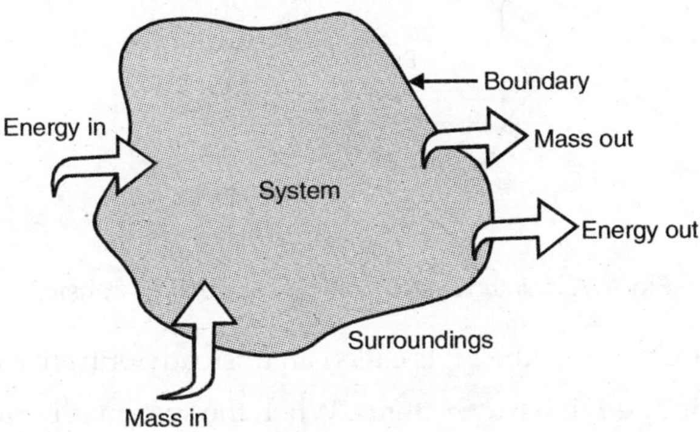

Fig. 1.5. An open system.

For example, consider a portion of steam plant as shown in Fig. 1.6. Mass of steam enters at section (1) and after expanding into the turbine, leaves at section (2). In this case work is produced when the turbine is driven. So, as shown, the mass, work and heat are crossing the system boundary.

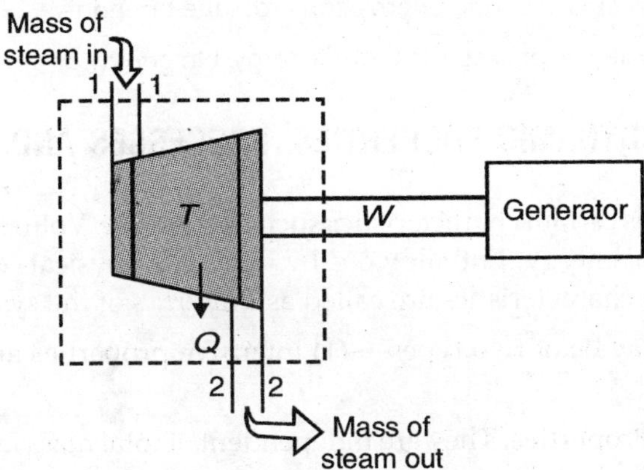

Fig. 1.6. Example of an open system.

(c) Isolated System. It is of fixed mass and energy and there is no mass or energy transfer across the system boundary. An isolated system is independent of all the changes that are taking place in the surroundings. (Fig. 1.7).

(d) Steady Flow and Unsteady Flow Systems. If the mass flow rate and energy flow rate at inlet and outlet are same, then the system is known as steady flow system.

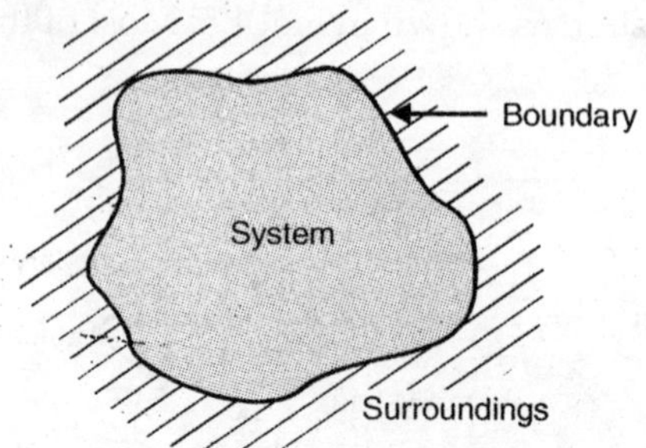

Fig. 1.7. Isolated system (No mass or energy transfer).

If the rates are varying, then it is called an unsteady or transient flow system.

(e) Adiabatic and Diabatic Systems. When the system is perfectly insulated, so that neither heat enters nor it leaves the system, then it is known as *adiabatic system*.

When heat energy, either enters or leaves the system, it is called as *diabatic system*. It is not insulated.

Points to Remember :

- Closed system is of fixed mass and energy exchange.
- In open system mass and energy can cross the boundary.
- Isolated system is of fixed mass and energy. No energy or mass transfer occurs.

1.3 THERMODYNAMIC PROPERTIES, PROCESSES AND CYCLES

Every system has certain characteristics such as Pressure, Volume. Temperature, Density, Internal energy, Enthalpy etc. by which its physical condition may be described. Such characteristics are called as *Properties* of the system.

Properties may be of two types— (1) Intensive properties and (2) Extensive properties.

1. Intensive Properties. They are independent of total mass in the system e.g., Pressure, Temperature, Density.

2. Extensive Properties. These are dependent on the total mass in the system. When the mass of the system changes, the values of the properties will change proportionately, e.g., Volume, Weight, Energy, Enthalpy, Entropy.

Properties may also be classified as —

1. Primary properties (P, V, T)

2. Secondary or derived properties (Internal energy u, Enthalpy h and Entropy s).

The Properties of unit mass are known as *Specific properties* for example specific volume, specific internal energy etc. Thus all the specific properties indicate the values for unit mass and they are independent of total mass, so *all the specific properties are intensive properties.*

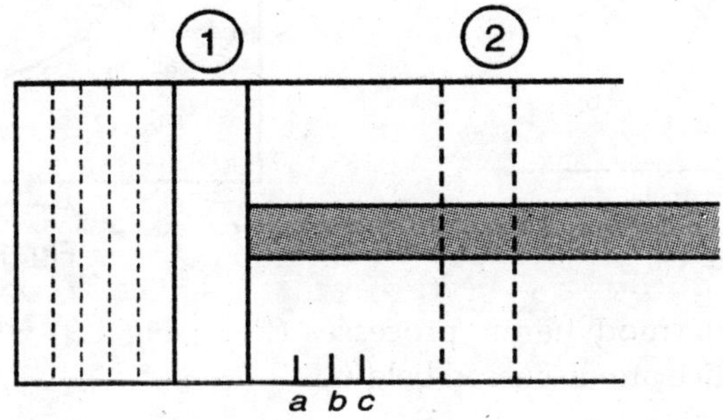

Fig. 1.8 (a)

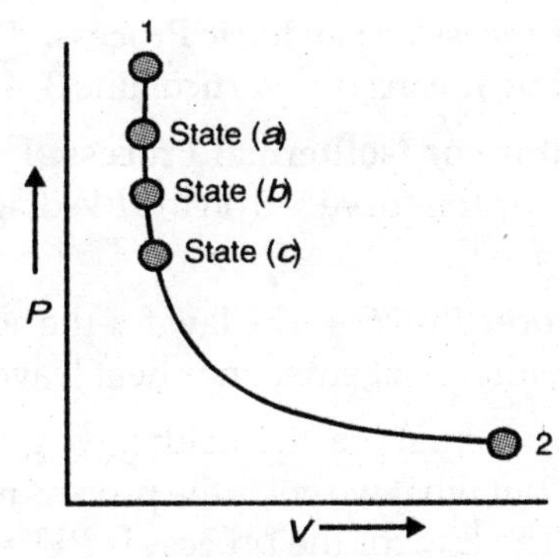

Fig. 1.8 (b)

When all the properties of a system have definite values (such as *P, V, T*) then the system is said to exist at a *Definite state* and the properties of the state are said to be *State functions or point functions.* Any operation in which one or more properties of the system changes, is called *Change of state,* and the series of states (and substates like *a, b, c,* etc. as shown in Fig. 1.9) passed during a change of state from state (1) to state (2) is called the *path* of change of state. The properties of the states and substances are called as *Path functions.*

When the path is completely specified, then the change of state is called a *Process.*

A *Process* is defined as the transformation of the system from one fixed state to another fixed state.

When any one of the properties changes, the working substance or system is said to have undergone a process.

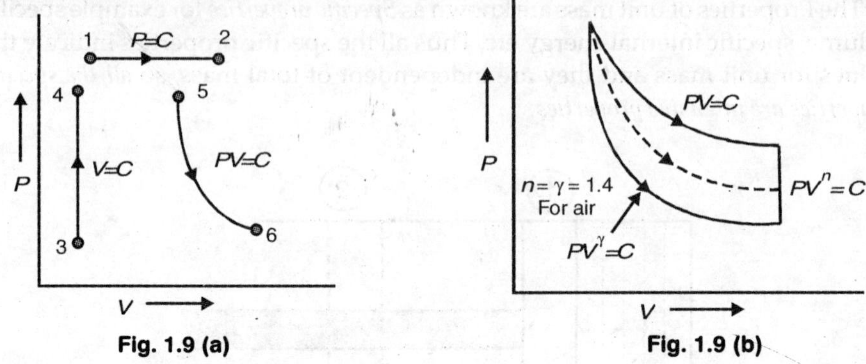

Fig. 1.9 (a) Fig. 1.9 (b)

The various thermodynamic processes (Refer Fig 1.9) which are used in engineering practice are discussed below:

1. Constant Pressure Process or Isobaric Process. The law for the process is ($P = C$) and is represented on the *PV*-diagram by means of a horizontal line 1–2.

2. Constant Volume Process or Isochoric Process. The law for the process is $V = C$ and is represented by means of a vertical line 3–4 on the *PV*-diagram.

3. Constant Temperature or Isothermal Process. The law for the process is $PV = C$ and is represented by the curve 5–6 on the *PV*-diagram. Also the law of the process may be given as $T = C$.

4. Adiabatic or Isentropic Process. The law for the adiabatic process is $PV^{\gamma} = C$. During this process neither heat enters nor heat leaves the system.

5. Polytropic Process. In actual practice neither we get isothermal process, nor we get adiabatic process, but what we get is the process in between these two and is known as polytropic. The law for the process is $PV^n = C$.

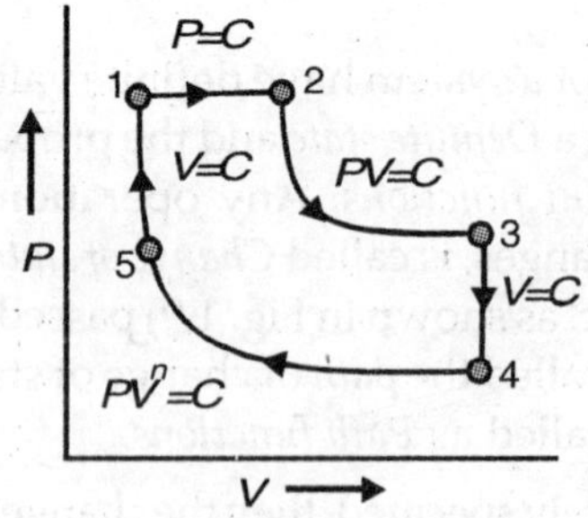

Fig. 1.10

Thermodynamic Cycle. 1. It is defined as the series of state changes such that the final identical with the initial state. 2. When the system undergoes the series of processes such that the end states are same, the system is said to have undergone a thermodynamic cycle. 3. Series of thermodynamic processes, the end states of which are identical, is called a *thermodynamic cycle.*

In a thermodynamic cycle, chemical composition of the working fluid during the process does not change. Thus all the properties in the initial and final states remain unchanged.

For example, water that circulates through the steam power plant. There is a change of phase during the process, but the end states are same.

Points to Remember :

- ⇨ Intensive property is independent of mass.
- ⇨ Extensive property depends on mass.

1.4 WORKING SUBSTANCE

"It is a fluid, in which energy can be stored and from which energy can be removed".

or

"It is a fluid which is used in thermodynamic systems, for converting one form of energy into the other".

The working substance may be a liquid, vapour or a gas. Generally air and water are used as working substances, as they are freely available. However NH_3 and Freon-12 are also used depending upon specific applications.

1.5 UNITS AND DIMENSIONS

Nowadays generally SI (System International) system of units is used. The basic units in this system are:

System. Basic Units

Sr. No.	Quantity	Unit	Symbols of Unit
1.	Length	Metre	m
2.	Mass	Kilogram	kg
3.	Time	Seconds	s
4.	Temperature	Celsius/Kelvin	ºC/K

The dimensions of all other quantities are derived from these basic units which are:

Sr. No.	Quantity	Unit	Symbol of Unit	Alternative Unit
1.	Force	Newton	N	—
2.	Energy (Work or Heat)	Joule	J	Nm
3.	Power	Watt	W	J/sec
4.	Pressure	Pascal	Pa	N/m^2
5.	Volume	Cubic Metres	m^3	Cu-m

A proper understanding of units of Force, Pressure, Temperature, Work, Heat and Power is necessary for solving numerical problems in Thermal Engineering.

1. Force. The unit of force in SI units is Newtons (N) A force of 1 N produces an acceleration of 1 m/sec² when applied to a mass of 1 kg.

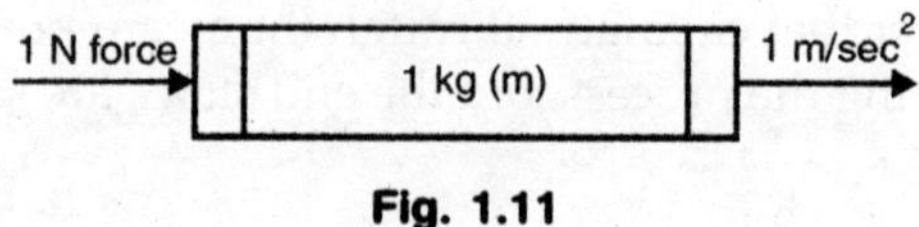

Fig. 1.11

We know that from Newton's Second law of motion $F \propto ma$

or $$F = \frac{ma}{g_c},$$ where g_c = constant of proportionality.

As 1 N force produces an acceleration of 1 m/sec² in 1 kg mass, we have,

$$1 \text{ Newton} = \frac{1}{g_c} \times 1 \text{ kg} \times 1 \text{ m/sec}^2$$

$\therefore$ $$g_c = 1 \text{ in SI units and units will be}$$

$$g_c = \frac{1 \text{ kg-m}}{\text{N-sec}^2}$$

When $g_c = 1$, the product of mass and acceleration becomes the magnitude of force.

If the mass of 1 kg is allowed to fall freely under the action of standard gravitational force, it is accelerated at the rate of 9.806 m/sec² (9.81 m/sec²) and we have,

$$\text{Force} = 1 \text{ kg} \times 9.81 \text{ m/sec}^2 = 9.81 \text{ N}$$

$\therefore$ It follows that the weight of 1 kg mass equals 9.81 N and since the weight is the force,

$$1 \text{ kgf} = 9.81 \text{ N}$$

2. Weight. Weight of a body is the force exerted on its mass due to gravity.

3. Density. Density is defined as mass per unit volume of the substance. It is denoted by ρ

$\therefore$ $$\rho = \frac{\text{mass}}{\text{volume}} = \frac{\text{kg}}{\text{m}^3}$$

Since unit of volume is m³ and mass is kg.

4. Work (mechanical): When a force F acts on a body and causes displacement through a distance dl in the direction of force. then the work is said to be done. This work is equal to the product of force and distance moved.

$$W = F \times dl$$

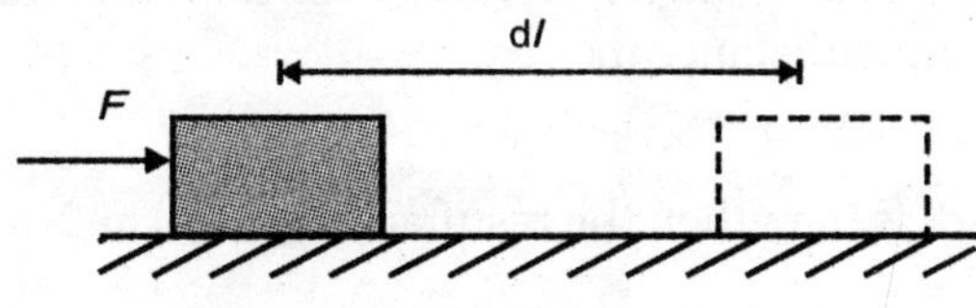

Fig. 1.12

If force F is in Newton and distance dl is in metre, then the resultant unit of W.D. will be Nm and 1 Nm = 1 joule.

5. Power. It is defined as the rate of doing work.

or $$\text{Power} = \frac{\text{W.D.}}{\text{Time taken}}$$

If the unit of W.D. is in joules and the time taken is in seconds, then the unit of Power is J/sec.

And the rate of doing work of 1 J/sec is called watt (W)

$$\therefore \quad \text{Power} = \frac{\text{J}}{\text{sec}} \text{ or W}$$

Earlier there used to be another unit of Power known as Horse Power (H.P.)

$$1 \text{ HP} = 746 \text{ watts} = 0.746 \text{ kW}$$

6. Pressure (P). Pressure is defined as the force per unit area. Thus if F is the force applied to an area A then,

$P = \dfrac{F}{A}$ and if F is in newton and A is in m^2 then the unit will be $\dfrac{\text{N}}{\text{m}^2}$.

This unit of pressure is called as pascal (Pa).

1.6 MECHANICAL AND THERMODYNAMIC WORK

Mechanical Work

$$\text{W.D.} = F \times dl$$

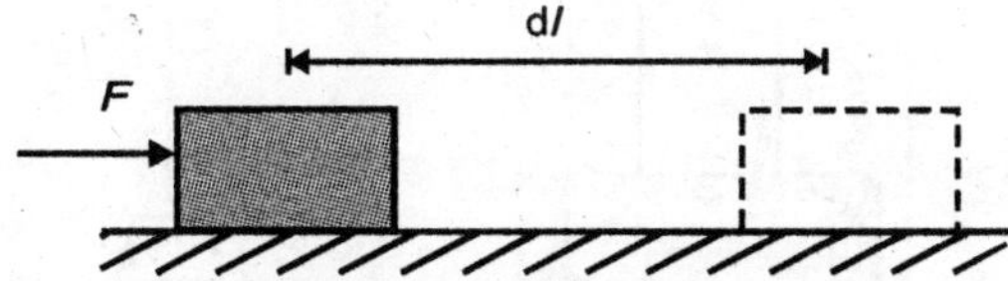

Fig. 1.13. Mechanical work.

When a force F acts on a body and causes a displacement through a distance dl in the direction of force, then the work is said to be done and this work is equal to the product of force and distance moved.

i.e. Work done $= F \times dl$

If F is in N, and dl is in m then the resultant unit will be Nm or Joule.

Thermodynamic Work

"It is an interaction between the system and the surroundings and the work is said to be done by the system on surroundings, if the complete external effect is lifting up of the body".

Let us consider a battery and motor system as shown in Fig. 1.14. The motor is driving a fan. The system is doing work on the surroundings.

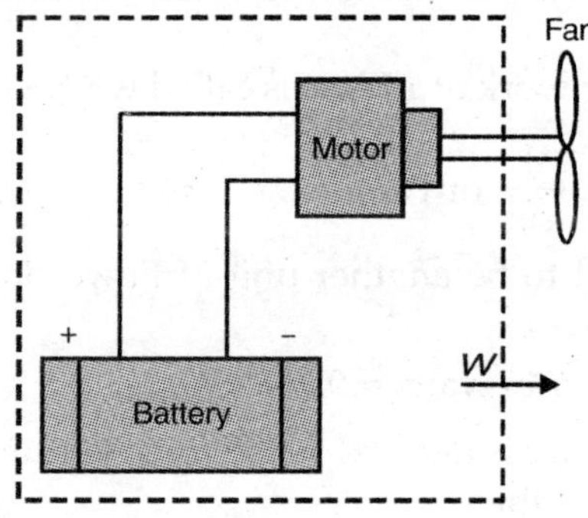

Fig. 1.14

When the fan is replaced by a pulley and weight as shown in Fig. 1.15, then the weight may be raised as the pulley is driven by the motor. Then the complete external effect is raising up the body, so the system is doing thermodynamic work.

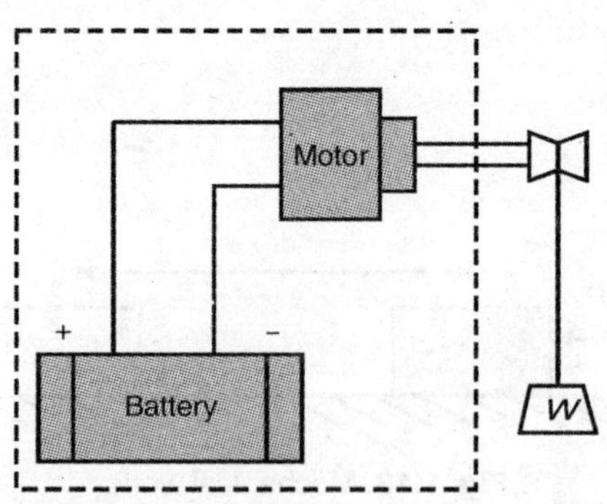

Fig. 1.15

When the work is done by the system, then it is taken as + ve, and when the work is done on the system, it is taken as –ve. (Refer Fig. 1. 16)

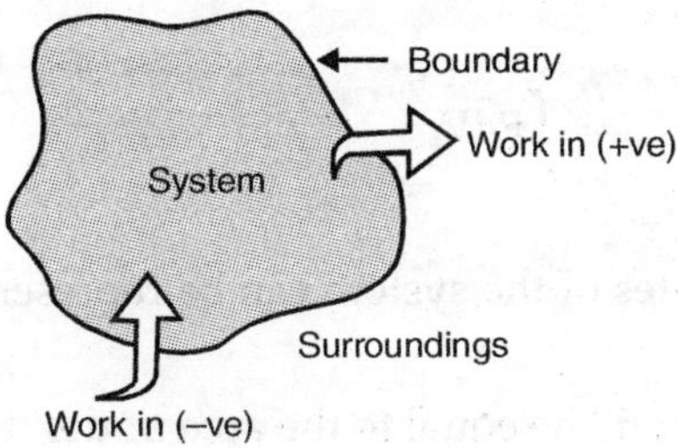

Fig. 1.16

1.7 *PdV*-WORK

Consider a system consisting of a gas in a cylinder fitted with a piston as shown in Fig. 1.17. During the initial condition of the piston i.e. when the piston is at position (1), the pressure inside the cylinder is P_1 and volume is V_1. Let the gas expands as the piston moves to position (2). Then the pressure falls to P_2 and volume will increase to V_2.

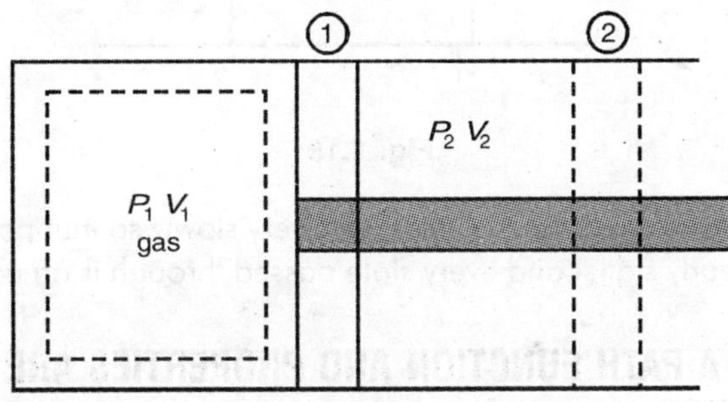

Fig. 1.17

Now, consider a small movement of the piston d*l* during which pressure *P* is assumed to be constant. If *a* is the cross sectional area in m^2, then the force acting on the piston is given by

$$F = P \times a$$

And the small amount of work done by the gas on the piston in causing the displacement d*l*, = force × displacement.

$$\delta W = F \times dl = P \times a \times dl \qquad (\because F = P \times a)$$

$$\delta W = PdV,$$

where $dV = a \times dl$ = small amount of displacement volume.

When the piston moves from position (1) to position (2) volume will change from V_1 to V_2 then the total amount of work done is given by,

$$W_{1\text{-}2} = \int_{v_1}^{v_2} PdV$$

The initial and final states of the system can be represented on *PV*-diagram as shown in Fig. 1.18.

The work done W_{1-2} will be equal to the area under the curve 1–2 on the *PV*-diagram.

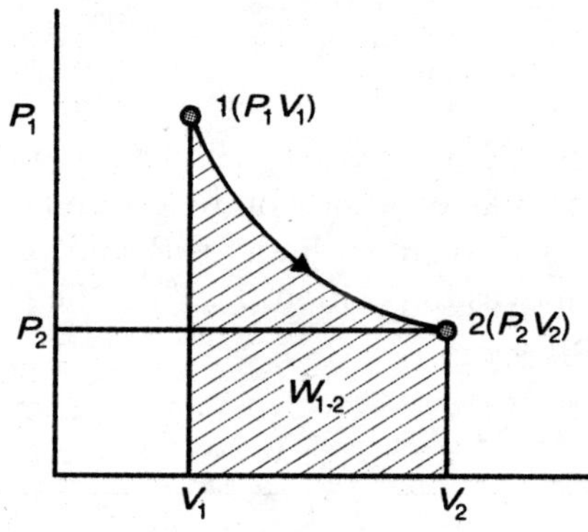

Fig. 1.18

Note: Here the piston is assumed to move very very slowly so that path 1–2 is Quasi-static (*i.e.* nearly static) and every state passed through is an equilibrium state.

1.8 WORK IS A PATH FUNCTION AND PROPERTIES ARE POINT FUNCTIONS

With reference to Fig. 1.19 it is possible to take a system from state (1) to state (2) by following many quasi-static paths such as *A, B, C* etc.

Now if we consider the path 1–*A*–2, then the work done during the process 1–*A*–2 will be equal to the area under the curve 1–*A*–2 and if we consider path 1–*B*–2, then the work done during the path 1–*B*–2 will be equal to the area under the curve 1–*B*–2.

So, the work done during the process directly depends upon the path and not on the end states. So, work is called Path function and δW is an inexact or imperfect differential.

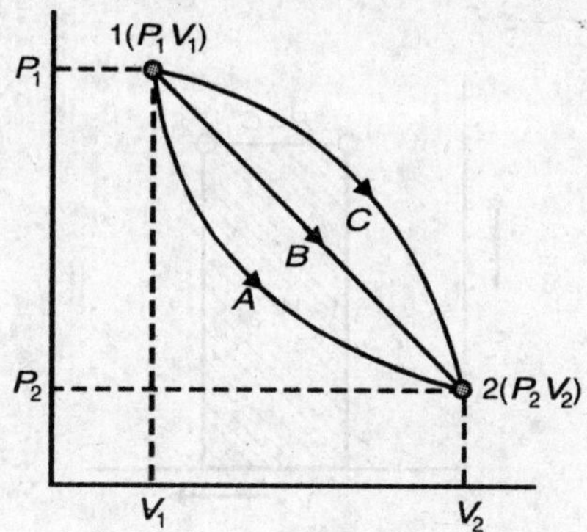

Fig. 1.19

i.e. $$\int_1^2 \delta W \neq W_2 - W_1$$

But $$\int_1^2 \delta W = W_{1-2} \text{ or } {}_1W_2$$

Thermodynamic properties are point functions, since for a given state, there is a definite value for each property. During change of states, the values of the properties will change and this change in values of the properties does not depend on path, but it depends upon end states and hence all the properties are Point functions.

The differentials of point functions are exact or perfect differentials and the integration is simply, e.g. $$\int_1^2 dV = V_2 - V_1$$

Operator δ is used to denote inexact differentials, operator d is used to denote exact differentials.

1.9 EQUATIONS FOR WORK DONE IN VARIOUS PROCESSES

PdV work in various quasi-static processes

1. Constant pressure process (Isobaric process). Here pressure is constant.

We know that Work Done is given by,

$$W_{1-2} = \int_{v_1}^{v_2} PdV = P\int_{v_1}^{v_2} dV$$

$$= P(V_2 - V_1)\text{Nm, if } P \text{ is N/m}^2 \times V \text{ is m}^3$$

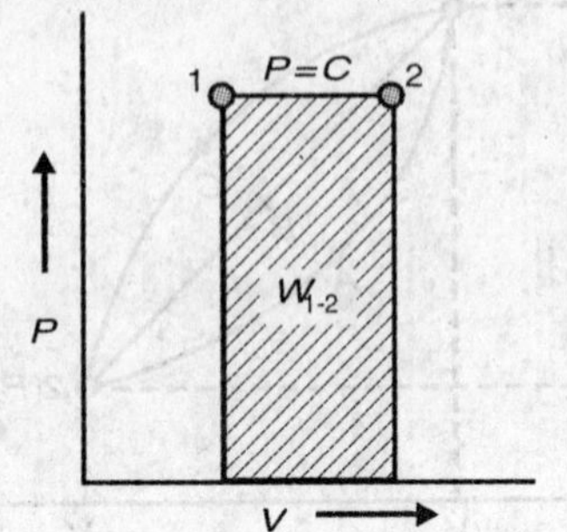

Fig. 1.20

2. Constant volume process. (Isochoric process)

$$W_{1-2} = \int PdV$$
$$= 0$$

During this process $V = C$ hence W.D. $= 0$

Since derivative of a constant is zero.
From PV-diagram area under the line 1–2 is zero.

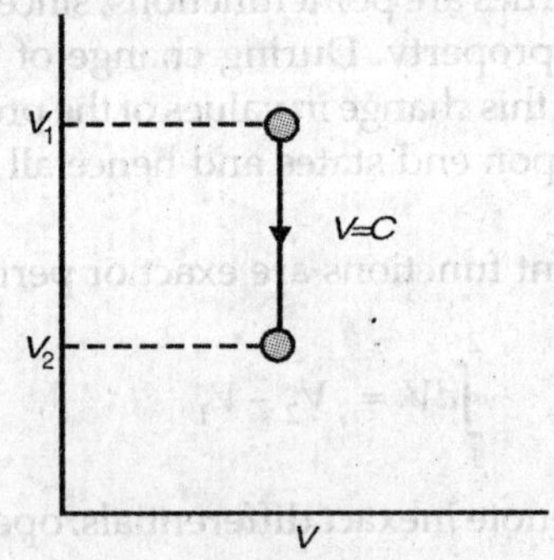

Fig. 1.21

3. Isothermal process. We know that work done in a non-flow system

$$W_{1-2} = \int_1^2 P.dV \qquad \text{...(1)}$$

The law for the process is,

$$PV = C$$

i.e. $$PV = P_1V_1 = C$$

$$\therefore \qquad P = \frac{P_1 V_1}{V}$$

Substituting this in Eq. (1) we get,

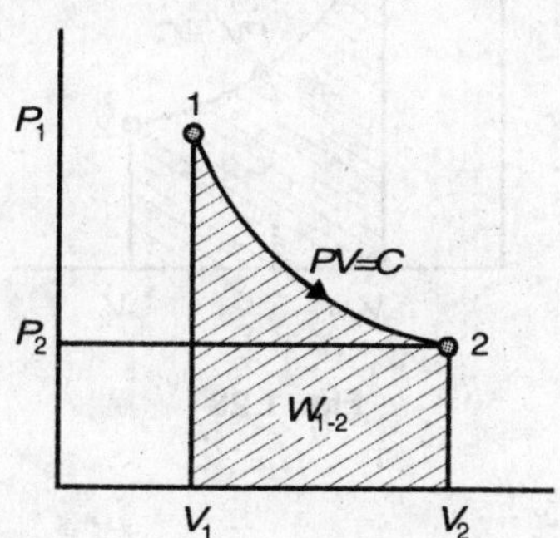

Fig. 1.22

$$\therefore \qquad W_{1-2} = \int_1^2 \frac{P_1 V_1}{V} \cdot dV = P_1 V_1 \int_1^2 \frac{dV}{V}$$

$$W_{1\text{-}2} = P_1 V_1 \ln \frac{V_2}{V_1}$$

$$\mathbf{W_{1\text{-}2} = P_1 V_1 \ln \frac{P_1}{P_2}}$$

4. Polytropic process. The law for the process is $PV^n = C$ where n is a constant.
We know that, the work down in non-flow system,

$$W_{1-2} = \int_1^2 PdV \qquad \ldots(2)$$

But for a polytropic process,

$$PV^n = P_1 V_1^n = P_2 V_2^n = C$$

$$\therefore \qquad P = \frac{C}{V^n}$$

Substituting in Eq. (2), we get,

$$W_{1-2} = \int_1^2 \frac{C}{V^n} dV$$

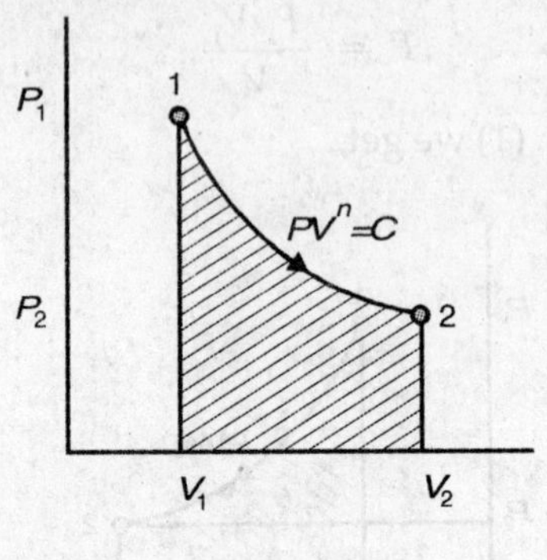

Fig. 1.23

$$W_{1-2} = \int_1^2 C\,V^{-n}\,dV$$

$$= C\left[\frac{V^{-n+1}}{-n+1}\right]_1^2$$

$$= C\left[\frac{V_2^{-n+1} - V_1^{-n+1}}{-n+1}\right] = \frac{CV_2^{-n+1} - CV_1^{-n+1}}{-n+1}$$

$$= \frac{CV_2^{-n+1}}{-n+1} - \frac{CV_1^{-n+1}}{-n+1}$$

Substituting for C

$$= \frac{P_2V_2^n \cdot V_2^{-n+1}}{-n+1} - \frac{P_1V_1^n \cdot V_1^{-n+1}}{-n+1}$$

$$= \frac{P_2V_2^{n-n+1} - P_1V_1^{n-n+1}}{1-n}$$

$$W_{1-2} = \frac{P_2V_2 - P_1V_1}{1-n} \text{ or } \frac{P_1V_1 - P_2V_2}{n-1} \qquad \ldots(3)$$

Since $P_1V_1 = mRT_1$ and $P_2V_2 = mRT_2$

$$W_{1-2} = \frac{mR(T_2 - T_1)}{1-r}$$

And for unit mass $$W_{1-2} = \frac{R(T_2 - T_1)}{1-r} \qquad \ldots(4)$$

1.10 HEAT

"Heat is a form of energy, which crosses the system boundary due to the temperature difference between the system and the surroundings."

When the two bodies, one hot and the other cold are kept in contact with each other, then the hot body loses heat and becomes colder and the cold body gains heat and becomes hotter and this process continues till the thermal equilibrium is reached. At this stage the two bodies will be at the same temperature.

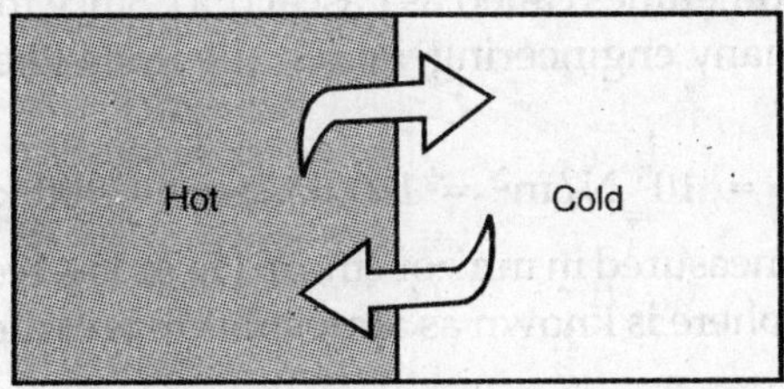

Fig. 1.24

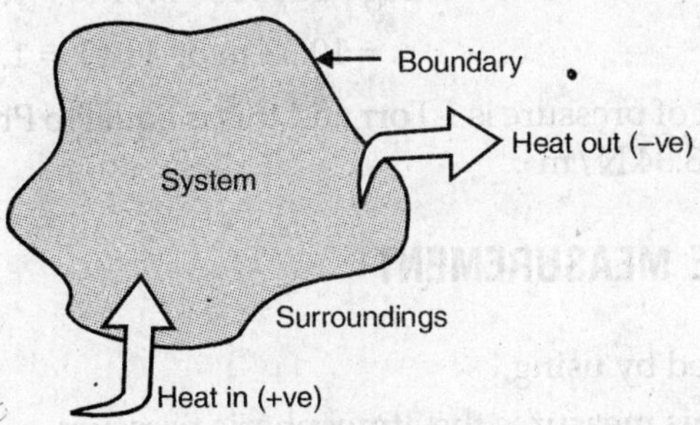

Fig. 1.25

Like work, heat is a path function and we know that the differentials of path functions are imperfect differentials. If Q is the heat transfer, then the magnitude of heat transfer during the process 1–2 is given by,

$$\int_1^2 \delta Q = Q_{1-2} \text{ or } Q_2$$

$$\int_1^2 \delta Q \neq Q_2 - Q_1$$

Note. When heat flows into the system then it is taken as +ve and when heat flows out of the system then it is taken as –ve.

1.11 PRESSURE

It is defined as force per unit area i.e.

$$P = \frac{F}{A}$$

$$F = \text{Force in newton (N)}$$

$$A = \text{Area is a m}^2$$

Then the unit of pressure will be N/m^2, which is the basic unit of pressure, in SI units. This is also sometimes called as Pascal (Pa). Since this unit is very small, when compared to many engineering values, the units like, KPa, MPa, bar are used.

$$1 \text{ bar} = 10^5 \text{ N/m}^2 = 100 \text{ kN/m}^2 = 100 \text{ kPa}$$

Pressures are also measured in mm, or cm, of Hg or H_2O column. The pressure exerted by the atmosphere is known as atmospheric pressure and is denoted by 1 atm.

In various units, the value of 1 atm pressure are given by,

$$1 \text{ atm} = 760 \text{ mm of Hg} = 101325 \text{ N/m}^2 = 1.01325 \text{ bar.}$$

$$= 10.33 \text{ m of } H_2O = 1.033 \text{ kg/cm}^2$$

Another useful unit of pressure is 1 Torr and this is equal to Pressure exerted due to 1 mm of Hg = 133.34 N/m^2.

1.12 PRESSURE MEASUREMENT

Pressure is measured by using,

1. **Barometer.** This measures the atmospheric pressure.
2. **Bourdon pressure gauge.** This is used to measure pressure inside the pipes / containers, etc.
3. **Manometers.** These are used to measure pressure inside the pipes / vessels, etc.

The atmospheric pressure acts both on the container and the gauge which measures the pressure inside the container.

When the pressure of the system is more (i.e. +ve pressure) than the atmosphere, then guage pressure is +ve and P_{abs} or P_{real} is given by,

$$P_{abs} = P_{aim} + P_{gauge}$$

If the pressure of the system is less (i.e. –ve pressure) than the atmospheric pressure, its gauge pressure is –ve and it is known as vacuum and,

$$P_{abs} = P_{atm} - P_{vac}$$

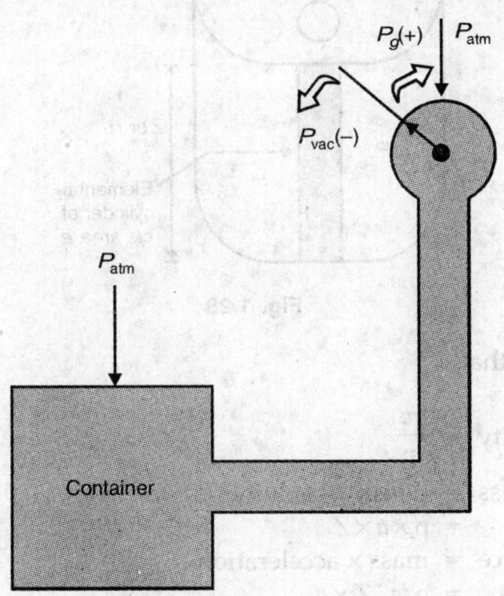

Fig. 1.26

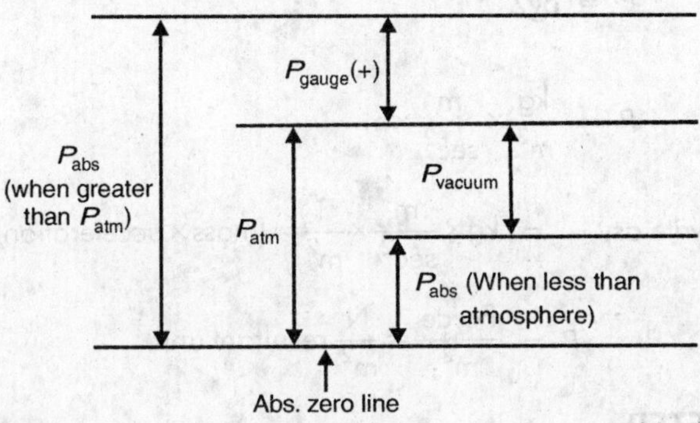

Fig. 1.27

Figure 1.27 shows the relationship between absolute and gauge pressure.

1.13 PRESSURE EXERTED DUE TO A COLUMN OF FLUID

Figure 1.28 shows a fluid of density ρ contained in container to a depth *z* or *h*. Now consider an elemental vertical cylinder of cross sectional area *a*.

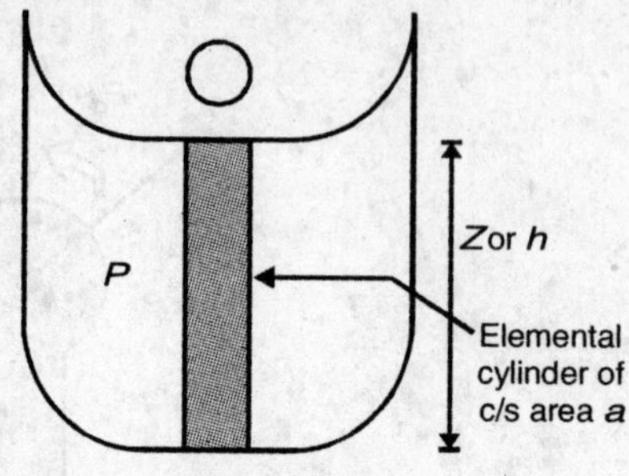

Fig. 1.28

Then we know that,

$$\text{Density} = \frac{m}{v}$$

$$\therefore \quad \text{Mass} = \text{density} \times \text{volume}$$

$$= \rho \times a \times Z$$

$$\text{But} \quad \text{Force} = \text{mass} \times \text{acceleration}$$

$$= \rho . a . Z \times g$$

But pressure exerted due to a column of fluid at a depth Z

$$= \frac{\text{Force}}{\text{Area}} = \frac{\rho \cdot a . \cdot Z \cdot g}{a}$$

$$P = \rho g Z$$

Note:

$$\text{Units} \quad P = \frac{\text{kg}}{\text{m}^3} \times \frac{\text{m}}{\text{sec}^2} \times \text{m}$$

This we can write as,

$$= \text{kg} \times \frac{\text{m}}{\text{sec}^2} \times \frac{1}{\text{m}^2} = [\text{mass} \times \text{acceleration}] \times \frac{1}{\text{m}^2}$$

$$P = \frac{\text{Force}}{\text{m}^2} = \frac{\text{N}}{\text{m}^2} \text{ resultant unit.}$$

1.14 BAROMETER

It is used to measure atmospheric pressure. Consider Hg in the container as shown in Fig. 1.29 (a) and (b). First of all, let a glass tube open at both the ends is immersed in the Hg as shown in Fig. 1.29 (a). Then the atmospheric pressure acting on the surface of Hg level is same as shown.

Now if the glass tube with one end sealed and evacuated to a state of vacuum is immersed as shown in Fig. 1.31 (b) then, due to atmospheric pressure acting on the surface of Hg, the Hg will rise in the glass tube and this will be equal to 760 mm at sea level.

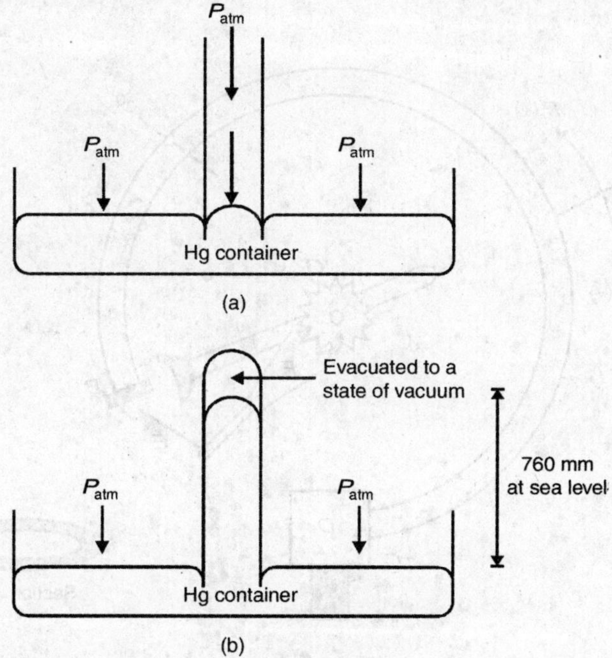

Fig. 1.29. Barometer.

1.15 BOURDON PRESSURE GAUGE

This is the most common type of pressure gauge. This is used to measure the pressure inside the pipes, containers, vessels etc.

The basic element of this gauge is the tube *A*. It is elliptical in cross section and is bent into an arc of circle as shown in Fig. 1.30. End *B* of this tube is sealed, whereas open end *C* is connected to the connecting union *D* through which it can be mounted over the pipes or containers of whose pressures are to be measured.

If the pressure of the system is more than the atmosphere (i.e. +ve pressure) then the tube will tend to curl out. Conversely if the pressure of the system is less than atmosphere (i.e. –ve pressure) then the tube will tend to curl in. The change in pressure will therefore appear as the movement of the end *B*. This end *B* is connected to the quadrant gear *F* through the link *E*. The quadrant gear *F* engages with die small gear *G*, on to which a pointer *H* is attached.

Thus any movement due to change in pressure of the end *B* will be transmitted to the quadrant gear which will rotate the gear *G* and hence the pointer. The pointer rotates on the calibrated scale and thus gives the pressure readings directly.

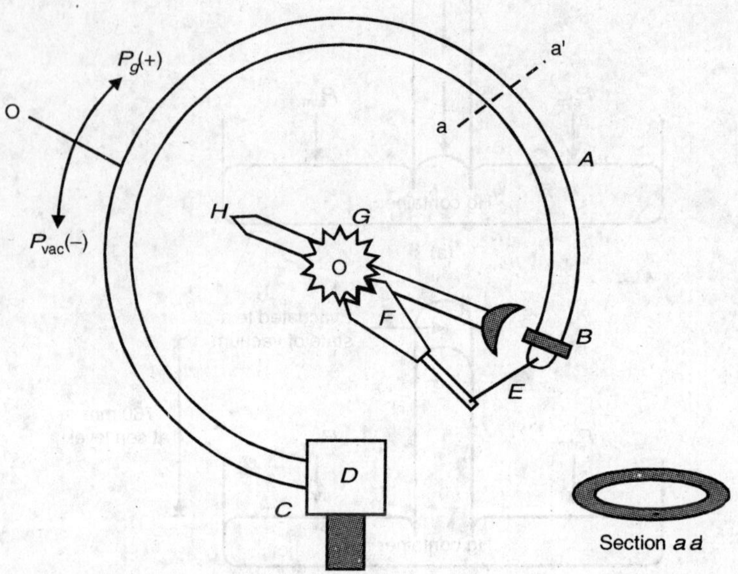

Fig. 1.30

1.16 MANOMETERS

These are used to measure the pressure inside the pipes, containers. It consists of a bent glass tube or a U-type glass tube which contains a liquid generally H_2O or Hg. One end of the manometer is connected to the container or pipe of whose pressure measurement is to be done and the other end is generally open to the atmosphere. The U-tube with one end closed and evacuated to state of vacuum indicates the absolute pressure directly as shown in Fig. 1.31(c).

Balancing the pressure values about the line 1–1', we get for Fig. 1.31 (a).

$$P_{abs} = P_{abs} + \rho g Z = P_{abs} + P_{gauge}$$

For Fig. 1.31 (b) $\quad P_{abs} + \rho g Z = P_{atm}$

i.e. $\quad P_{abs} = P_{atm} - \rho g Z = P_{atm} - P_{gauge} = P_{atm} - P_{vac}$

For Fig. 1.31 (c) $\quad P_{abs} = \rho g Z$

1.17 ZEROTH LAW OF THERMODYNAMICS (THERMAL EQUILIBRIUM)

When the two bodies one hot and the other cold, are placed in contact with each other, then the hot body loses heat and becomes colder and the cold body gains heat and becomes hotter, and this process continues till the thermal equilibrium is reached. In this state the two bodies will be at the same temperature.

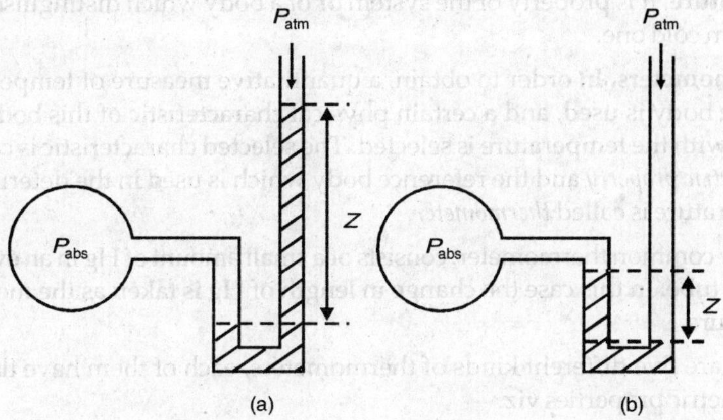

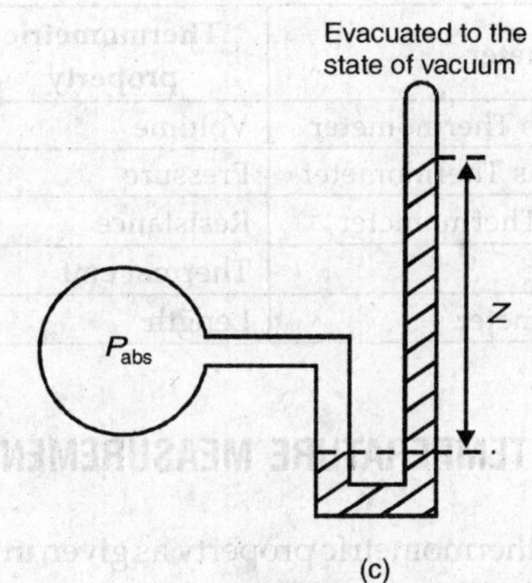

Fig. 1.32

From this a law is deduced known as *Zeroth law of thermodynamics*. It states that, "When a body *A* is in thermal equilibrium with the body *B* and also separately with the body *C*, then the bodies *B* and *C* will be in thermal equilibrium with each other. On the basis of this law, it is possible to compare the temperatures of *B* and *C* and it can be said that the temperatures of *B* and *C* are equal without actually making any contact between them. This law thus forms the basis of temperature measurement: by comparison method.

Thermal Equilibrium	**Thermal Equilibrium**	**Thermal Equilibrium**
A | *B*	*A* | *C*	*B* | *C*

If *A* is thermal equilibrium separately with *B* and *C* then *B* and *C* will also be in thermal equilibrium.

1.18 THERMOMETRIC PROPERTY AND THERMOMETERS

Temperature. It is property of the system or of a body which distinguishes a hot body from cold one.

Thermometers. In order to obtain, a quantitative measure of temperature, a reference body is used, and a certain physical characteristic of this body which changes with the temperature is selected. The selected characteristic is called the *thermometric property* and the reference body which is used in the determination of temperature is called *thermometer*.

A very common thermometer, consists of a small amount of Hg in an evacuated capillary tube. In this case the change in length of Hg is taken as the measure of temperature.

There are five different kinds of thermometers, each of them have their own thermometric properties viz.

Thermometer	Thermometric property	Thermometric substance
1. Constant Volume gas Thermometer	Volume	Gas
2. Constant Pressure gas Thermometer	Pressure	Gas
3. Electrical resistance Thermometer	Resistance	Wire
4. Thermo couple	Thermal emf	Wire
5. Hg in Glass Thermometer	Length	Hg

1.19 PRINCIPLE OF TEMPERATURE MEASUREMENT

Let x be the value of any thermometric property as given in the table and θ be the temperature corresponding to x.

When the temperature changes, the values of the thermometric properties change proportionately. Therefore the value of temperature is proportional to the values of thermometric properties.

So we have, $\theta \propto x$

i.e. $\theta = ax$ ...(1)

For Eqs (1) and (2) $\theta_1 = ax_1$

$\theta_1 = ax_2$ ...(2)

where a is the constant of proportionality.

Dividing we get, $\dfrac{\theta_1}{\theta_1} = \dfrac{x_1}{x_2}$ ...(3)

Thus the ratio of temperature is equal to the ratio of values of properties, such as Length of Hg, volume, pressure etc.

Now, to construct a scale we need a reference point. The internationally accepted reference point is the *Triple point of water*, which is a temperature, at which all the 3-phases of water, i.e. ice, water and vapour, exists in equilibrium. And the value of triple point temperature is 273.16 K (or 0.01ºC) at a pressure of 0.006 bar, where *K* stands for degree Kelvin.

∴ Equation (1) becomes,

$$273.16 = ax_t \Rightarrow \frac{273.16}{x_t} = a$$

where x_t is the value of the property at triple point.

Now susbtitute this value of *a* in Eq. (1)

i.e. $\theta = ax$ we get, $\theta = \frac{273.16}{x_t} \cdot x$

$$\theta = 273.16 \times \frac{x}{x_t}$$

Thus for any change in value of *x* (i.e. Thermometric property) the temperature θ can be found out. Thus from this equation. we can construct a scale and calibrate it to give us direct temperature readings. Here it is to be noted that x_t is the value of the property at triple point and is constant.

1.20 SCALE OF TEMPERATURE

A scale of temperature is based on some reference or fixed point to which an arbitrary value of temperature is assigned. Thus on a thermometer with some fixed or reference point, any temperature has a unique numerical value w. r. t. the fixed reference point.

Scales of Temperature

1. (a) Celsius scale, (b) Fahrenheit scale.
2. Absolute Temperature scale.
3. Ideal Gas temperature scale.

1. (a) Celsius Scale. This scale is named after Anders the Celsius, who devised this scale. There are two fixed points on this scale. One is ice point and the other is the steam point. These two points are numbered as 0 and 100 on Celsius scale. The interval between them is divided into 100 equal parts and each division, is known as one degree Celsius and degree Celsius denoted as ºC.

(b) Fahrenheit Scale. On this scale the ice point is numbered as 32 and steam point is numbered as 212. The interval between them is divided into 180 equal parts and each division, is known as 1 ºF.

The temperature readings on one scale can be converted into a reading on the other scale by the following formula.

$$\frac{C}{100} = \frac{F-32}{180}$$

$$\therefore \quad C = \frac{100}{180}(F-32) = \frac{5}{9}(F-32)$$

and $$F = \frac{9}{5}(C)+32$$

2. Absolute Temperature Scale or Kelvin Temperature Scale or Thermodynamic Temperature Scale.

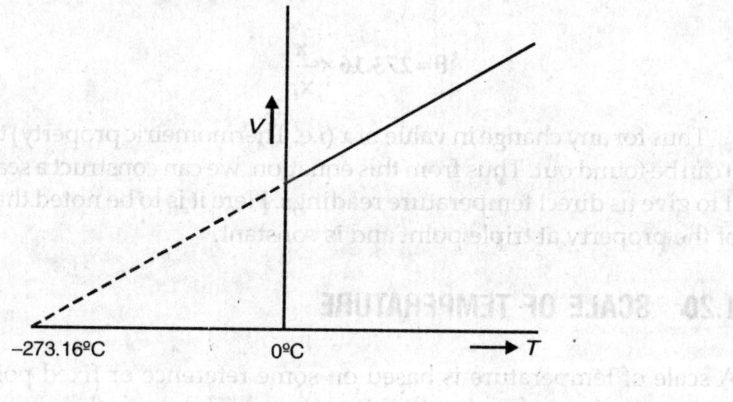

Fig. 1.32

This is an internationally accepted scale. In this scale the temperature is measured from absolute zero temperature.

Absolute zero is the temperature at which all the vibratory, translatory and rotational motion of the molecules of a gas, is supposed to cease. A gas on cooling contracts in volume. In case of perfect gases, 'Charle's found that decreases in volume per degree centigrade decrease in temperature is 1/273. 16th of its volume at 0°C and pressure remaining constant. Thus the volume of gas will be zero at temperature –273.16°C. This temperature 273.16°C below or behind 0ºC (or –273.16°C) is called the *absolute zero temperature.*

In this scale there is one more reference point taken, the Triple point of water. This is a temperature at which all the three phases of water exists in equilibrium, and the value of triple point temperature is 0.01 °C or 273.16 K at a pressure of 0.006 bar. Here it is to be noted that, the ice point temperature on this scale is 273.15 K, and steam point temperature is 373.15 K and therefore to convert the temperature given in °C to K, just add 273.15.

$K = °C + 273.15$ or in general, $K = °C + 273$. Here it is to be noted that the value of Absolute zero on this scale is $-273.16°C$.

3. Ideal Gas-temperature Scale. This is similar to absolute temperature scale. This scale is constructed by using constant volume gas thermometer.

In a constant volume gas thermometer, gas will be used as the working substance and as all the gases behave as ideal gases at low pressure, thus scale is known as Ideal gas temperature scale. The value of absolute zero on this scale is same as that given by Absolute temperature scale i.e. – 273.16ºC.

1.21 MICROSCOPIC AND MACROSCOPIC POINT OF VIEW

There are two points of views.

1. Microscopic or Statistical Thermodynamics.
2. Macroscopic or Classical Thermodynamics. From which the behaviour of matter or working of a system can be studied.

From microscopic point of view, it is considered that, the matter is not continuous, but it is made up of a large number of identical particles called molecules. For example, consider a gas in a cylinder as a system. Then this gas is made up of number of molecules. Each molecule of a gas, at a given instant, has certain position, energy, etc., and for each molecule these change very rapidly because of the collisions (strikings). The behaviour of the gas is described by summing up the behaviours of each molecule. And for summing up the behaviour of molecules statistical methods are employed and hence it is also called as *statistical thermodynamics*.

In macroscopic point of view a certain quantity of matter is considered, and the events that are taking place at the molecular level are not taken into account. For example, let us consider a system of an IC Engine consisting of a charge in the engine cylinder. At any instant the system has certain volume depending upon the position of the piston, this volume is easily measurable.

Another quantity to describe the system is the pressure of the gas inside the cylinder. A pressure gauge can be used to measure the same. Similarly temperature, chemical composition etc. may be described. Thus, in macroscopic point of view the system will be described by large scale properties.

Though the approach based on microscopic point of view seems to be different, from that based on macroscopic point of view, but there exists a relationship between them. The relationship between macroscopic and microscopic point of view lies in the fact that the macroscopic properties are in fact the average properties of a large number of microscopic characteristics.

Hence, when both the methods are applied to a particular system, they give the same result.

Points to Remember:

⇨ The microscopic point of view postulates the existence of molecules, their movement and collisions and hence it is constantly being changed. Whereas the macroscopic properties can be measured and their existence can be felt by our senses.

1.22 QUASI-STATIC PROCESS

Quasi means nearly or almost. So quasi-static process means nearly static process or nearly stationary process, or a process which proceeds with extreme slowness.

Quasi-static process proceeds from one equilibrium state to another equilibrium state till the end of the process i.e. all the states passed during the process are all equilibrium states.

In each state the process deviates from the equilibrium state by a very small amount and immediately attains its equilibrium.

Let us consider a system consisting of a gas in a cylinder fitted with a piston as shown in Fig. 1.33 (a). Let *W* be the weight kept on the piston which just balances the upward force exerted by the gas. The system is initially in the equilibrium state. During initial condition, let the properties be P_1 V_1 and t_1. If the weight is removed then there will be an unbalanced force between the system and the surroundings and the piston will move up till it hits the stops. Then again the system comes to another equilibrium state described by the properties P_2, V_2, t_2 but the intermediate states passed through by the system are non-equilibrium states. In Fig. 1.33 (b) points (1) and (2) are initial and final equilibrium states and since the system is passing through the non equilibrium states, it is joined by dotted line. Such a process is called as *Irreversible process*.

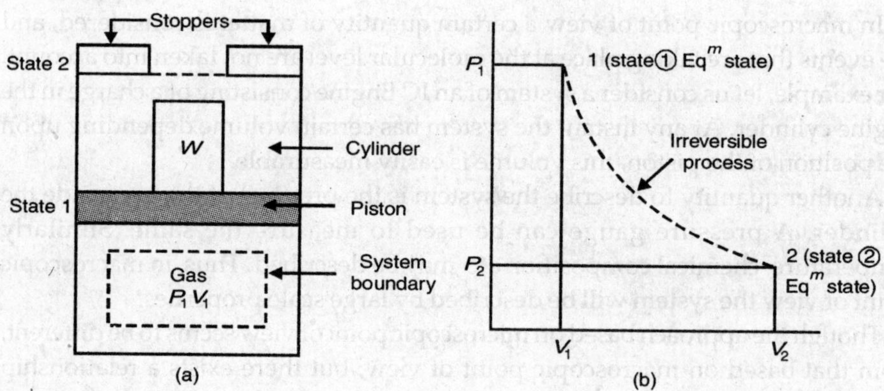

Fig. 1.33

Now if the single weight on the piston is made up of many very small pieces of weights as shown in the Fig. 1.33 (c) and these weights are removed one by one, then the departure of the state of the system from the thermodynamic equilibrium state will be very small. So, every state passed through by the system will be an equilibrium state.

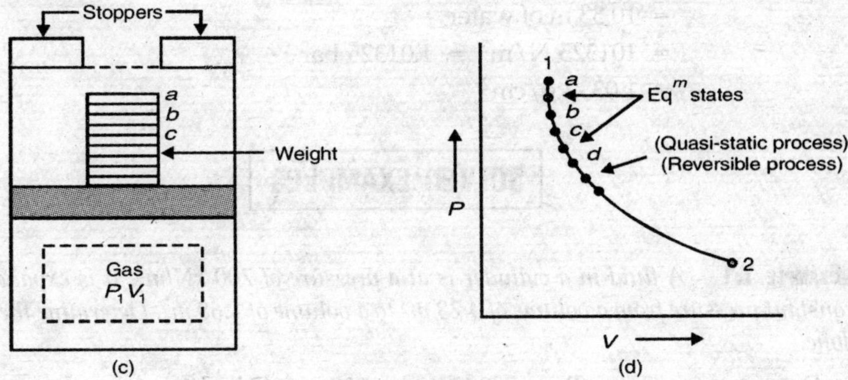

Fig. 1.33

A Quasi-static process is also called as *Reversible process* and is represented on the *PV*-diagram by a continuous line as shown in Fig. 1.33 (d).

In other words suppose when one weight say *a* is removed then the system will deviate from its equilibrium by a very small amount and immediately comes to an equilibrium state *a* (as shown in *PV*-diagram (d) and when weight *b* is removed, the system again deviates by a small amount and immediately comes to another equilibrium state *b*. Thus when these small weights are removed one by one the system proceeds from one equilibrium state to another equilibrium state till end, such a process which passes through all the equilibrium states is known as quasistatic process.

1.23 ENERGY IN TRANSIT

When energy flows from one system (or one body) to another system (other body), then it is referred as energy in transit.

In general heat and work are called energies in transit. The energy transfer is due to some energy potential, e.g.

(i) Mechanical energy or work energy are transferred due to difference in pressure potential.
(ii) Electrical energy is transferred due to different in voltage potential.
(iii) Heat energy is transferred due to difference in temperature potential.

LIST OF FORMULAE

Note: 1. Unit of work in SI units is Nm or joules
2. Unit of pressure in SI units is N/m^2 or Pa

3. 1 bar $= 10^5$ N/m^2 = 100 kN/m^2 = 100 kPa
4. 1 atm = 760 mm of Hg
= 10.33 m of water
= 101325 N/m^2 = 1.01325 bar
= 1.033 kg/cm^2

SOLVED EXAMPLES

EXAMPLE 1.1 *A fluid in a cylinder is at a pressure of 700 kN/m². It is expanded at constant pressure from a volume of 0.28 m³ to a volume of 1.68 m³. Determine the work done.*

Data: $P = 700 \text{ kN/m}^2 = 700 \times 10^3 \text{ N/m}^2$

Solution

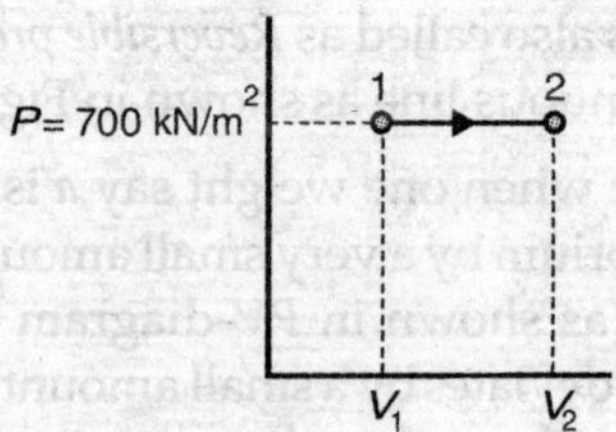

Fig. Ex. 1.1

Expanded at constant pressure,

$$V_1 = 0.28 \text{ m}^3 \quad V_2 = 1.68 \text{ m}^3$$

We know that, work done $= \int_{V_1}^{V_2} P dV$

i.e. Work done $= W_{1-2} = P\,(V_2 - V_1)$ since $(P = C)$

Note : Units $\frac{\text{N}}{\text{m}^2} \times \text{m}^3 = \text{Nm}$

$$= 700 \times 10^3 (1.68 - 0.28) \text{ Nms}$$

$$= 700 \times 10^3 \times 1.4 \text{ Nm or Joules}$$

$$= 980000 \text{ J} = 9.8 \times 10^5 \text{ J}$$

$$\mathbf{W_{1\text{-}2} = 980\,kJ = 0.98\,MJ}$$

EXAMPLE 1.2. *0.112 m^3 of gas has a pressure of 138 kN/m^2. It is compressed to 690 kN/m^2, according to the law $PV^{1.4} = C$. Determine the new volume of the gas.*

Data: The gas is compressed according to the $PV^{1.4} = C$

$P_1 = 138$ kN/m^2 $\quad P_2 = 690$ kN/m^2

$V_1 = 0.112$ m^3 $\quad V_2 = ?$

Solution

We know that polytropic process is given by $PV^n = C$

i.e. $$PV^n = P_1 V_1^n = P_2 V_2^n = C$$

$\therefore$ $$P_1 V_1^{1.4} = P_2 V_2^{1.4}$$

$$\frac{P_1}{P_2} = \left[\frac{V_2}{V_1}\right]^{1.4}$$

or $$\frac{V_2}{V_1} = \left[\frac{P_1}{P_2}\right]^{\frac{1}{1.4}}$$

or $$V_2 = V_1 \times \left[\frac{P_1}{P_2}\right]^{\frac{1}{1.4}}$$

$$= 0.112 \text{ m}^3 \times \left[\frac{138 \text{ kN/m}^2}{690 \text{ kN/m}^2}\right]^{1/1.4}$$

$$= 0.112 \text{ m}^3 \times (0.2)^{1/1.4} = 0.112 \times 0.3168$$

$$\mathbf{V_2 = 0.03555 \text{ m}^3}$$

EXAMPLE 1.3 *0.014 m^3 gas at a pressure of 2070 kN/m^2 expands to a pressure of 207 kN/m^2, according to the law $PV^{1.35} = C$. Determine the work done by the gas during the expansion.*

Data: Gas expands according to the law $PV^{1.35} = C$

$V_1 = 0.14$ m^3 $\quad P_1 = 2070$ kN/m^2

$P_2 = 207$ kN/m^2 $\quad$ *Work done = ?*

Solution

The work done during a polytropic process is given by,

$$W_{1-2} = \frac{P_1V_1 - P_2V_2}{n-1} = \frac{P_2V_2 - P_1V_1}{1-n}$$

Since V_2 is not given, to determine V_2,
We know that since it is a polytropic process,

$$P_1V_1^n = P_2\,V_2^n$$

$$\frac{P_1}{P_2} = \left[\frac{V_2}{V_1}\right]^n$$

$$V_2 = V_1\times\left[\frac{P_1}{P_2}\right]^{\frac{1}{n}} = 0.014\,\text{m}^3\left[\frac{2070\ \text{kN/m}^2}{207\ \text{kN/m}^2}\right]^{\frac{1}{1.35}}$$

$$= 0.014\times5.5048\ \text{m}^3$$

$$\mathbf{V_2 = 0.0771\ m^3}$$

$$\therefore\quad \text{Work done} = W_{1-2} = \frac{P_1V_1 - P_2V_2}{n-1}$$

$$= \frac{2070\times10^3\text{N/m}^2\times0.014\text{m}^3 - 207\times10^3\text{N/m}^2\times0.077\text{m}^3}{1.35-1}$$

$$= \frac{28980-15959.7}{0.35} = \frac{13020.3}{0.35} = 37200.857\ \text{Nm}$$

$$= 37.2\times10^3\ \text{Nm or Joules}$$

$$\mathbf{W_{1-2} = 37.2\ kJ}$$

EXAMPLE 1.4 *A gas is compressed hyperbolically (Isothermally) from a pressure and volume of 100 kN/m² and 0.056 m³ respectively to a volume of 0.007 m³. Determine the final pressure and the work done on the gas.*

Data: Since the gas is compressed Isothermally, $PV = C$ i.e. $P_1V_1 = P_2V_2$

$$P_1 = 100\ kN/m^2 = 100\times10^3\ N/m^2$$

$P_2 = ?$ *Work done* $= ?$

$V_1 = 0.056\ m^3$ $V_2 = 0.007\ m^3$

Solution

Since it is an Isothermal process $P_1V_1 = P_2V_2$

$$P_2 = \frac{P_1V_1}{V_2}$$

$$\therefore \quad P_2 = \frac{100\times1000\ \text{N/m}^2\times0.056\ \text{m}^3}{0.007\ \text{m}^3}$$

$$P_2 = \frac{5600}{0.007}$$

$$P_2 = 8\times10^5\ \text{N/m}^2$$

$$\mathbf{P_2 = 8\times10^2\ kN/m^2 = 800\ kN/m^2}$$

We know that, Work done $= P_1V_1 \ln \dfrac{V_2}{V_1}$

$$= 100\times1000\times0.056\times\ln\left(\frac{0.007}{0.056}\right)$$

$$= 100\times1000\times0.056\times-2.079$$

$$W_{1-2} = -11644.973\ \text{Nm} = -11.65\ \text{kNm or kJ}$$

∴ The work done on the gas = 11.65 kJ

EXAMPLE 1.5 *A non-flow reversible, quasi static process occurs for which pressure $P = [V^2 + 8/V]$ bar. Determine the work done if volume changes from 1 m³ to 3 m³.*

Data: $P = [V^2 + 8/V]\ \text{bar} = [V^2 + 8/V]\times10^5\ \text{N/m}^2$

$V_1 = 1\ m^3 \quad V_2 = 3\ m^2, \quad$ *work done = ?*

Solution

We know that, work done in a closed system or displacement work done

$$W_{1-2} = \int PdV$$

$$W_{1-2} = \int_{V_1=1}^{V_2=3} PdV, \text{ subtitute for } P,$$

$$= \int_1^3 \left[V^2 + \frac{8}{V}\right]\times10^5\,dV = 10^5\int_1^3\left[V^2 + \frac{8}{V}\right]dV$$

$$= 10^5\left[\frac{V^3}{3} + 8\ln V\right]_1^3 = 10^5\left[\frac{3^3-1^3}{3} + 8\ln\frac{3}{1}\right]$$

$$= 10^5\left[\frac{27-1}{3}+8\times1.0986\right]=10^5\left[\frac{26}{3}+8.789\right]$$

$$\mathbf{W_{1-2} = 1745556.5\ Nm = 17.455\times10^5\ Nm\ or\ Joules}$$

EXAMPLE 1.6 *In a certain thermodynamic process of an ideal gas, the volume changes from 0.2 m^3 to 0.5 m^3 while pressure change occurs according to the law P = 1500 $\left[\frac{V}{100}+1\right]$. Where P is in N/$m^2$ and V is in m^3, then find out the work done by the gas in kJ.*

Data: $V_1 = 0.2\,m^3$; $V_2 = 0.5\,m^3$ and $P = 1500\left[\frac{V}{100}+1\right] N/m^2$

Solution

We know that, work done

$$= W_{1-2} = \int PdV; \text{ substitute for } P$$

$$= \int_{0.2}^{0.5} 1500\left[\frac{V}{100}+1\right]dV = 1500\times\int_{0.2}^{0.5}\left[\frac{V}{100}+1\right]dV$$

$$= 1500\times\left[\frac{V^2}{2\times100}+V\right]_{0.2}^{0.5} = 1500\times\left[\frac{0.5^2-0.2^2}{200}+(0.5-0.2)\right]$$

$$= 1500\times(0.00105+0.3) = 451.575\ Nm$$

$$\mathbf{W_{1-2} = 0.4515\ kJ}$$

EXAMPLE 1.7 *Show that for Van-der-Waals equation,*

$$\left[P+\frac{a}{V^2}\right](V-b) = mRT$$

The work done is given by, $mRT\log_e\frac{V_2-b}{V_1-b} - a\left[\frac{1}{V_1}-\frac{1}{V_2}\right]$*, where* V_1 *and* V_2 *are initial and final volumes.*

Solution

From the given equation i.e.

$$\left[P+\frac{a}{V^2}\right](V-b)=mRT$$

$$P = \frac{mRT}{V-b}-\frac{a}{V^2}$$

$$\therefore \text{Work done} = W_{1-2} = \int PdV = \int_1^2\left[\frac{mRT}{V-b}-\frac{a}{V^2}\right]dV$$

$$= mRT\int_1^2 \frac{1}{V-b}dV - \int_1^2 \frac{a}{V^2}dV$$

$$\mathbf{W_{1-2} = mRT \log_e \frac{V_2 - b}{V_1 - b} - a\left[\frac{1}{V_1}-\frac{1}{V_2}\right]}$$

EXAMPLE 1.8 *A gas in a cylinder and piston arrangement comprise the system. Gas expands frictionlessly from 1.5 m³ to 2 m³ while receiving 2 Nm of work from a Paddle wheel. Pressure of gas remains constant at 6 N/m². Determine the net work done by the system.*

Solution

Work done by the gas on the piston,

W_{gas} = Force × Displacement = $F \times dl = P \times A \times dl = P \times dV$

The total work done by the gas, when the volume changes from V_1 to V_2 and the pressure remaining constant = $\int PdV$

i.e. $$W_{gas} = \int_1^2 PdV$$

$$= P\int_{1.5}^2 dV \quad \text{since } P \text{ is constant}$$

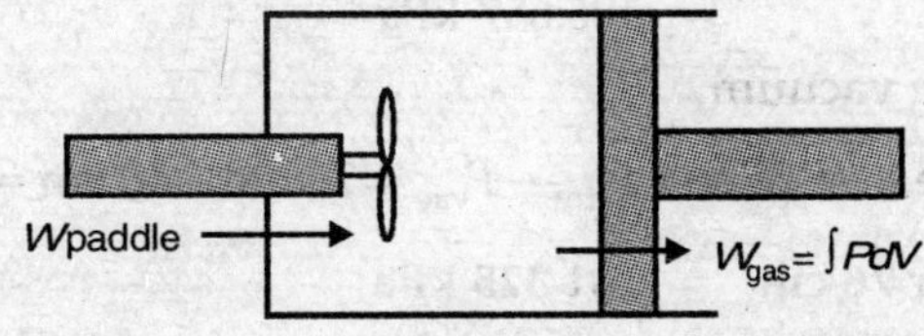

Fig. Ex. 1.8

$$= 6[V]_{1.5}^{2}$$

$$= 6[2-1.5] = 3 \text{ Nm}$$

But gas receives 2 Nm of work from the paddle wheel,

∴ **Net work done by the system = 3–2 = 1 Nm**

EXAMPLE 1.9 *Convert the following readings of pressure to kPa assuming that the barometer reads 760 mm of Hg*

(a) 80 cm of Hg

(b) 40 cm of Hg vac

(c) 1.5 mt of water

(d) 5.2 bar

Solution

Let us assume ρ of Hg $= 13596 \text{ kg/m}^3$

and $g = 9.806 \text{ m/sec}^2$, if the values are not given

Let Z_o = Barometric pressure = 760 mm of Hg

∴ Pressure exerted due to 760 mm of Hg column in the barometer,

$$P_{atm} = \rho gh \text{ or } \rho g Z_o = 13596 \times 9.806 \times \frac{760}{1000}$$

$$= 101325.01 \text{ N/m}^2 \text{ or Pa} = 101.325 \text{ kN/m}^2 \text{ or kPa}$$

(a) Pressure of 80 cm of Hg

For 76 cm = 101.325 kPa

∴ 80 cm = ?

$$\mathbf{\frac{80}{76} \times 101.325 = 106.6578 \text{ kPa}}$$

or

We know that, pressure exerted due to a column of fluid = $\rho.g.Z$

$$= 13596 \times 9.806 \times 0.8 = 106657.8 \text{ N/m}^2$$

$$= 106.657 \text{ kPa}$$

(b) 40 cm of Hg vacuum

In this case, $P_{abs} = P_{atm} - P_{vac} = 76 \text{ cm} - 40 \text{ cm} = 36 \text{ cm}$

Since for 76 cm – 101.325 kPa

∴ 36 cm – ?

$$\frac{36}{76} \times 101.325 = 47.996 \text{ kPa}$$

(c) Pressure due to 1.5 mt of H_2O column on kPa

$$= \rho_w \cdot g \cdot Z_w = 1000 \times 9.806 \times 1.5$$

$$= \mathbf{14709 \text{ N/m}^2 \text{ or Pa} = 14709 \text{ kPa}}$$

(d)

$$5.2 \text{ bar} = 5.2 \times 10^5 \text{N / m}^2 \text{or Pa}$$

$$= \mathbf{5.2 \times 10^2 \text{ kPa} = 520 \text{ kPa}}$$

Note: Pressure of 1 atm. equals 760 mm of Hg or 101325 N/m^2. These are standard values. To get this valve of 101325 N/m^2, we have to use ρ of Hg = 13596 kg/m^3 and g = 9.806 m/sec^2.
If we use ρ of Hg = 13.6 × 10^3 kg/m^3 and g = 9.81 m/sec^2 then we get P_{atm} = 101396 N/m^2.

EXAMPLE 1.10 *Convert the following readings of pressure into kPa (Abs) assuming barometric pressure of 750 mm of Hg.*

(i) 55 cm of Hg (Abs)
(ii) 3.3 Atm (Gauge)
(iii) 3.2 m of H_2O (Gauge)
(iv) 4.6 bar (Abs)

Solution

For

$$P_{ptm} = \rho_{Hg} \cdot g \cdot Z_o = \frac{13596 \times 9.81 \times 0.75}{1000}$$

$$= 100.32 \text{ kPa}$$

(i) 55 cm of Hg (Abs)

$$P_{abs} = 13596 \times 9.81 \times 0.55 = 73357 \text{ N/m}^2$$

$$= 73.35 \text{ kPa}$$

(ii) 3.3 atm (Gauge)

$$\text{Since 1 atm} = 101325 \text{ N/m}^2$$

$$\therefore \quad 3.3 \text{ atm} = 3.3 \times 101325 \text{ N/m}^2$$

$$\therefore \quad P_{abs} = P_{atm} + P_{gauge} = 100.032 + \frac{3.3 \times 101325}{1000}$$

$$= 434.405 \text{ kPa}$$

(iii) 3.2 m of H_2O (Gauge)

$$P_{abs} = P_{atm} + P_{gauge} = 100.032 + \frac{100 \times 9.81 \times 3.2}{1000}$$

$$= 131.42 \text{ kPa}$$

(iv) 4.6 bar (Abs)

Since $\quad 1 \text{ bar} = 10^5 \text{N/m}^2 = 10^2 \text{ kN/m}^2$

$\therefore \quad 4.6 \text{ bar} = 4.6 \times 100 \text{ kPa}$

$$= \mathbf{460 \text{ kPa}}$$

EXAMPLE 1.11 *A manometer shows a reading of 50 cm of water. If atmospheric pressure is 763 mm of Hg. Find absolute pressure in kPa. Take ρ of Hg = 13.6 × 10³ kg/m³, g = 9.81 m/sec².*

Solution

$$P_{abs} = P_{atm} + P_{gauge}$$

$$P_{atm} = \rho \cdot g \cdot Z_o = 13.6 \times 10^3 \frac{\text{kg}}{\text{m}^3} \times 9.81 \frac{\text{m}}{\text{sec}^2} \times 0.765 \text{ m}$$

$$= 102063.24 \text{ N/m}^2 \text{ or Pa}$$

$$= 102.063 \text{ kPa}$$

and $P_{manometer}$ or $P_{gauge} = \rho_w \cdot g \cdot Z_w = 1000 \times 9.81 \times 0.50 = 4905 \text{ kPa} = 4.905 \text{ kPa}$

$$P_{abs} = P_{atm} + P_{gauge}$$

$$P_{abs} = 102.063 + 4.905 = 106.968 \text{ kPa}$$

EXAMPLE 1.12 *A vacuum gauge mounted on the condenser reads 70 cm of Hg. If atmospheric pressure is 101.325 kPa. Find absolute pressure in kPa.*

Take $\rho_{Hg} = 13.6 \times 10^3$ kg/m³ and g = 9.81 m/sec²

Solution

We know that $\quad P_{abs} = P_{atm} - P_{vac}$

To find $\quad P_{vac} = \rho \cdot g \cdot Z = 13.6 \times 10^3 \times 9.81 \times 0.70$

$$= 93391.2 \text{ Pa} = 93.3912 \text{ kPa}$$

$$\mathbf{P_{abs} = 101.325 - 93.3912 = 7.9338 \text{ kPa}}$$

EXAMPLE 1.13 A vacuum gauge connected to the tank reads 30 kPa at a location where a barometric reading is 755 mm of Hg. Determine the absolute pressure in the tank. Take density of Hg = 13590 kg / m³.

Solution

$$P_{vac} = 30 \text{ kPa}$$

$$Z_o = 755 \text{ mm of Hg}$$

$$P_{abs} = ?$$

$$P_{abs} = P_{atm} - P_{vac} \quad \ldots(1)$$

To find P_{atm}

$$P_{atm} = \rho_{Hg} \cdot g \cdot Z_o = 13590 \times 9.81 \times 0.755$$

$$P_{atm} = 100655.01 \text{ N/m}^2$$

$$P_{atm} = 100.655 \text{ kPa} \quad \ldots(2)$$

∴ From (1)

$$P_{abs} = P_{atm} - P_{vac} = 100.655 - 30$$

$$\mathbf{P_{abs} = 70.655 \text{ kPa}}$$

EXAMPLE 1.14 *The pressure of a gas in a pipe line measured with Hg–manometer having one limb open to the atmosphere. The level in the open limb is 562 mm higher than the arm connected to the gas pipe. Calculate the gas pressure. The barometer reads 761 mm of Hg, the acceleration due to gravity is 9.79 m/sec² and the density of Hg is 13640 kg /m³.*

Data: $z = 0.526 \ m \qquad Z_o = 0.761 \ m$

$g = 9.79 \ m/sec^2 \qquad \rho_{Hg} = 13640 \ kg/m^2$

Solution

Note: Since the plane 1–1′ is horizontal plane, and is in the same continuous static mass of liquid, the pressure heads at the marked (*) points in figure will be equal.

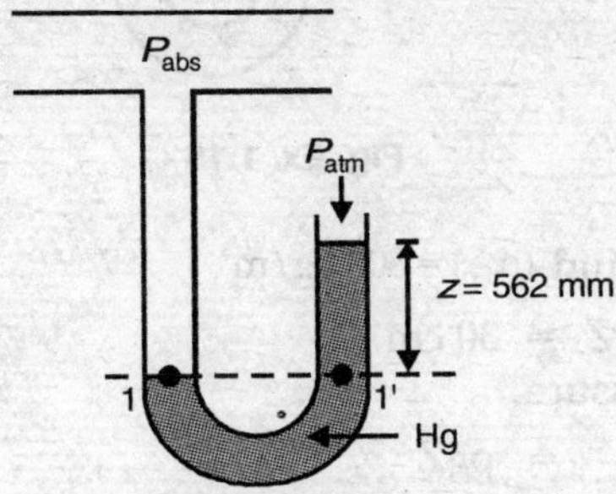

Fig. Ex. 1.14

Therefore the pressure along a horizontal line 1–1' is same. Balancing the pressure values about the line 1–1' we get,

$$P_{abs} = P_{atm} + \rho \cdot g \cdot Z \quad \text{...(1)}$$

But $\quad P_{atm} = \rho \times g \times Z_o$

$$= 13640 \times 9.79 \times 0.761$$

$$= 101620.59 \text{ N/m}^2$$

$$= 101.621 \text{ kN/m}^2$$

The pressure exerted due to the column of Hg of height $Z = \rho \times g \times Z$

$$= 13640 \times 9.79 \times 0.562 = 75047 \text{ N/m}^2$$

$$= 75.047 \text{ kN/m}^2$$

∴ From (1), (2) and (3)

$$P_{abs} = \mathbf{101.621 + 75.047 = 176.668 \text{ kN/m}^2}$$

EXAMPLE 1.15 *A U-tube manometer contains a liquid having a density of 800 kg/m³. When the manometer is connected to a gas pipe, the level in the open arm is 30 cm higher than the level in the column connected to the gas pipe. Find the absolute pressure of the gas in bar, if the barometric pressure is 75 cm of Hg.*

Solution

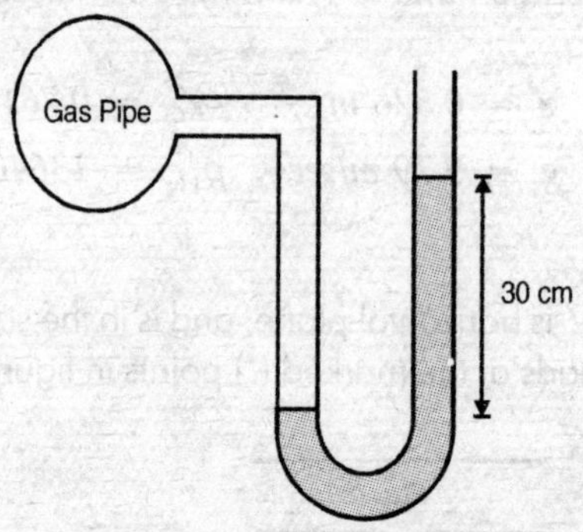

Fig. Ex. 1.15

We have, Density of liquid $(\rho_{liq}) = 800 \text{ kg/m}^3$

$$Z = 30 \text{ cm}$$

∴ We have, gauge pressure

$$= \rho g Z$$

$$= 800 \times 9.81 \times 0.3$$

$$= 2354.4 \text{ N/m}^2 = 0.023544 \text{ bar}$$

Now $\quad P_{atm} = \rho . g . Z = 75 \text{ cm of Hg}$

$$= 13.596\times10^3\times9.81\times0.75$$

$$= 99991.78\ \text{N/m}^2$$

$$= 0.9999\ \text{bar}$$

Also, $P_{abs} = P_{gauge} + P_{atm} = 0.023544 + 0.9999$

$$P_{abs} = \mathbf{1.234618\ bar}$$

EXAMPLE 1.16 *A U-tube manometer is connected to a gas pipe. The level of the liquid in the manometer arm open to the atm. is 17 cm lower than the level of liquid in the arm connected to the gas pipe. The liquid in the manometer has specific gravity of 0.8. Find the absolute pressure of the gas if the barometer reads760 mm of Hg. Take $\rho_{Hg} = 13596$ and $g = 9.806$.*

Data: $Z = 0.17\ m$ $\quad P_{abs} = ?$

$Z_o = 0.76$ $\quad$ *Sp. gr.* $= 0.8$

Solution

Note: Specific gravity $= \dfrac{\rho \text{ of given liquid}}{\rho \text{ of water}}$

$\therefore$ ρ of given liquid $= 0.8\times10^3\ \text{kg/m}^3$

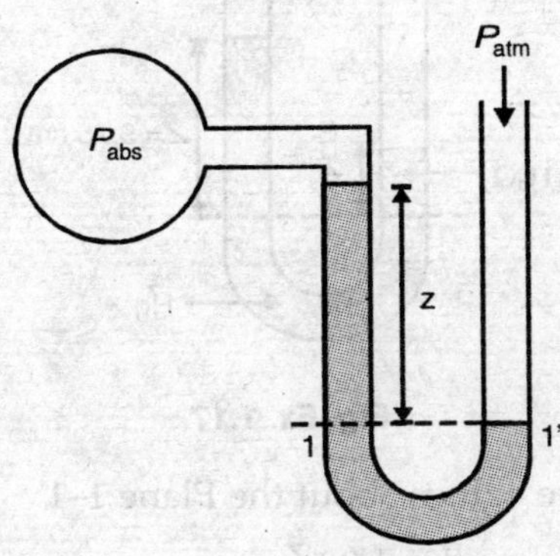

Fig. Ex. 1.16

Balancing the pressure values about the plane 1–1'

$$P_{abs} + \rho g Z = P_{atm}$$

$$P_{abs} = P_{atm} - \rho g Z$$

(a) $\quad P_{atm} = \rho g Z_o \quad$ ...(1)

$$= 13596\ \text{kg/m}^3\times9.806\ \text{m/sec}^2\times0.7\ \text{m}$$

$$= 101325\ \text{N/m}^2$$

and $\rho g Z = 0.8\times10^3\times9.806\times0.17 = 1333.616\ \text{N/m}^2$

$\therefore$ $P_{abs} = 101325 - 1333.616 = 99991.384\ \text{N/m}^2$

$$\mathbf{P_{abs} = 0.999913\ bar}$$

EXAMPLE 1.17 *A U-tube manometer with one arm open to atmosphere is used to measure the pressure of steam flowing through a pipe. Hg is used in the manometer. The height of the column open to the atmosphere is 9.75 cm greater than the Hg level in the arm connected to the steam pipe. Steam condenses in the manometer on the steam side.*

The column of condensate is 3.4 cm. The atmospheric pressure is 76 cm of Hg. Find the absolute pressure of steam. Take,

$$g = 9.806\ m/sec^2 \quad and \quad \rho_{Hg} = 13596\ kg/m^3$$

$$P_{abs} = P\ of\ steam = ?$$

Data: $P_{abs} = ?$ $\quad Z = 9.75\ cm$

Solution

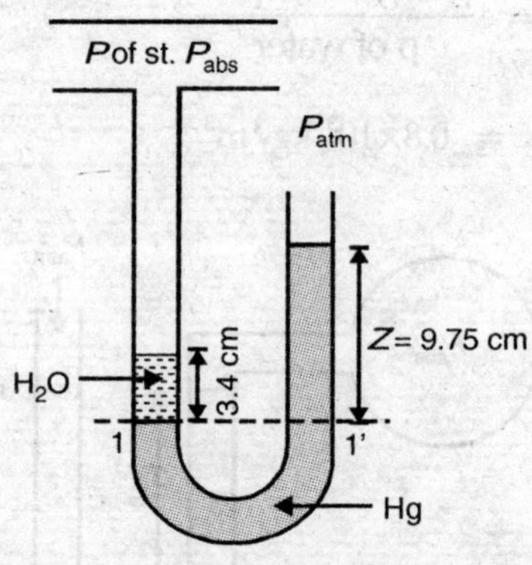

Fig. Ex. 1.17

Balancing the pressure values about the Plane 1–1'

$$P_{abs} + \rho_w g Z_w = P_{atm} + \rho g Z$$

Now to find,

(a) $\rho_w g Z_w = 1000\times9.806\times0.034$

$$= 333.404\ \text{N/m}^2 \quad \text{...(2)}$$

(b) $P_{atm} = \rho\, g Z_o = 13596\times9.806\times0.76$

$$= 101325\ \text{N/m}^2 \quad \text{...(3)}$$

(c) $$\rho g Z = 13596 \times 9.806 \times 0.0975$$

$$= 12998.932 \text{ N/m}^2 \quad \text{...(4)}$$

∴ Now substituting the values of Eqs (2), (3) and (4) in Eq. (1) we get,

$$P_{abs} + 333.404 = 101325 + 12998.932$$

$$P_{abs} = 114323.93 - 333.404 = 113990.53 \text{ N/m}^2$$

$$\mathbf{P_{abs} = 1.139905 \text{ bar}}$$

EXAMPLE 1.18 *A Hg manometer is used to measure the pressure of steam in the mains. (i.e. main steam pipe). Its one arm having higher level of Hg is open to atmosphere at 98 kPa The other arm is connected to steam pipe. The manometer records a level difference of 60 cm of Hg. On the lower level of Hg, 4 cm of water column accumulates due to condensation of steam. Find the absolute pressure of steam in bar.*

Take $g = 9.7 \text{ m/sec}^2$ *and* $\rho_{Hg} = 13.69 \text{ gm/cc}$

$\rho_w = 1000 \text{ kg/m}^3$

Data: $g = 9.7 \text{ m/sec}^2$

$\rho_{Hg} = 13.69 \text{ gm/cc} = 13.69 \times 10^3 \text{ kg/m}^3$

$\rho_{water} = 1000 \text{ kg/m}^3$

Solution

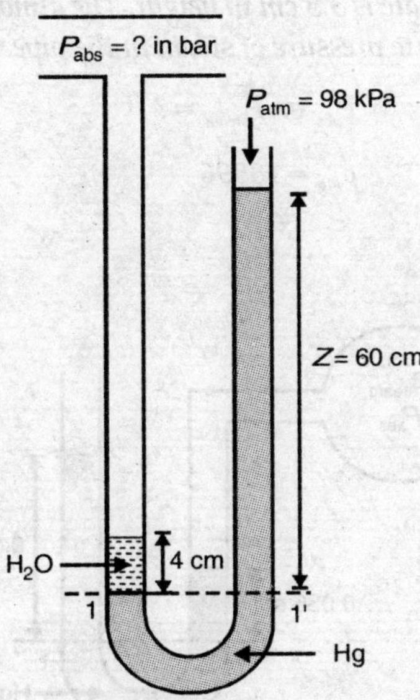

Fig. Ex. 1.18

Now balancing pressure values the plane 1–1'

$$P_{abs} + \rho_w g Z_w = P_{atm} + \rho g Z \quad \text{...(1)}$$

Now to find,

(a) $\rho_w g Z_w = 1000 \times 9.7 \times 0.04$

$= 388 \text{ N/m}^2$

$= 0.388 \text{ kN/m}^2$ or kPa ...(2)

(b) $\rho g Z = 13.69 \times 10^3 \times 9.7 \times 0.6$

$= 79675.8 \text{ N/m}^2 = 79.6758 \text{ kPa}$

∴ From (1) $P_{abs} = P_{atm} + \rho g Z - \rho_w g Z_w$

$= 98 + 79.68 - 0.388$

$= 177.68 - 0.388$

$= 177.292$ kPa

Since 1 bar $= 10^5 \text{ N/m}^2 = 100$ kPa

$\mathbf{P_{abs} = 1.77292}$ **bar**

EXAMPLE 1.19 *A U-tube manometer with one arm open to atmosphere is used to measure the pressure of steam flowing through a pipe. Mercury is used as a manomeric fluid. The height of the column of mercury in the open end is 10 cm greater than the mercury level in the arm connected to the steam pipe. Steam condenses in the manometer on the steam side. The column of condensate is 3.5 cm in height. The atmospheric pressure is 760 mm mercury. What is the absolute pressure of steam in the pipe in bar ?*

Data: $Z_o = 0.76\ m$ P_{abs} *or* $P_{steam} = ?$

$g = 9.806$ $\rho_{Hg} = 13596$

Solution

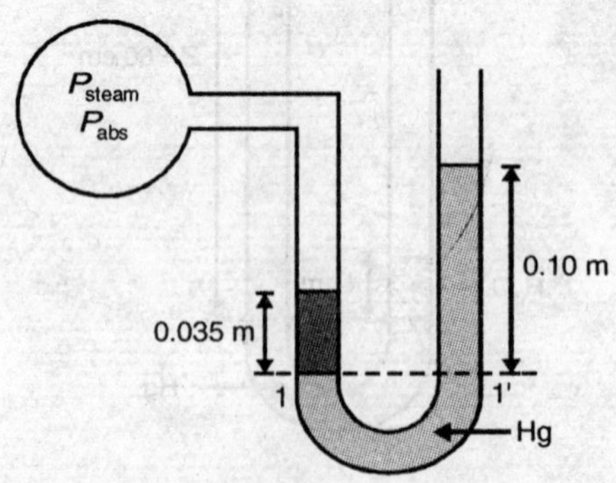

Fig. Ex. 1.19

$$P_{abs} + \rho_w g Z_w = P_{atm} + \rho_{Hg}\ g\ Z_{Hg}$$

$$P_{atm} = \rho_{Hg}\ g\ Z_o$$

$$= 13596 \times 9.806 \times 0.76$$

$$= 101325 \frac{N}{m^2}$$

$$\rho_w g Z_w = 1000 \times 9.806 \times 0.035$$

$$= 333.404 \text{ Pa}$$

$$\rho_{Hg}\ g\ Z_{Hg} = 13596 \times 9.806 \times 0.10$$

$$= 12998.93 \text{ Pa}$$

$\therefore$ From (1) $\quad P_{abs} = P_{atm} + \rho_{Hg} g Z_{Hg} - \rho_w g Z_w$

$$= 101325 + 12998.9 - 333.404$$

[From (1), (2), (3) and (4)]

$$= 113990.52 \text{ Pa}$$

$$= 1.139 \text{ bar}$$

$$\mathbf{P_{abs} = 1.14 \text{ bar}}$$

EXAMPLE 1.20 *A U-tube manometer with one arm open to atmosphere is used to measure the pressure of kerosene vapour passing in a pipe. Hg level in the open arm is 10 cm greater than that in the arm connected to the kerosene pipe. There is a condensation of kerosene vapour on the Hg column of 5.1 cm height. If the atmospheric pressure is 75.5 cm of Hg. Find the absolute pressure of vapour in kPa. Assume specific gravity of kerosene as 0.8 and that of Hg as 13.6.*

Solution

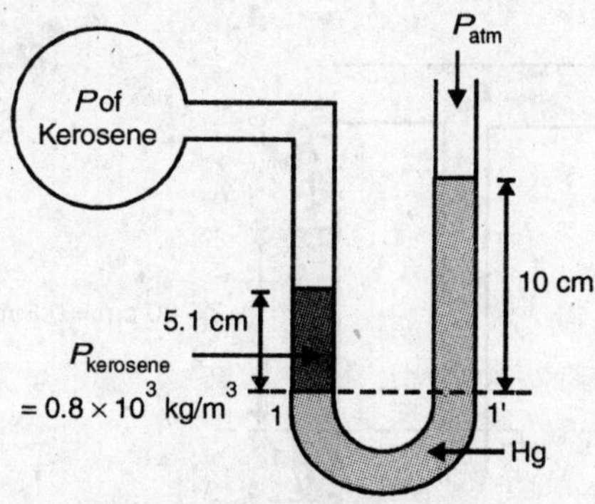

Fig. Ex. 1.20

Balancing the pressure values @ the plane 1–1'

$$P_{\text{abs of Kero. vap}} + \rho_{\text{kero}} g Z_{\text{kero}} = P_{\text{atm}} + \rho_{\text{Hg}} g Z_{\text{Hg}} \quad ...(1)$$

Now to find

(a) $\rho_{\text{kero}} \cdot g \cdot Z_{\text{kero.}} = 800 \times 9.81 \times 0.051 \text{ m}$

$= 400.02 \text{ N/m}^2 \quad ...(2)$

(b) $P_{\text{atm}} = \rho_{\text{Hg}} g Z_o$

$= 13600 \times 9.81 \times 0.755 \text{ m}$

$= 100726.08 \text{ N/m}^2 \quad ...(3)$

(c) $\rho_{\text{Hg}} g Z_{\text{Hg}} = 13600 \times 9.81 \times 0.1 \text{ m}$

$= 13341.6 \text{ N/m}^2 \quad ...(4)$

Sustituting Eqs (2), (3) and (4) in Eq. (1) we get,

$$P_{\text{abs}} + 400.02 = 100729.08 + 13341.6$$

$$P_{\text{abs}} = 113670.68 \text{ N/m}^2$$

$$\mathbf{P_{abs} = 113.67068 \ kP}$$

EXAMPLE 1.21 *A U-tube manometer having equal arms with one arm open to atmosphere is used to measure the pressure of a gas passing through a pipe line. The fluid used in the manometer has a specific gravity of 0.8, while the height of the fluid column in the arm open to the atmosphere is 30 cm greater than the height of fluid column in the arm connected to the gas pipe line. Determine, the absolute pressure of the gas in pipe line.*

If the fluid used were Hg, what would be the difference of height of Hg-column in the 2-arms.

Take $\rho_{Hg} = 13596$ *and* $g = 9.806 \text{ m/sec}^2$

Solution

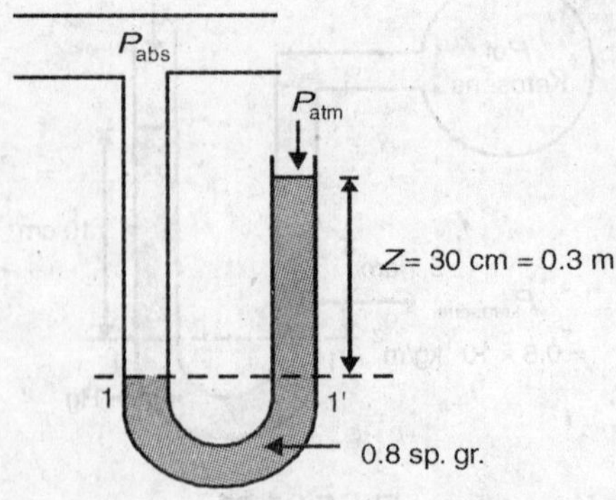

Fig. Ex. 1.21

ρ of fluid used in the manometer = $0.8 \times 1000 \text{ kg/m}^3$

$$P_{abs} = P_{atm} + \rho g Z$$

Let the barometer reads 760 cm of Hg

$$P_{atm} = \rho g Z_o$$
$$= 13596 \times 9.806 \times 0.76$$
$$= 101325 \text{ N/m}^2$$

and $$\rho g Z = 0.8 \times 10^3 \times 9.806 \times 0.3$$
$$= 2353.44 \text{ N/m}^2$$

∴ $$P_{abs} = 101325 + 2353.44$$
$$= 1013678.44 \text{ N/m}^2$$

$$\mathbf{P_{abs} = 1.0367844 \ \ bar}$$

If Hg is used, then to find Z,

(i.e. to produce the same absolute pressure, we have to find the equivalent column of Hg)

$$P_{abs} = P_{atm} + rgZ$$
$$103678.44 = 101325 + 13596 \times 9.806 \times Z$$
$$Z = \frac{103678.44 - 101325}{13596 \times 9.806}$$
$$= 0.01765 \text{ mt of Hg}$$

$$\mathbf{Z = 1.765. \text{ cm of Hg}}$$

EXAMPLE 1.22 *A vertical composite liquid column with its upper end exposed to atmospheric pressure, comprise of 45 cm of Hg (sp. gr. 13.6); 65 cm of water and 80 cm of oil (sp. gr. 0.8). Calculate the pressure in bar,*

(i) At the bottom of the column.

(ii) At the inter surface of oil and water.

(iii) At the inter surface of water and Hg.

Assume atmospheric pressure as
100 kPa and g = 9.81 m/sec²

Solution

Since it is given P_{atm} = 100 kPa i.e. equal to 1 bar

(Since $10^5 \text{ N/m}^2 = 1 \text{ bar} = 100 \text{ kPa}$)

(i) $$P_{\text{at bottom}} = P_{atm} + P_{oil} + P_{water} + P_{Hg}$$

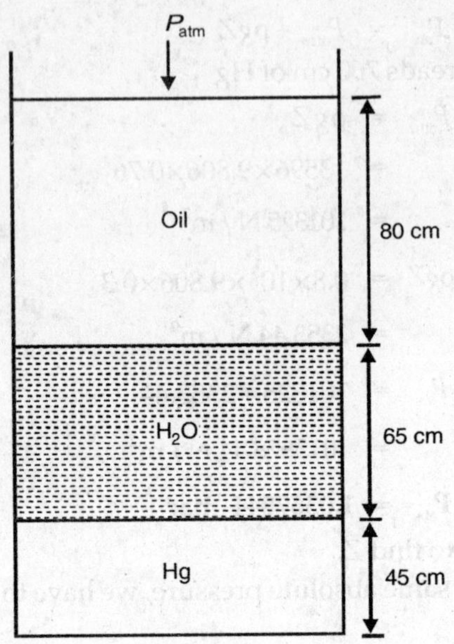

Fig. Ex. 1.22

$$= 1+\rho_{oil}gZ_{oil}+\rho_w gZ_w+\rho gZ$$

$$= 1+\frac{0.8\times1000\times9.81\times0.8}{10^5}+\frac{1000\times9.81\times0.65}{10^5}+\frac{13.6\times10^3\times9.81\times0.45}{10^5}$$

Note: If ρ is in kg/m^3, g is in m/sec^2, Z is in m, then the resultant unit will be N/m^2.

$\therefore$ To get in bar divide by 10^5

(i) $P_{\text{at bottom}} = 1+0.062784+0.063765+0.6$

$\mathbf{P_{\text{at bottom}} = 1.72692\ bar}$

(ii) Pressure at the interface of oil and water

$= 1+0.062784 = 1.062784$ bar

(iii) Pressure at the intersurface of water and Hg $= 1+0.062784+0.063765$

$= 1.1265$ bar

EXAMPLE 1.23 *An air tank has three manometers connected to it. The fluids in them are oil (Specific Gravity = 0.8) water and mercury (Specific Gravity = 13.6). If the absolute pressure in the tank is 1.2 bar and the barometer reads 760 mm of Hg. Estimate the height of fluid in each of manometer.*

Data: Fluids in the manometers are oil (Sp. Gr. = 0.8) water and mercury (Sp. Gr. = 13.6)

$$P_{abs} = 1.2 \text{ bar}$$

$$\text{Barometric reading} = 760 \text{ mm of Hg}$$

Height of fluid in each of manometer = ?

Solution

We know that, $P_{abs} = P_{atm} + P_{gauge}$

$$P_{atm} = \rho_{Hg} g Z_o = 13.6 \times 10^3 \times 9.81 \times 0.76$$

$$= 101396.16 \text{ N}$$

$$P_{abs} = 1.2 \text{ bar} = 1.2 \times 10^5 \text{ N/m}^2$$

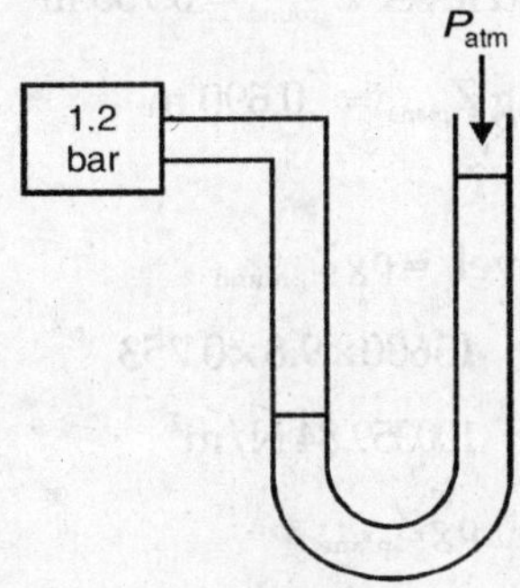

Fig. Ex. 1.23

Case I: For Oil Manometer

(a) $$\text{Sp. Gravity} = \frac{\rho \text{ of given fluid}}{\rho \text{ of water}}$$

$$0.8 = \frac{\rho_{oil}}{1000} \qquad \therefore \rho_{oil} = 800 \text{ kg/m}^3$$

$$\therefore \quad P_{gauge} = \rho_{oil} g Z_{oil} = 800 \times 9.81 \times Z_{oil}$$

Substituting these values in Eq. (1) we get,

$$1.2 \times 10^5 = 101396.6 + 800 \times 9.81 \times Z_{oil}$$

$$\therefore \quad \mathbf{Z_{oil} = 2.37 \text{ m}}$$

Case II: For Water Manometer

$$P_{abs} = P_{atm} + \rho_w g Z_w$$

$$1.2 \times 10^5 = 101396.6 + 1000 \times 9.81 \times Z_w$$

$$Z_w = 1.89 \text{ m}$$

Case III: For Hg Manometer

$$P_{abs} = P_{atm} + \rho_{Hg} g Z_{Hg}$$

$$1.2 \times 10^5 = 101396.6 + 13.6 \times 10^3 \times 9.81 \times Z_{Hg}$$

$$Z_{Hg} = 0.139 \text{ Hg}$$

EXAMPLE 1.24 *A basic barometer can be used as an altitude measuring device in aeroplanes. The ground control reports a barometric reading of 753 mm of Hg, while the pilot's reading is 690 mm of Hg. Estimate the altitude of the plane from the ground level if the average air density is 1.25 kg/m³ and g = 9.8 m/sec².*

Solution

Barometer reading at ground level $Z_{ground} = 0.753$ m

Pilot's Barometric reading $Z_{plane} = 0.690$ m

Altitude = ?

Pressure at the ground level $= \rho g Z_{ground}$

$$= 13600 \times 9.8 \times 0.753$$

$$= 100359.84 \text{ N/m}^2 \qquad ...(1)$$

Pressure at plane level $= \rho g Z_{plane}$

$$= 13600 \times 9.8 \times 0.690$$

$$= 91963.2 \text{ N/m}^2 \qquad ...(2)$$

∴ Change of pressure at the ground level and that of plane level,

$$\Delta P = 100359.84 - 91963.2$$

$$= 8396.64 \text{ N/m}^2$$

∴ This is the difference of pressure, now to find altitude for this much difference of pressure,

$$\Delta P = \rho_{air} g Z_{altitude}$$

$$8396.64 = 1.25 \times 9.8 \times Z_{altitude}$$

$$Z_{altitude} = 684.44 \text{ m}$$

EXAMPLE 1.25 *A manometer connected between 2 pipes as shown in Fig. Ex. 1.25. Take,*

$\rho_{water} = 1000$ kg/m³ $\quad \rho_{Hg} = 13590$ kg/m³

Pressure at $A = 400$ kPa $\quad g = 9.8$ m/sec².

Find Pressure at B = ?

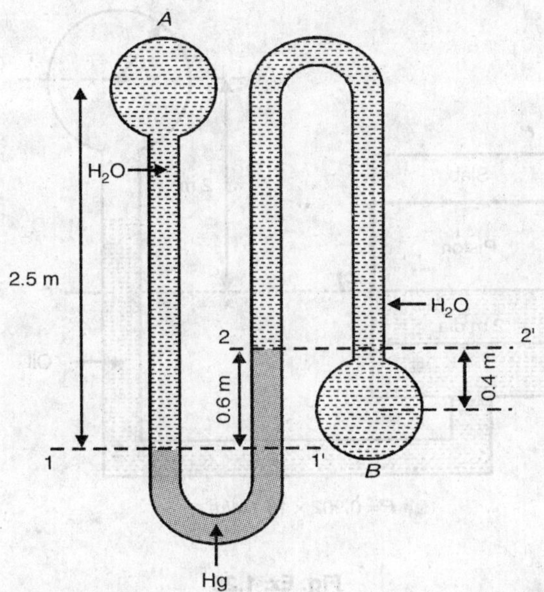

Fig. Ex. 1.25

Solution

Note. If there are more than one U-turns, then start from point and as we go downwards take it as positive, and as we go upwards take it as negative. Then equate this pressure equation to pressure at the second point.

So, writing the pressure equation,

$$P_A + P_w \, g \, Z_w - \rho_{Hg} \, g \, Z_{Hg} + \rho_w \, g \, Z_w = P_B$$

$$400 \times 1000 + 1000 \times 9.8 \times 2.5 - 13590 \times 9.8 \times 0.6 + 1000 \times 9.8 \times 0.4 = P_B$$

$$400000 + 24500 - 79409.2 + 3920 = P_B$$

$$P_B = 348510.8 \text{ Pascals}$$

$$\mathbf{P_B = 348.5108 \ kPa}$$

EXAMPLE 1.26 *The cylinder and tubing shown in Fig. Ex. 1.26, contain oil of density 0.902×10^3 kg/m^3 for gauge reading of 2 bar. What is the total weight of piston and slab placed on it?*

Solution

We know that, Pressure $= \dfrac{\text{Force}}{\text{Area}} = \dfrac{\text{Weight}}{\text{Area}}$

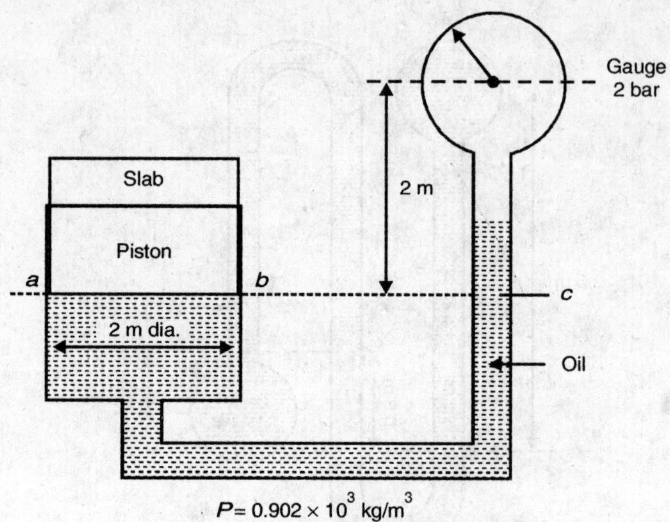

Fig. Ex. 1.26

And from Fig. Ex. 1.26 it can be seen that the pressure at $a-b-c$ level will be same.

$$\therefore \quad \frac{W}{A} = P_{gauge} + \rho_{oil} g h_{oil}$$

$$\therefore \quad \frac{W}{\frac{\pi}{4}(2)^2} = 2\times10^5 \frac{\text{N}}{\text{m}^2} + 0.902\times10^3 \times 9.81\times2\text{ m}$$

$$\mathbf{W = 683916.05\ N\ \text{(i.e. weight of piston and slab)}}$$

EXAMPLE 1.27 *The piston shown in Fig. Ex. 1.27 is held in equilibrium by pressure of gas flowing through the pipe. The piston has a mass of 21 kg.* $P_I = 600$ *kPa;* $P_{II} = 170$ *kPa. Determine the pressure in the gas* P_{III}.

Solution

Total downward force acting will be:

(i) Due to self weight.

(ii) Pressure P_I acting on 10 cm dia. piston.

(iii) Pressure P_{II} acting on 20 cm dia. piston.

And this is balanced by pressure P_{III}.

$\therefore$ Force due to mass $= 21\times9.81 = 206.01$ N $= 0.20601$ kN

Then force I acting on 10 cm dia piston =

$$\text{Pr.} \times \text{Area} = 600 \times \frac{\pi}{4}(0.1)^2 = 4.712389 \text{ kN}$$

And force II acting on 20 cm dia. piston =

$$170 \times \left(\frac{\pi}{4}(0.2^2 - 0.1^2)\right) = 4.0055306 \text{ kN}$$

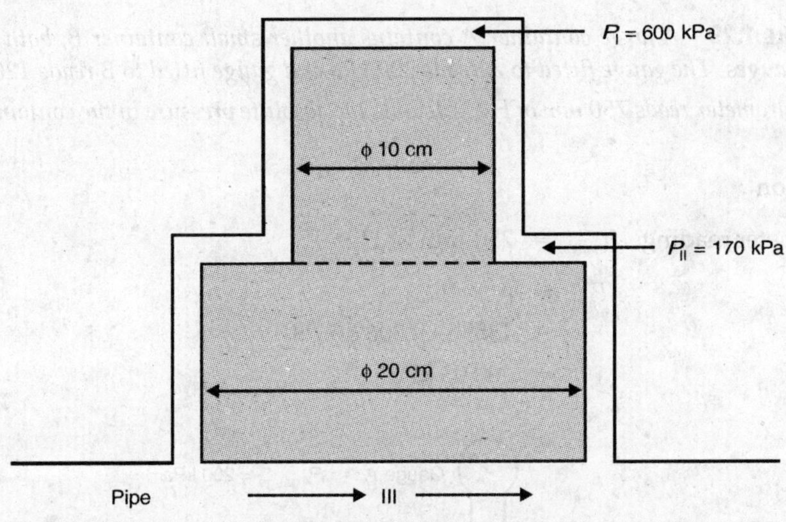

Fig. Ex. 1.27

$$\therefore \text{Total downward load/force} = 0.20601 + 4.712389 + 4.005306$$
$$= 8.9239296 \text{ kN}$$

This is balanced by P_{III}

$$\therefore \quad 8.9239296 = P_{III} \times \text{Area} = P_{III} \times \frac{\pi}{4}(0.2)^2$$
$$= P_{III} \times 0.0314$$
$$\mathbf{P_{III} = 284.0575 \text{ kPa}}$$

EXAMPLE 1.28 *The basic barometer can be used to measure the height of a building. If the barometric readings at top and bottom of a building are 730 and 755 mm of Hg respectively. Determine the height of the building. Assume an average air density of 1.18 kg/m³.*

Solution

The difference between pressure at top and bottom is equal to the air head of height equal to that of the building.

$$\therefore \qquad P_1 - P_2 = \rho_{air} g h$$

$$(0.755 - 0.73) \times 13600 \times 9.81 = 1.18 \times 9.81 \times h$$

$$h = \frac{13600 \times 0.025}{1.18}$$

$$\mathbf{h = 288.14\ m}$$

EXAMPLE 1.29 *A large container A contains another small container B, both fitted with gauges. The gauge fitted to A reads 200 kPa and gauge fitted to B reads 120 kPa.*

If barometer reads 750 mm of Hg, calculate the absolute pressure in the containers A and B.

Solution

$$\text{Barometer reading } h_{baro} = 750 \text{ mm of Hg} = Z_o$$

$$P_{atm} = \rho \cdot g \cdot h_{baro}$$

$$= 13596 \times 9.806 \times 0.750$$

$$= 100.03 \text{ kPa.}$$

Fig. Ex. 1.29

$$P_{A\,abs} = P_{A\,gauge} + P_{atm}$$

$$= 200 + 100.03$$

$$= 300.03 \text{ kPa}$$

Note that gauge *B* reads pressure relative to its surrounding presure i.e. gauge *B* is surrounded by pressure of 300.03 kPa.

$$P_{B\,abs} = P_{B\,gauge} + P_{atms} \text{ of } B$$

$$= 120 + 300.03$$

$$\mathbf{P_{B\,abs} = 420.03\ kPa}$$

EXAMPLE 1.30 *Convert the following temperatures into degree Fahrenheit.*

(i) 40 °C

(ii) – 20 °C

Solution

(i) Using the relation $\frac{C}{100} = \frac{F-32}{180}$

$$\frac{40}{100} = \frac{F-32}{180}$$

$$F-32 = \frac{40\times180}{100} = 72$$

$$\mathbf{F = 72+32 = 104\ °F}$$

(ii) $\frac{-20}{100} = \frac{F-32}{180}$

$$F-32 = \frac{-20\times180}{100} = -36$$

$$\mathbf{F = 32-36 = -4\ °F}$$

EXAMPLE 1.31 *Find the temperature which has the same value on both the centigrade and fahrenheit scales.*

Solution

Using the relation $\frac{C}{100} = \frac{F-32}{180}$

Putting C = F, $\frac{C}{100} = \frac{C-32}{180}$

$\therefore$ 180 C = 100 C – 3200

$\therefore$ C = – 40

$\therefore$ **– 40 ºC = – 40 ºF**

EXAMPLE 1.32 *A temperature scale of a certain thermometer is given by the relation t = a $\log_e$ P + b, where a and b constants and P is the thermometric property of the fluid in the thermometer. If at the ice point and steam point, thermometric properties are found to be 1.5 and 7.5 respectively. What would be the temperature corresponding to the thermometric property 3.5 on Celsius Scale.*

Solution

$$t = a\log_e P + b$$

$$100 = a \log_e 7.5 + b$$

$$0 = a \log_e 1.5 + b$$

Solving theses two equations we get,

$$\therefore \quad a = 62.1336 \text{ and } b = -25.168$$

$\therefore$ For $P = 3.5,\ t = 62.1336 \log_e 3.5 - 25.168$

$$\mathbf{t = 52.670676\ ^{\circ}C}$$

EXAMPLE 1.33 *The temperature scale of a certain thermometer is given by $t = a \ln P + b$, when a and b are constants and P is the thermometric property. The temperature of ice point and steam point are 100 and 300 respectively on this scale. Experiment gives values of thermometric properties of $P_1 = 1.86$ and $P_2 = 6.81$ at ice point and steam point respectively. Evaluate the temperature corresponding to a reading of thermometric property at $P = 2.5$.*

Solution

$$t = a \ln P + b \qquad ...(1)$$

$$300 = a \ln 6.8 + b$$

$$100 = a \ln 1.86 + b \qquad ...(2)$$

On subtracting we get, $200 = a \ln\left(\dfrac{6.81}{1.86}\right)$

$$\therefore \quad a = \frac{200}{1.3} = 154 \qquad ...(3)$$

Substituting this value of a in (2) we get,

$$100 = 154 \ln 1.86 + b$$

$$\therefore \quad b = 100 - 154 \ln 1.86 = 4.43$$

$\therefore$ For $P = 2.5,\ t = 154 \ln 2.5 + 4.43$

$$\mathbf{t = 145.5\ ^{\circ}}$$

EXAMPLE 1.34 *The pressure in a constant volume gas thermometer is measured as 32 mm of Hg above the atmospheric pressure at triple point of water, determine the temperature in ºC, when pressure is 76 mm of Hg above atmospheric pressure. Barometric pressure is 752 mm of Hg.*

Solution

Note: In a constant volume gas thermometer, volume remains constant but as the temperature changes, pressure change occurs.

$$\frac{\theta_1}{\theta_2} = \left(\frac{P_2}{P_1}\right) \text{constant volume} \quad ...(1)$$

Given that $\theta_1 = 273.16$ K (triple point of water)

$P_1 = P_{abs} + P_{gauge} = 752 + 32$

$P_1 = 784$ mm of Hg

and $P_2 = P_{atm} + P_{gauge} = 752 + 76 = 828$ mm of Hg

Substituting these in Eq. (1) we get

$$\frac{273.16}{\theta_2} = \frac{784}{828}$$

$$\theta_2 = 288.49 \text{ K}$$

$$\theta_2 = 15.49°\text{C}$$

EXAMPLE 1.35 *The temperature T on a thermometric scale is defined in terms of a property P by the relation,*

T = a ln P + b where a and b are constants.

The temperatures of the ice point and steam points are assigned the numbers 32 and 212 respectively. Experiment gives values of P of 1.86 and 6.81 at the ice point and steam point respectively. Evaluate the temperature corresponding to a reading of P = 2.5 on the thermometer.

Solution

We have $T = a \ln P + b$

Substituting the values for ice point and steam point,

$32 = a \ln 1.86 + b$

and $212 = a \ln 6.81 + b$

Solving we get $a = 138.7, b = -54.07$

∴ Equation becomes $T = 138.7 \ln P - 54.07$

Now for $P = 2.5$, $T = 138.7 \ln 2.5 - 54.07$

$= 73$

THEORY QUESTIONS

1. What is Thermodynamics?
2. Define Thermodynamic system? Explain its types with suitable examples.
3. Discuss :
   (1) System,
   (2) Surrounding,
   (3) Boundary,
   (4) Universe.
4. What do you understand by thermodynamic system ? How these are classified ? Explain each with suitable sketch.
   Explain the following
   1. Open system
   2. Closed system
   3. Isolated system
   and state the examples of each.
5. Distinguish between: Open system and Closed system.
6. State with reasoning whether the following systems are closed, open or isolated.
   1. IC engine
   2. Refrigerator
   3. Pressure cooker.
7. Explain Thermodynamic properties, processes and cycles.
8. Discuss :
   1. State
   2. Property
9. Explain : Thermodynamic properties.
10. Compare: Intensive and extensive properties.
11. What do you understand by state function and path function ?
12. Compare state function and path function.
13. Discuss – Unit extensive property is an intensive property.
14. Define the term Process.
15. Explain – 1. Cycle; 2. Process.
16. Define property and classify it with proper examples.
17. What is meant by Working Substance? Explain with an example.
18. Write a short note on Units and Dimensions.
19. What is Mechanical and Thermodynamic Work ? Explain.

20. What do you mean by thermodynamic work and mechanical work ?
21. What is *PdV*-work ? or Work done at the moving boundary ? or Displacement work.
22. Define heat and work and explain why neither is a property?
23. Prove that work is a path function and properties are point functions.
24. Define heat and work and explain why neither is a property?
25. 'Property is an exact differential'. Explain.
26. Derive the equations for work done in various quasi-static processes.
27. Write a note on Heat Energy.
28. Define the terms : Work and heat.
29. Explain the similarities and differences between heat and work.
30. Define the property — Pressure. State the various units used in practice.
31. How Pressure measurement is done ?
32. Write a note on Pressure exerted due to a column of fluid.
33. Write a short note on Barometer.
34. Briefly explain why you cannot use a mercury barometer to measure pressure in a gravity free environment.
35. Write a note on Bourdon Pressure gauge.
36. Explain — Bourdon's Pressure gauge with sketch.
37. Write a note on Manometers.
38. Define Thermal Equilibrium. State and explain the Zeroth law of Thermodynamics.
39. State and explain zeroth law of thermodynamics and explain how it leads to the concept of temperature.
40. What is meant by thermometric property. Enlist the temperature measuring devices.
41. Discuss various thermodynamic properties used for temperature measurement.
42. What is the principle of Temperature Measurement ?
43. What are the different Scales of Temperature ?
44. Write a note on Microscopic and Macroscopic point of view.
45. What do you understand by microscopic and macroscopic approach of thermodynamic study ? Explain.
46. What is the difference between the classical and statistical approaches to Thermodynamics.
47. Write a note on Energy in transit.
48. Explain the term — Energy in transit.
49. What is a Quasi–static Process? Explain.

50. What do you mean by Quasi-static Process?
51. Explain why a non-quasi equilibrium compression process requires a larger work input than the corresponding quasi equilibrium one, and, why a non-quasi equilibrium expansion process delivers less work than the corresponding quasi-equilibrium one.

PROBLEMS FOR PRACTICE

1. The pressure of a fluid of specific gravity 0.8 flowing in a horizontal pipe line in determined with a simple mercury U-tube manometer. The level of Hg surface in the right limb which is open to the atmosphere is 90 mm above the centre of the pipes. The level of Hg in the left limb which is connected to the pipe is 60 mm below the centre of pipe. Determine absolute pressure of liquid in the pipe is N/m². Assume standard P_{atm} to be equal to 10 m of water.

   (*Ans.* 117641.52 N/m²)
2. An inverted U-tube is connected across two pipes *A* and *B* carrying fluid as shown in Fig. P2. Calculate the pressure difference between the centre lines of the pipes.

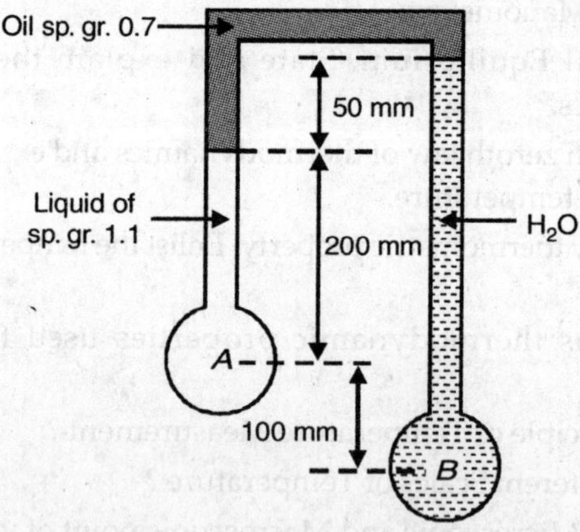

Fig. P2

(*Ans.* $P_A - P_B = -931.95$ N/m²)
3. In the arrangement shown in Fig. P3. Find the difference of pressure in the pipe *A* and *B*.

   (*Ans.* $P_A - P_B = 29135.7$ N/m²)

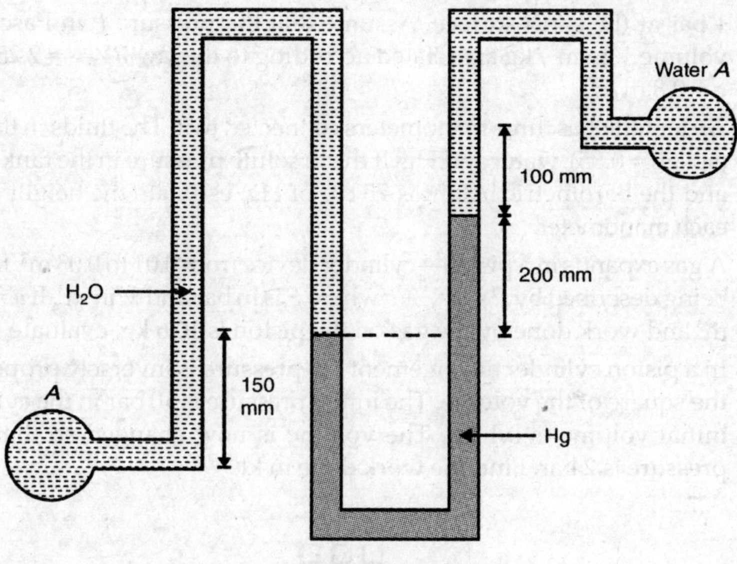

Fig. P3

4. A pressure gauge is attached to closed tank which contains some gas. The gauge reads 1.3 bar when barometer reads 75 cm of Hg. If the barometer changes to 72 cm of Hg, what will be the reading on the gauge for the same absolute pressure.

5. A U-tube manometer with one arm open to atmosphere is used to measure the pressure of the steam flowing through a pipe using mercury as manometric fluid. The level of the Hg column in the arm connected to the pipe is 125 mm below the level of Hg in the open arm. The column of condensate collected is 35 mm. Taking $g = 9.81 \text{ m/s}^2$, $\rho_{Hg} = 13560 \text{ kg/m}^3$ and atm. pressure as 760 mm of Hg, estimate the absolute pressure of steam in the pipe in bar.

6. Both a Bourdon gauge and manometer are attached to a gas tank. If the reading on the pressure gauge is 80 kPa, determine the distance between the two fluid levels of the manometer if the fluid is (i) Hg having sp. gravity of 13.6, (ii) water.

7. A U-tube manometer with one arm open to atmosphere is used to measure the pressure of steam flowing through a pipe. The height of Hg column open to atmosphere is 15 cm greater than the Hg level in the arm connected to steam pipe. The column of condensate is 5 cm. The barometer reading is 70 cm of Hg. Find the absolute pressure of steam in kPa.

8. A water pump is required to supply water at 900 lit./min into a tank at a height of 20 m. Find the power required to drive the pump in kW.

9. Estimate the height of the atmospheric column to produce a pressure of 1 bar at the earth surface. Assume that the pressure P in Pascal and sp. volume V in m^3/kg are related according to the law $PV^{1.39} = 2.25 \times 10^5$ and $g = 9.8\ m/s^2$.
10. An air tank has three manometers connected to it. The fluids in them are oil (sp. gr. = 0.85), water and Hg. If the absolute pressure in the tank is 1.24 bar and the barometric height is 75 cm of Hg, estimate the height of fluid in each manometer.
11. A gas expands in a piston—cylinder device from 0.01 to 0.03 m^3, the process being described by $P = aV^{-1} + b$ where P is in bar and V in m^3. If $a = 0.06$ bar/m^3 and work done by the gas on the piston is 10.6 kg, evaluate b.
12. In a piston cylinder arrangement the pressure is inversely proportional to the square of the volume. The initial pressure is 10 bar in the cylinder and initial volume is 0.1 m^3. The volume is now changed so that the final pressure is 2 bar. Find the work done in kJ.

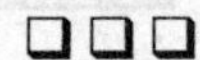

CHAPTER 2

FIRST LAW OF THERMODYNAMICS

2.1 INTRODUCTION

Energy is inherent in all matters. Energy may appear in many different forms. Conversion can be made from one form of energy to another. We are unable to define the general term energy in a simple way, but we can define with precision the various forms in which it appears.

One of the consequences of Einstein's theory of Relativity is that mass may be converted into energy and energy into mass, the relation being given by the famous equation,

$$\text{Energy } (E) = m \text{ (mass)} \times C^2 \qquad \text{(where } C - \text{velocity of light)}$$

Energy is a scalar quantity, not a vector quantity.

Except the nuclear reaction, (where mass is converted into energy) total energy of the universe is constant. For this matter, First Law Law of Thermodynamics can be expressed as,

Energy can neither be created nor destroyed (except in nuclear reactions). This is the law of conservation of energy.

2.2 JOULE'S EXPERIMENT

Joule carried out experiments in 1843, which led to the formulation of the First Law of Thermodynamics. He took a known quantity of water in a rigid vessel, which was insulated adiabatically from the surroundings. The vessel was fitted with a paddle wheel and a thermometer as shown in Fig. 2.1.

Now let a certain amount of work W_{1-2} be done upon the closed system by the paddle wheel. The quantity of work can be measured by the fall of weight W which drives the paddle wheel.

Let t_1 be the initial temperature of water before work transfer and after work transfer let the temperature rise to t_2. So, the system is changing its state from state (1) to state (2) and the process 1–2 undergone by the system is shown in Fig. 2.2.

Now let the insulation be removed. Then the system will interact by heat transfer (i.e. heat will be dissipated from the system to the surroundings) till the system comes to original temperature t_1. The amount of heat transfer Q_{2-1} from the system during the process 2–1 can be estimated: ($Q_{2-1} = mC_p \Delta T$).

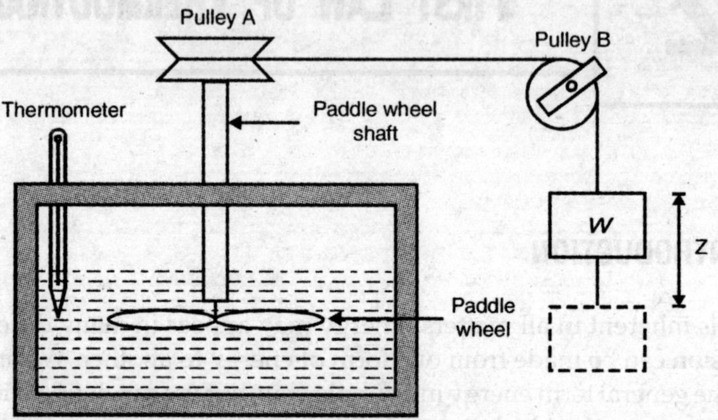

Fig. 2.1. Joules experiment.

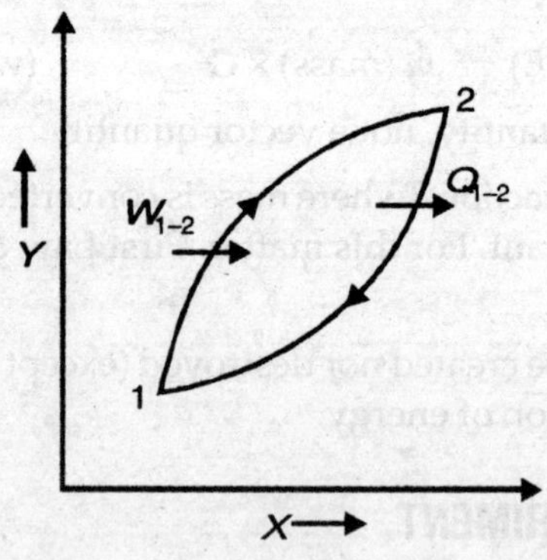

Fig. 2.2

The system thus undergoes a cycle, which consists of a definite amount of work input W_{1-2} to the system, followed by a heat transfer Q_{2-1} from the system.

Joule, repeated this experiment for different weights and different heights and in each case he found that, net work transfer during a cycle is proportional to net heat transfer,

i.e. $$\Sigma \delta W \propto \Sigma \delta Q$$

Or in other words *"when a closed system undergoes any cyclic process, the cyclic integral of work is proportional to the cyclic integral of heat"*. This is known as the first law of thermodynamics for a closed system undergoing a cycle.

i.e.
$$\oint \delta W \propto \oint \delta Q$$
$$\oint \delta W = J\oint \delta Q \text{; Since in S.I. units } J = 1$$
$$\oint \delta W = \oint \delta Q$$

Thus, the First Law states the general principle of conservation of energy i.e. *"Energy can neither be created nor be destroyed, but energy can be converted from one form to the other."*

2.3 FIRST LAW FOR A CLOSED SYSTEM UNDERGOING A PROCESS

If a closed system undergoes a change of state or a process and during which, both work transfer and heat transfer are involved, then the net energy transfer will be stored within the system. If Q is the amount of heat transferred to the system and W is the amount of work transferred from the system, during the process as shown in Fig. 2.3, then the net energy transfer ($Q - W$) will be stored in the system.

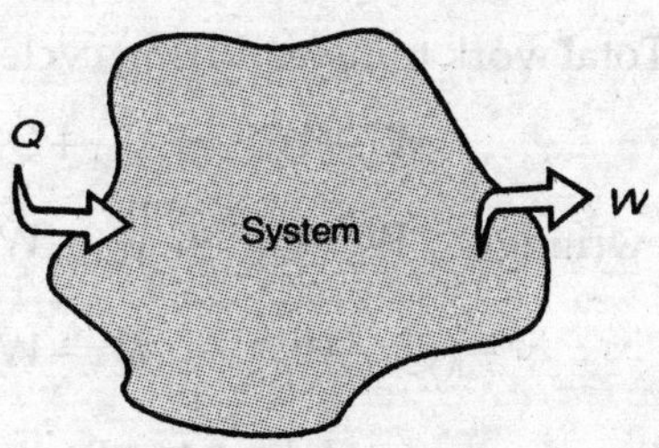

Fig. 2.3

Energy in storage is neither heat nor work, but is called as Internal energy or simply energy of the system.

$$\therefore \quad Q - W = \Delta E$$

or
$$Q = \Delta E + W$$

where ΔE is the change in energy. Here Q, W and ΔE all are expressed in joules.

2.4 ENERGY — A PROPERTY OF THE SYSTEM

Consider a system which changes its state from state (1) to state (2) by following the path A, and returns from state (2) to state (1) by following the path B as shown in Fig. 2.4. So the system undergoes a cycle. Now writing the First Law for the path A,

$$Q_A - W_A = \Delta E_A \qquad \text{...(1)}$$

and for path B.

$$Q_B - W_B = \Delta E_B \qquad \text{...(2)}$$

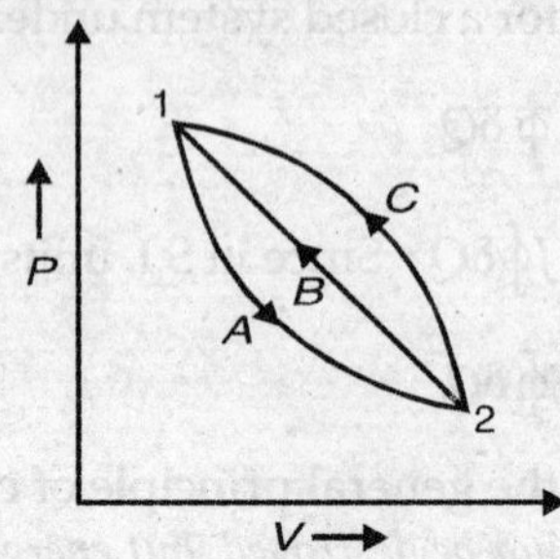

Fig. 2.4

The processes 1–*A*–2 and 2–*B*–1 together will constitute a cycle for which,

$$\oint \delta W = \oint \delta Q$$

i.e. Total work transfer during cycle = Total heat transfer during the cycle.

$$W_A + W_B = Q_A + Q_B$$

Rewriting $\quad W_B - Q_B = Q_A - W_A \qquad \text{...(3)}$

$$\therefore \quad -(Q_B - W_B) = Q_A - W_A$$

$$-\Delta E_B = \Delta E_A$$

i.e. $\quad \Delta E_A = -\Delta E_B \qquad \text{...(4)}$

Similarly if the system returns from point (2) to state point (1) by following the path *C* instead of path *B* then,

$$\Delta E_A = -\Delta E_C \qquad \text{...(5)}$$

From Eqs (4) and (5),

$$\mathbf{\Delta E_B = \Delta E_C} \qquad \text{...(6)}$$

Thus the change in energy for the path *B* and *C* are same. Hence change in energy does not depend upon path, so it depends on end states. Hence, it is a point function and since properties are point functions, *Energy is a property of the system.*

2.5 DIFFERENT FORMS OF STORED ENERGIES

The total energy *E* is made up of Kinetic Energy (K.E.), Potential Energy (P.E.) and Internal Energy (I.E.).

$$\therefore \quad E = \text{K.E.} + \text{P.E.} + \text{I.E.}$$

Kinetic Energy is due to the motion of the fluid in the system. Potential Energy is the energy due to the gravitational force.

A part of the total energy which is stored in the molecular and atomic structure is known as *Internal energy* and is denoted by U.

In the absence of Kinetic Energy and Potential Energy i.e. when

$$KE = 0 \text{ and } PE = 0$$

$$E = U, \text{ and the I law for process,}$$

$$Q = \Delta E + W$$

becomes

$$Q = \Delta U + W$$

In the differential form, Eqs (1) and (2) become

$$\delta Q = dE + \delta W$$

i.e.

$$\delta Q = dE + PdV \,(\text{as } \delta W = PdV)$$

Also,

$$\delta Q = dU + \delta W \,(\text{when } E = U)$$

i.e.

$$\delta Q = dU + PdV$$

When K.E. and P.E. are considered, then the I law for a process will be

$$\delta Q - \delta W = dU + d(\text{KE}) + d(\text{PE})$$

In the integrated form, Eq. (5) becomes

$$\mathbf{Q_{1\text{-}2} - W_{1\text{-}2} = U_2 - U_1 + \frac{m\left(C_2^2 - C_1^2\right)}{2} + mg\left(Z_2 - Z_1\right)}$$

where C is the velocity and Z is the height.

For unit mass, Eq. (6) will be,

$$\mathbf{q_{1\text{-}2} - w_{1\text{-}2} = u_2 - u_1 + \frac{\left(C_2^2 - C_1^2\right)}{2} + g\left(Z_2 - Z_1\right)} \quad \ldots(7)$$

2.6 INTERNAL ENERGY (U)

"It is the energy stored in the molecular structure due to heat and work interactions".

or

In case of gases we know that the gas is made up of a number of molecules, which are moving continuously. Internal Energy (I.E.) is the energy which arises from the motion of these molecules.

If the temperature of the gas is increased, the molecular activity increases, therefore the I.E. also increases. Thus I.E. is a function of temperature and its value can be increased or decreased by adding or removing heat to or from the system.

Due to practical difficulties it is very very difficult to determine the absolute value of Internal energy. Fortunately, in most of the thermodynamic applications change in I.E. is used, since changes in the states of a system is considered. Change in I.E. is denoted by ΔU and $\Delta U = U_2 - U_1$.

For unit mass, $\Delta \mathbf{u} = \mathbf{u}_2 - \mathbf{u}_1$

2.7 ENTHALPY (H)

It is an extensive property, since its value depends on mass. Enthalpy of a substance is given by the sum of Internal energy and Pressure–Volume product.

i.e. $\mathbf{H = U + PV}$ and specific enthalpy, $\mathbf{h = u + Pv}$

2.8 SPECIFIC HEAT AT CONSTANT PRESSURE AND SPECIFIC HEAT AT CONSTANT VOLUME

"Specific heat of a gas is defined as the amount of heat required to rise the temperature of unit mass of a gas through one degree".

Let

$$C = \text{General specific heat}$$
$$Q = \text{Heat transfer (J)}$$
$$\Delta T = \text{Change in temperature K}$$
$$m = \text{mass (kg)}$$

Then from the definition, $$C = \frac{Q}{m \cdot \Delta T}\left[\text{units will be } \frac{\text{J}}{\text{kg-K}}\right] \quad ...(1)$$

or $$Q = m \cdot C \cdot \Delta T \quad ...(2)$$

In the differential form, $$\delta Q = m \cdot C \cdot dT \quad ...(3)$$

And for unit mass, $$\delta q = C \cdot dT \quad ...(4)$$

or $$C = \frac{\delta q}{dT} \quad ...(5)$$

There are 2-types of specific heats,

(1) Specific heat at constant volume C_v

(2) Specific heat at constant pressure C_p

1. Specific Heat at Constant Volume C_v

"It is the amount of heat required to rise the temperature of unit mass of a gas through one degree when the volume is constant".

If a unit mass of a gas is taken in a closed vessel and is heated, the volume of the gas remains constant, but the temperature increases. As the volume remains constant, there is no external work done by the gas and as temperature of the gas increases, there is increase in internal energy of the gas.

Therefore heat supplied to the gas is completely utilised in increasing the I.E. of the gas,

and $$C_V = \left(\frac{\delta q}{dT}\right)_V = \left(\frac{du}{dT}\right)_V$$

Therefore C_v is also defined as, *"the rate of change of specifc internal energy with respect to temperature when the volume is kept constant."*

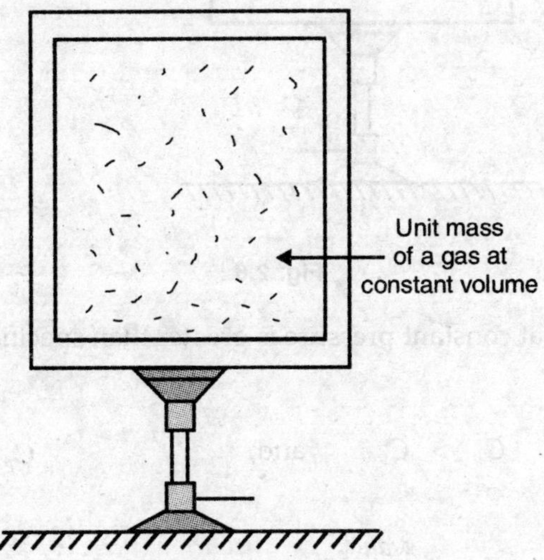

Fig. 2.5

2. Specific Heat at Constant Pressure C_p

"It is the amount of heat required to rise the temperature of unit mass of a gas through one degree when the pressure is kept constant".

Consider a unit mass of a gas in a cylinder fitted with a frictionless piston as shown in Fig. 2.6. When the gas is heated, the piston moves up, maintaining the same pressure. But the volume and temperature of the gas increases during heating. As there is increase in volume, there is external work done by the gas and as there is increase in temperature, there is increase in I.E.

Thus heat supplied when the pressure is constant, is utilised for two purposes,

(a) To do some external work.

(b) To increase the I.E. of the gas.

Whereas in case of constant volume heating, the heat supplied is completely utilised for increasing the I.E.

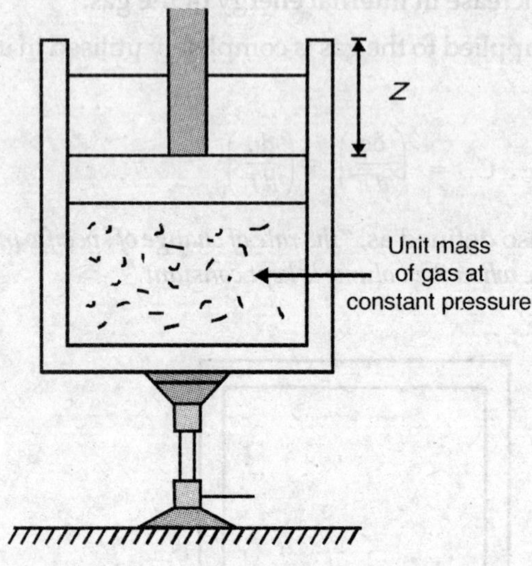

Fig. 2.6

$\therefore$ Specific heat at constant pressure is greater than specific heat at constant volume.

i.e. $C_P > C_V$ and, $C_P = \left[\dfrac{\delta q}{dT}\right]_P$

Also $C_P = \left[\dfrac{dh}{dT}\right]_P$

Therefore, C_p is also defined as, *"the rate of change of specific enthalpy with respect to temperature when the pressure is kept constant."*

2.9 ADIABATIC INDEX

It is the ratio of specific heat at constant pressure to the specific heat at constant volume and is given by,

$$K \text{ or } \gamma = \frac{C_p}{C_v}$$

Note: For air, $C_p = 1.005 \dfrac{\text{kJ}}{\text{kg-K}}$ and $C_v = 0.718 \dfrac{\text{kJ}}{\text{kg-K}}$ and $\boldsymbol{\gamma} = 1.4$

2.10 PERPETUAL MOTION MACHINE OF FIRST KIND [PMM – 1]

First law states the general principle of conservation of energy, i.e. energy can neither be created nor be destroyed but it can be transformed from one form to the other.

But PMM–1 is defined as a machine which will produce continuous work output without receiving any energy from any other system or the surroundings. (Fig. 2.7).

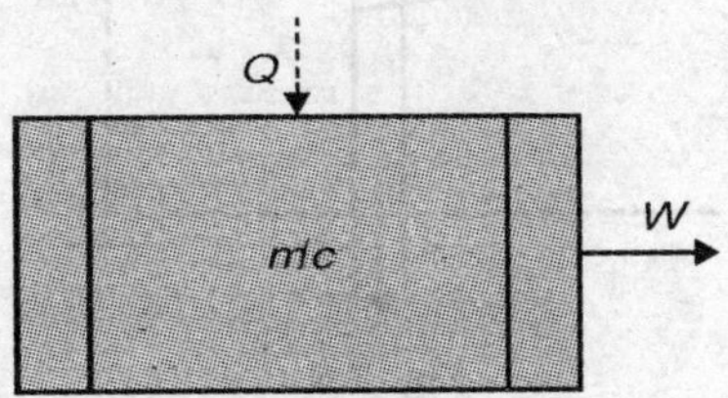

Fig. 2.7

It will create energy and thus violates First Law of Thermodynamics. All the attempts made so far to make such a machine have failed, thus they show the validity of I–law. Thus PMM–1 is just a conceptual machine.

Converse of PMM–1 is also true, i.e. there can be no machine which would continuously consume work, without producing some other form of energy. (Fig. 2.8)

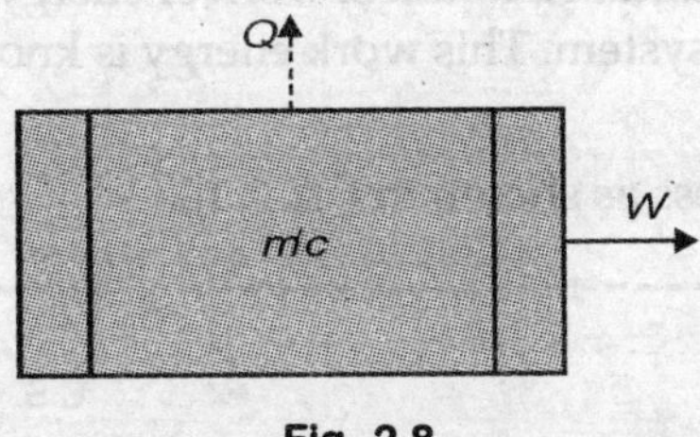

Fig. 2.8

2.11 FLOW PROCESS; CONTROL VOLUME; CONTROL SURFACE

Flow Process. Processes performed in the open systems are called as Flow-processes.

Figure 2.9 shows a portion of steam power plant. High pressure steam enters the steam turbines at section (1), expands and produces work output and then it leaves the turbine at section (2).

For analysing the expansion process, two methods are used.

In the first method certain mass of the fluid Marked *A* is considered and this is suppose to flow through the turbine.

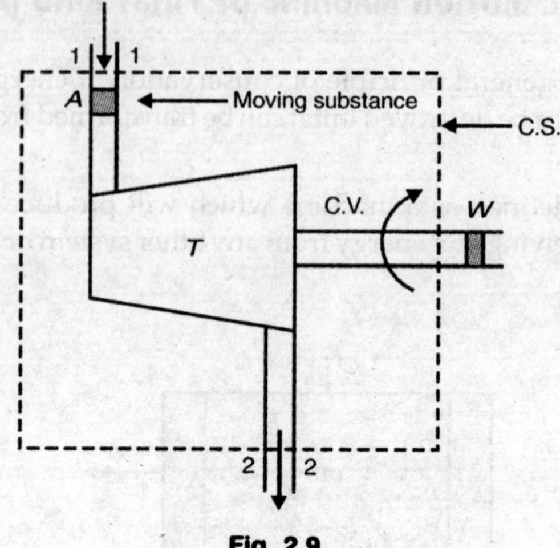

Fig. 2.9

In the second method, certain fixed volume of the system known as *control volume* (*CV*) is considered.

The moving substance flows through this control volume. The surface of the control volume is known as *control surface* (*CS*).

2.12 FLOW WORK OR FLOW ENERGY

In case of flow processes certain amount of work or energy is required to push the fluid into and out of the system. This work energy is known as *flow work or flow energy*.

Consider a flow process as shown in Fig. 2.10.

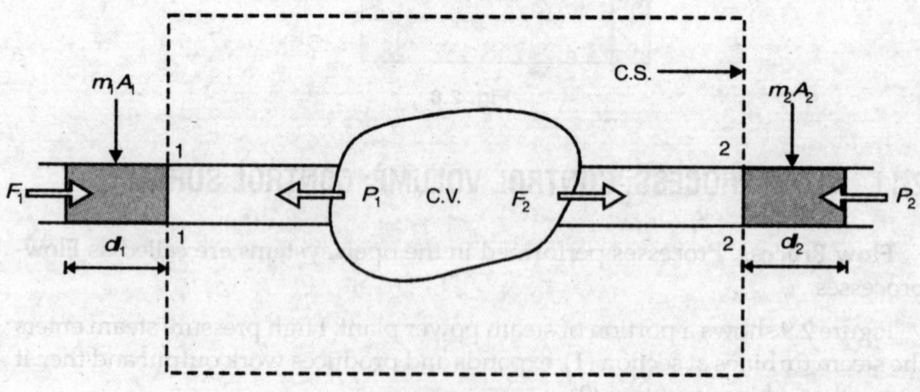

Fig. 2.10

Let F_1 be the force which forces the mass m_1 of cross sectional area A_1 into the system against the pressure P_1 of the system. Now consider a small amount of work done on the system in causing the displacement dl_1 for the mass m_1.

Thus the small amount of work done on the system in causing the displacement dl_1,

$$\delta W = -F_1 \times dl_1$$

[**Note:** –ve sign, since the work is done on the system.]

But $\quad F_1 = P_1 \times A_1$

$\therefore \quad \delta W = -P_1 \times A_1 \times dl_1$

Since $\quad A_1 \times dl_1 = dV_1 =$ small amount of displacement volume.

$\therefore \quad \delta W = -P_1 \times dV_1$

$\therefore$ Total flow work at sec. (1) $= -P_1 V_1$

Similarly work done by the system to force the fluid out of the system at sec. (2) $= P_2 V_2$

$\therefore$ **Net flow work** $= \mathbf{P_2V_2 - P_1V_1}$

and **for unit mass** $= \mathbf{P_2V_2 - P_1V_1}$

Note: Since the flow work is entirely expressed in terms of properties of the system, the net flow work depends on the end states and it is a *Property*. Whereas the other forms of work are path functions so in problems involving flow processes, flow work is to be calculated separately.

2.13 CONDITIONS OF STEADY FLOW SYSTEM

A steady flow process should satisfy the following conditions,

(i) The mass flow rate into and out of the system are equal and do not vary with time i.e. mass in the system does not change.
(If the mass flow rate at the inlet is 10 kg / sec., then the mass flow rate at the exit will be 10 kg / sec. and it does not vary with time i.e. if we note the mass flow rate say, when we come to the college and while going home also it will be same.)

(ii) The energy of the fluid at the entrance and at the exit are same and do not vary with time.

(iii) The rates of heat and work transfer into and out of the system do not vary with time.

2.14 STEADY FLOW ENERGY EQUATION (S.F.E.E.)

Steady flow energy equation is obtained by applying the first law of thermodynamics to a steady flow system.

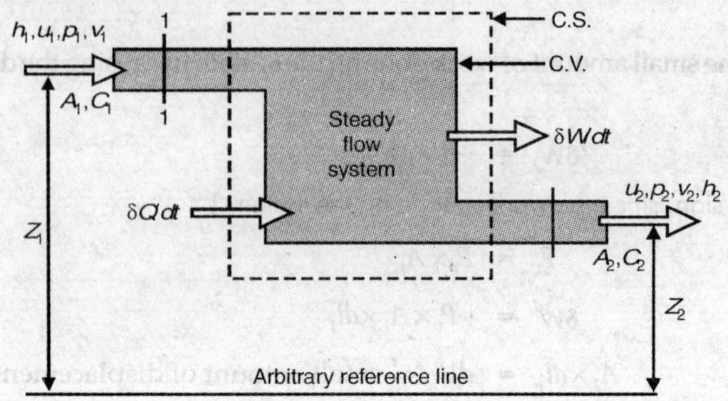

Fig. 2.11

Let,

$\dot{m}$ = Mass flow rate in kg/sec

$A_1\ A_2$ = Cross sectional azreas at sec 1–1 and 2–2 respectively in m^2

$C_1\ C_2$ = Velocities in m/sec

$P_1\ P_2$ = Absolute pressures in N/m^2

$v_1\ v_2$ = Specific volumes in m^3/kg

$u_1\ u_2$ = Specific internal energies in J/kg

$h_1\ h_2$ = Specific enthalpies in J/kg

$Z_1\ Z_2$ = Elevations above arbitrary reference datum line in m

Q = Net heat transfer in Joules

$\frac{\delta Q}{dt}$ = Net rate of heat transfer in J/sec

$\frac{\delta Q}{dm}$ = q = Net rate of heat transfer in J/kg

W = Net work transfer in joules

$\frac{\delta W}{dt}$ = Net rate of work transfer in J/sec

$\frac{\delta W}{dm}$ = w = Net rate of work transfer in J/kg

t = time in seconds

Figure 2.11 shows a steady flow system. Fluid entes at sec. 1–1 and leaves at sec. 2–2.

Since it is a steady flow system. According to 2nd condition.

Total energy at the entrance = Total energy at the exit

∴ We can write,

$$\dot{m}\begin{bmatrix}\text{Energy carried}\\ \text{into the system}\end{bmatrix}+\dot{m}\begin{bmatrix}\text{Specific enthalpy}\\ \text{of entering fluid}\end{bmatrix}+\begin{bmatrix}\text{Net rate of heat}\\ \text{transfer into}\\ \text{system}\end{bmatrix}$$

$$=\dot{m}\begin{bmatrix}\text{Energy carried}\\ \text{out of the system}\end{bmatrix}+\dot{m}\begin{bmatrix}\text{Specific enthalpy}\\ \text{of the fluid}\\ \text{at the exit}\end{bmatrix}+\begin{bmatrix}\text{Net rate of work}\\ \text{transfer by the}\\ \text{system}\end{bmatrix}$$

i.e. $$\dot{m}[KE_1+PE_1]+\dot{m}(h_1)+\frac{\delta Q}{dt} = \dot{m}[KE_2+PE_2]+\dot{m}(h_2)+\frac{\delta W}{dt}$$

i.e. $$\dot{m}\left[\frac{C_1^2}{2}+g\,Z_1\right]+\dot{m}(h_1)+\frac{\delta Q}{dt} = \dot{m}\left[\frac{C_2^2}{2}+g\,Z_2\right]+\dot{m}(h_2)+\frac{\delta W}{dt}$$

Note: $m$of KE and PE will be unity, since we are deriving the SFEE for unit mass.

$$\therefore \dot{\mathbf{m}}\left[\frac{\mathbf{C}_1^2}{2}+\mathbf{g}\,\mathbf{Z}_1+\mathbf{h}_1\right]+\frac{\delta\mathbf{Q}}{\mathbf{dt}} = \dot{\mathbf{m}}\left[\frac{\mathbf{C}_2^2}{2}+\mathbf{g}\,\mathbf{Z}_2+\mathbf{h}_2\right]+\frac{\delta\mathbf{W}}{\mathbf{dt}} \qquad \ldots(1)$$

This is steady flow energy equation on *time basis*.

Also, since $h_1 = u_1 + P_1v_1$ and $h_2 = u_2 + P_2v_2$

∴ Eq. (1) becomes,

$$\therefore\ \dot{m}\left[\frac{C_1^2}{2}+g\,Z_1+u_1+P_1v_1\right]+\frac{\delta Q}{dt} = \dot{m}\left[\frac{C_2^2}{2}+g\,Z_2+u_2+P_2v_2\right]+\frac{\delta W}{dt}$$

or $$\dot{\mathbf{m}}[\mathbf{KE}_1+\mathbf{PE}_1+\mathbf{IE}_1+\mathbf{FE}_1]+\frac{\delta\mathbf{W}}{\mathbf{dt}} = \dot{\mathbf{m}}[\mathbf{KE}_2+\mathbf{PE}_2+\mathbf{IE}_2+\mathbf{FE}_2]+\frac{\delta\mathbf{W}}{\mathbf{dt}} \qquad \ldots(2)$$

2.15 STEADY FLOW ENERGY EQUATION ON MASS BASIS

For deriving this, we have to consider $\dot{m} = 1$ kg/sec and all other quantities will be for per kg mass such as $\delta W/dm$ and $\delta Q/dm$.

∴ Eq. (1) becomes,

$$\left[\frac{C_1^2}{2}+gZ_1+h_1\right]+\frac{\delta Q}{dm} = \left[\frac{C_2^2}{2}+gZ_2+h_2\right]+\frac{\delta W}{dm}$$

$$\therefore \qquad \left[\frac{C_1^2}{2}+gZ_1+h_1\right]+q = \left[\frac{C_2^2}{2}+gZ_2+h_2\right]+w$$

This is the SFEE on mass basis.

2.16 IN A STEADY FLOW SYSTEM W.D.= $-\int v.dP$ WHEN CHANGES IN KE AND PE ARE NEGLECTED

We know that, SFEE on mass basis. From Eq. (3)

$$\left[\frac{C_1^2}{2}+gZ_1+h_1\right]+q = \left[\frac{C_2^2}{2}+gZ_2+h_2\right]+w \qquad \ldots(1)$$

$$\text{i.e.} \qquad \left[\frac{C_1^2}{2}+gZ_1+u_1+P_1v_1\right]+q = \left[\frac{C_2^2}{2}+gZ_2+u_2+P_2v_2\right]+w$$

$$\therefore \qquad q = u_2-u_1+P_2v_2-P_1v_1+\frac{C_2^2-C_1^2}{2}+g(Z_2-Z_1)+w$$

$$\therefore \qquad q = \Delta u+\Delta(Pv)+\Delta(\text{KE})+d(\text{PE})+w$$

In the differential form,

$$\delta q = du+d(Pv)+d(\text{KE})+d(\text{PE})+\delta w \qquad \ldots(2)$$

But we also know that I law for a closed system undergoing a process is,

$$Q-W = \Delta E = \Delta U$$

For unit mass, $q-w = \Delta u$

i.e. $q = \Delta u+w$

In the differential form, $\delta q = du+\delta w$

$$\text{i.e.} \qquad \delta q = du+Pdv \qquad \ldots(3)$$

Assuming that δq is same for closed system and for the steady flow system, equating Eqs (2) and (3)

$$du+Pdv = du+d(Pv)+d(\text{KE})+d(\text{PE})+\delta w \qquad \ldots(4)$$

$$\text{i.e.} \qquad du+Pdv = du+Pdv+vdP+d(\text{KE})+d(\text{PE})+\delta w$$

$\therefore$ We can write, $-v\,dP = \delta w+d(\text{KE})+d(\text{PE})$

$\therefore$ For total mass, $\qquad -\int vdP = w+\Delta(\text{KE})+\Delta(\text{PE}) \qquad \ldots(5)$

When change in KE and PE are neglected,

$$-\int \mathbf{vdP} = \mathrm{w}$$

This work done is given by the area behind the curve 1–2 as shown in Fig. 2.12.

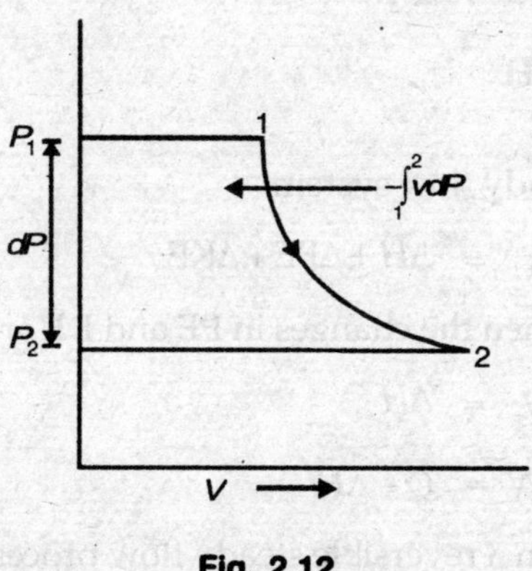

Fig. 2.12

2.17 SIGNIFICANCE OF $\int Pdv$ IN CASE OF STEADY FLOW PROCESS AND NON-FLOW PROCESS

Upto Eq. (4) derivation is same as in the above article.

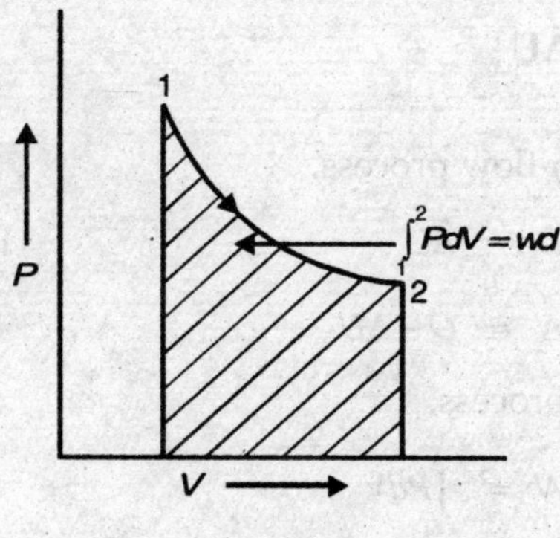

Fig. 2.13

From Eq. (4), $du + Pdv = du + d(Pv) + d(\mathrm{KE}) + d(\mathrm{PE}) + \delta w$

$$Pdv = d(Pv) + d(\mathrm{KE}) + d(\mathrm{PE}) + \delta w$$

$\therefore$ For total mass, $\int_1^2 \mathbf{Pdv} = \mathbf{\Delta(Pv) + \Delta(KE) + \Delta(PE) + w}$

For a non-flow process (i.e. for a closed system),

$$\int P dv = \text{work done and is given by the area under the curve 1–2 as shown in Fig. 2.13.}$$

2.18 PROOFS

2.18.1 $-\int \mathbf{v dP = Q - \Delta H}$

We know that, for a steady flow system,

$$Q - W = \Delta H + \Delta PE + \Delta KE \quad \ldots(1)$$

Consider the case when the changes in PE and KE are neglected,

$$Q - W = \Delta H$$

$$\therefore \quad W = Q - \Delta H \quad \ldots(2)$$

Also we know that, in a reversible steady flow process,

Work done $= -\int vdP$ when changes in KE and PE are neglected. ...(3)

Hence from (2) and (3) we can write,

$$\therefore \quad -\int \mathbf{v dP} = \mathbf{Q - \Delta H}$$

2.18.2 $\int \mathbf{P dV = Q - \Delta U}$

We know that, for a non-flow process,

$$Q - W = \Delta U$$

$$W = Q - \Delta U \quad \ldots(1)$$

And for a Non-flow process,

$$W = \int PdV \quad \ldots(2)$$

$\therefore$ From (1) and (2) $\quad \int \mathbf{PdV} = \mathbf{Q - \Delta U}$

2.18.3 $-\int \mathbf{v dP} - \int \mathbf{P dv = -\Delta Pv}$

We know that,

$$-\int vdP = w+\Delta\text{KE} + \Delta\text{PE}$$

Also we know that, significance of $\int Pdv$ in case of steady flow process and non-flow process,

$$\int Pdv = w+\Delta(\text{PE})+\Delta(\text{KE})+\Delta Pv \qquad ...(2)$$

$\therefore$ Substracting Eq. (2) from (1) we get,

$$-\int \mathbf{v\,dP} - \int \mathbf{P\,dv} = \mathbf{-\Delta Pv}$$

2.19 APPLICATIONS OF ENERGY EQUATIONS

(a) Boiler is a steam generator

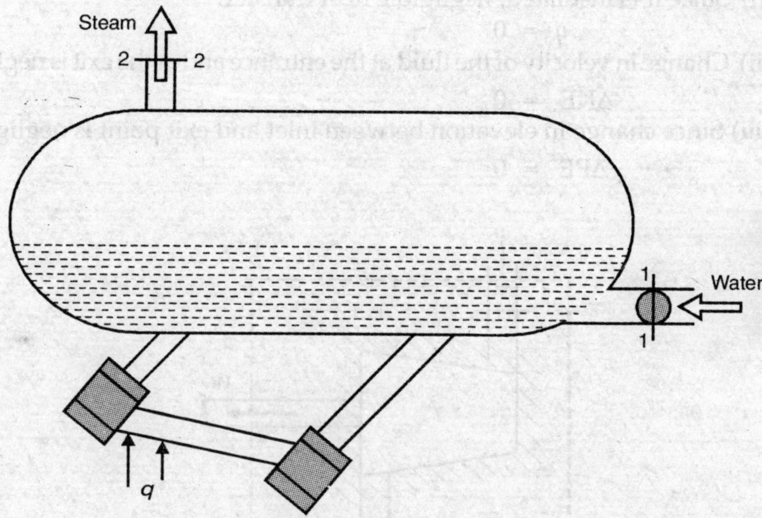

Fig. 2.14

Characteristics of the boiler are,

(1) Change in velocity of the fluid at the entrance and exit is very small, and hence ΔKE can be neglected, i.e. $\Delta\text{KE} = 0$

(2) Change in elevation between the inlet and exit point is very small so change in PE may be neglected, i.e. $\Delta\text{PE} = 0$

(3) Since no work is done in a boiler $w = 0$

$\therefore$ From SFEE on mass basis,

$$\left[\frac{C_1^2}{2} + gZ_1 + h_1\right] + q = \left[\frac{C_2^2}{2} + gZ_2 + h_2\right] + w$$

i.e.
$$q = \left[\frac{\left(C_2^2 - C_1^2\right)}{2} + g\left(Z_2 - Z_1\right) + \left(h_2 - h_1\right)\right] + w \quad \ldots(1)$$

On substituting the above conditions Eq. (1) becomes,

$$q = \left[0 + 0 + \left(h_2 - h_1\right)\right] + 0$$

i.e.
$$\mathbf{q = h_2 - h_1}$$

(b) Turbine (Engine) / Compressor (or pump)

The turbines and engines are power producing devices whereas Compressors, Pumps, Blowers etc. are power consuming devices.

Charcteristics of a turbine are,

(i) Since it is insulated, negligible heat transfer.

$\therefore \quad q = 0$

(ii) Change in velocity of the fluid at the entrance and at the exit is negligible.

$\therefore \quad \Delta KE = 0$

(iii) Since change in elevation between inlet and exit point is negligible,

$\Delta PE = 0$

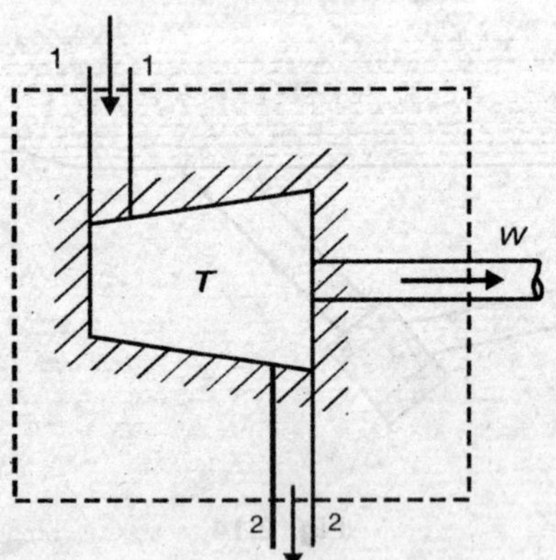

Fig. 2.15

$\therefore$ From SFEE, i.e. from Eq. (1),

$$q = \left[\frac{\left(C_2^2 - C_1^2\right)}{2} + g\left(Z_2 - Z_1\right) + \left(h_2 - h_1\right)\right] + w$$

∴ The equation reduces to,

$$0 = [0+0+(h_2-h_1)]+w$$

or $$\mathbf{w = h_1 - h_2}$$

Here work is produced at the expense of enthalpy. Similarly for an Adiabatic Pump / Compressor, work is done on the system, so w is – ve.

∴ SFEE becomes,

$$\mathbf{w = h_2 - h_1} \quad (\text{since } w \text{ is } -\text{ve})$$

(c) Throttling Process

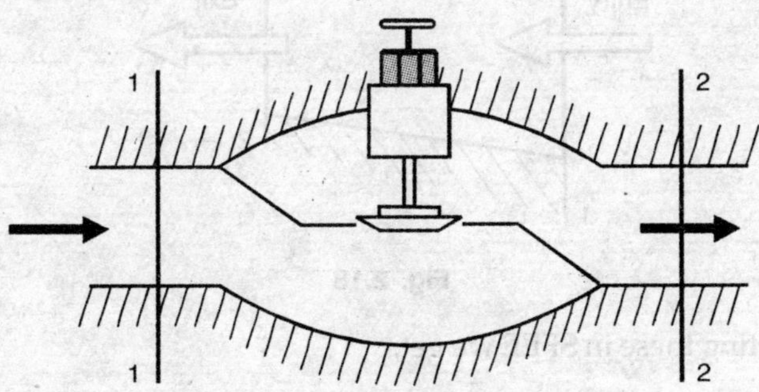

Fig. 2.16

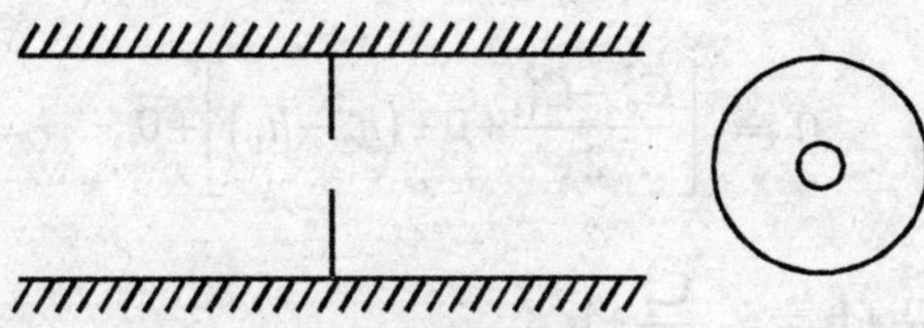

Fig. 2.17

When a fluid flows through a partially opened valve or an orifice then there is an appreciable pressure drop and the flow is said to be throttled.

Characteristics are,

(i) $q = 0$

(ii) $w = 0$

(iii) $\Delta PE = 0$

Substituting these in SFEE we get

$$q = \left[\frac{C_2^2 - C_1^2}{2} + g(Z_2 - Z_1) + (h_2 - h_1)\right] + w$$

$$0 = \left[\frac{(C_2^2 - C_1^2)}{2} + 0 + (h_2 - h_1)\right] + 0$$

i.e. $$\frac{C_1^2}{2}+h_1 = \frac{C_2^2}{2}+h_2$$

(d) Nozzle. It is a device which is used to increase the KE of flowing fluids

(i) $q = 0$

(ii) $w = 0$

(iii) $\Delta PE = 0$

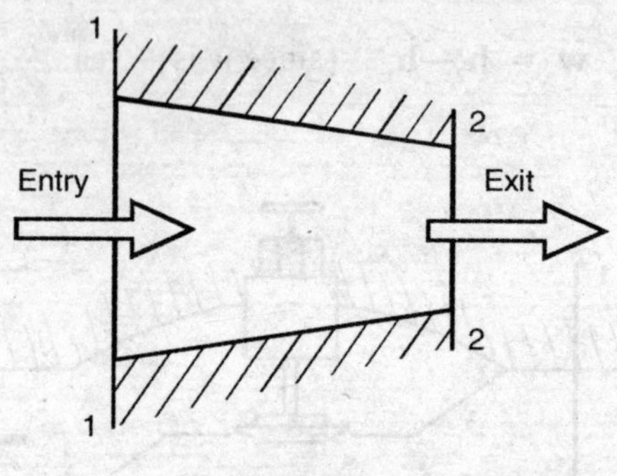

Fig. 2.18

∴ Substituting these in SFEE we get,

$$q = \left[\frac{C_2^2 - C_1^2}{2} + g(Z_2 - Z_1) + (h_2 - h_1)\right] + w$$

$$0 = \left[\frac{C_2^2 - C_1^2}{2} + 0 + (h_2 - h_1)\right] + 0$$

i.e. $$\frac{C_1^2}{2}+h_1 = \frac{C_2^2}{2}+h_2$$

(e) Condenser or to show in a steady flow process with multiple streams of fluid without mixing.

$$\frac{\dot{m}_1}{\dot{m}_3} = \frac{h_4 - h_3}{h_1 - h_2}$$

Figure 2.19 shows a surface condenser. In this case, cooling water enters at Section 1– 1 passes through the tubes and leaves the condenser at Section 2 – 2.

Steam to be condensed enters the condenser at Section 3–3 and when it comes in contact with the cold surface of water tubes, it gets condensed and the condensate leaves the condenser at Section 4–4.

So, in this case working fluids are water and steam i.e. two fluids flow through the system and since they do not mix with each other, it may be considered as a steady flow system with multiple streams of fluid without mixing.

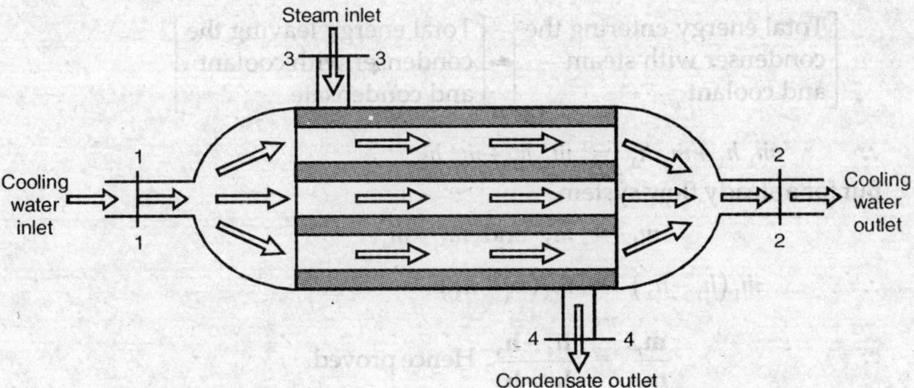

Fig. 2.19

Let, $\dot{m}_1\ \dot{m}_2$ — Mass flow rates of cooling water at sec. 1 – 1 and 2 – 2 in kg / sec.

$\dot{m}_3\ \dot{m}_4$ — Mass flow rates of steam and condensate sec. 3 – 3 and 4 – 4 in kg / sec.

$h_1\ h_2$ — Specific enthalpies of cooling water at sec. 1 – 1 and 2 – 2 respectively in J / kg

$h_3\ h_4$ — Specific enthalpies of steam and condensate at sec. 3 – 3 and 4 – 4 respectively in J / kg

Since it is steady flow system, writing SFEE for the coolant.

$$\dot{m}_1\left[\frac{C_1^2}{2}+gZ_1+h_1\right]+\frac{\delta Q}{dt} = \dot{m}_2\left[\frac{C_2^2}{2}+gZ_2+h_2\right]+\frac{\delta W}{dt} \qquad \text{...(1)}$$

And for steam,

$$\dot{m}_3\left[\frac{C_3^2}{2}+gZ_3+h_3\right]+\frac{\delta Q}{dt} = \dot{m}_4\left[\frac{C_4^2}{2}+gZ_4+h_4\right]+\frac{\delta W}{dt} \qquad \text{...(2)}$$

Neglecting the changes in KE and PE, since they are very small and as the condenser is insulated $\delta Q / dt = 0$ and as no work is done in the condenser $\delta W / dt = 0$.

So, Eqs (1) and (2) reduce to,

$$\dot{m}_1\ h_1 = \dot{m}_2\ h_2 \qquad \text{...(3)}$$

$$\dot{m}_3\ h_3 = \dot{m}_4\ h_4 \qquad \text{...(4)}$$

Now since it is a steady flow system,

$$\begin{bmatrix}\text{Total energy entering the}\\ \text{condenser with steam}\\ \text{and coolant}\end{bmatrix} = \begin{bmatrix}\text{Total energy leaving the}\\ \text{condenser with coolant}\\ \text{and condensate}\end{bmatrix}$$

$$\therefore \quad \dot{m}_1 h_1 + \dot{m}_3 h_3 = \dot{m}_2 h_2 + \dot{m}_4 h_4$$

But for a steady flow system

$$\dot{m}_1 = \dot{m}_2 \text{ and } \dot{m}_3 = \dot{m}_4$$

$$\therefore \quad \dot{m}_1 (h_1 - h_2) = \dot{m}_3 (h_4 - h_3)$$

$$\therefore \quad \frac{\dot{\mathbf{m}}_1}{\dot{\mathbf{m}}_3} = \frac{\mathbf{h}_4 - \mathbf{h}_3}{\mathbf{h}_1 - \mathbf{h}_2}. \text{ Hence proved.}$$

2.20 CONTINUITY EQUATION OR LAW OF CONSERVATION OF MASS

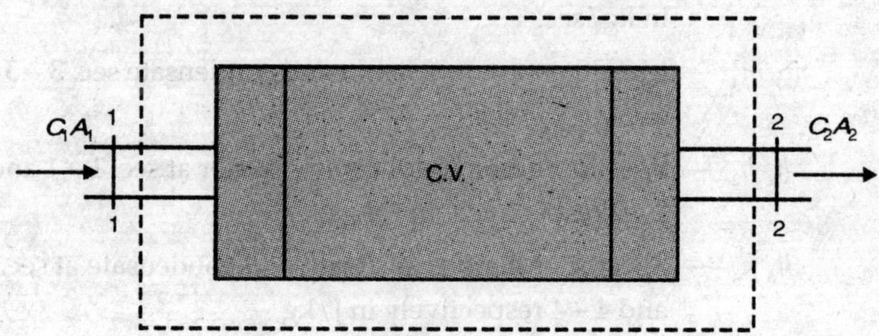

Fig. 2.20

Continuity equation is based on the principle of conservation of mass

Let $\rho_1 \rho_2$ — Density of the fluid at sec 1 – 1 and 2 – 2 respectively in kg/m^3

$A_1 A_2$ — Cross sectional areas at sec 1 – 1 and 2 – 2 respectively in m^2

$C_1 C_2$ — Velocities of the fluid at sec 1 – 1 and 2 – 2 respecytively in m/sec

Then the volume of the fluid entering at sec 1–1 per unit time $= A_1 \times C_1$

($\therefore$ A_1 is in m^2 and C_1 is in m / sec then resultant unit will be m^3 / sec)

$\therefore$ Mass of the fluid entering at sec 1–1 per unit time,

$$= \text{Density} \times \text{Volume}$$

$$= \rho_1 \times A_1 \times C_1$$

Simlarly mass of fluid leaving at sec 2–2 per unit $= \rho_2 A_2 C_2$

Since it is steady flow system,

Mass flow rate at sec 1 – 1 = Mass flow rate at sec 2 – 2

$$\therefore \quad \rho_1 A_1 C_1 = \rho_2 A_2 C_2$$

$$\text{Since} \quad \rho = \frac{1}{v}$$

$$\therefore \quad \frac{A_1 C_1}{v_1} = \frac{A_2 C_2}{v_2}$$

$$\text{or} \quad \rho AC = \text{constant}$$

$$\frac{AC}{v} = \text{constant}$$

These are the continuity equations.

2.21 LIMITATIONS OF FIRST LAW OF THERMODYNAMICS

We have seen that first law of thermodynamics is simply a law of conservation of energy and has its own limitations. These limitations are:

1. The law considers all forms of energies equivalent i.e. the first law of thermodynamics is a law of energy equivalence. This law is necessary condition so far as the account of energy balance is concerned with the possibility of transformation of one kind of energy to another.
2. The first law does not consider the direction of energy transformation.
3. The first law of thermodynamics does not consider the grade of the energy or energy reservoirs. It, assumes that all energy reservoirs are identical.

The second law of thermodynamics relates to the direction of energy exchange processes.

One of the statements of the second law of thermodynamics is Clausius statement and states that no cyclic process is prossible whose result is the flow of heat out of a heat reservoir at one temperature and the flow of an equal quantity of heat into a second reservoir at a higher temperature, without external work. Or, heat can not flow from one body at a lower temperature to the another body at higher temperature without any external work.

This principle is used in a refrigerator. For refrigerator external power is supplied to the compressor.

Without proof or discussion we here define the third law of thermodynamics.

The third law states that the entropy of a pure substance is zero at absolute zero temperature.

SOLVED EXAMPLES

EXAMPLE 2.1 *During a cycle consisting of 4-processes, the heat transfers are 60 kJ, – 8kJ, – 34 kJ and 6 kJ. Determine the net work for the cycle.*

Solution

We know that from the first law for a closed system undergoing a cycle,

Total heat transfer in a cycle = Total work transfer in a cycle

i.e. $$\oint \delta Q = \oint \delta W$$

As given in the problem, the cycle consists of 4-processes and heat transfers for the processes are given.

$$\therefore \quad \oint \delta Q = 60\text{ kJ} - 8\text{ kJ} - 34\text{ kJ} + 6\text{ kJ}$$

$$\oint \delta Q = 24\text{ kJ} = \oint \delta W$$

∴ Net work transfer in a cycle = 24 kJ

EXAMPLE 2.2 *In a non-flow reversible process 300 kJ of heat leaves the system consisting of a gas. The Internal Energy of the gas remains the same. Calculate the work done.*

Data: $Q = -300$ *kJ (Since heat flows out of the system)*

$U =$ *Constant* $\quad \therefore \quad \Delta U = 0$

$W = ?$

Solution

We know that first law for a closed system undergoing a process $Q - W = \Delta U$. As given in problem $Q - W = 0$, since $\Delta U = 0$.

$$\mathbf{Q = W = -300\text{ kJ}}$$

i.e. 300 kJ of work is done on the gas.

EXAMPLE 2.3 *In a non-flow reversible process, the pressure and volume are related by*

$$P = \left[\frac{3.5}{V} + 3\right] \text{bar}$$

where P is in bar and V is in m³. During the process the volume changes from 1.5 m³ to 4.5 m³ and heat added is 2000 kJ. Determine change in Internal Energy (I.E.)?

Data: $$P = \left[\frac{3.5}{V} + 3\right] \times 10^2 \text{ kPa}$$

$$V_1 = 1.5\ m^3$$

$$V_2 = 4.5\ m^3$$
$$Q = 2000\ kJ$$
$$\Delta U = ?$$

Solution

We know that first law for a closed system undergoing a process,

$$Q - W = \Delta U$$

Q is given, ΔU is to be found

∴ To find W,

Also we know that,

$$\text{Work done} = \int P\,dV \text{ in closed system}$$

Substituting $$P = \left[\frac{3.5}{V} + 3\right] \times 10^2 \text{ kPa}$$

We get $$\text{W.D.} = \int \left[\frac{3.5}{V} + 3\right] \times 10^2\ dV$$

$$= 100 \int_{1.5}^{4.5} \left(\frac{3.5}{V} + 3\right) dV \text{ kJ}$$

$$= 100\left[(3.5 \ln V + 3V)\right]_{1.5}^{4.5}$$

$$= 100\left[3.5 \ln \frac{4.5}{1.5} + 3(4.5 - 1.5)\right]$$

$$= 100\left[3.5 \ln 3 + 3 \times (3)\right]$$

$$= 100[3.845 + 9] = 100[12.845]$$

$$\text{Work done} = 1284.5143 \text{ kJ}$$

$$\therefore \quad Q - W = \Delta U$$

$$2000 - 1284.5143 = \Delta U$$

$$\therefore \quad \mathbf{\Delta U = 715.4857}$$

EXAMPLE 2.4 *In a boiler water enters with an enthalpy of 35 kJ/kg and steam leaves with the enthalpy of 705 kJ/kg. Find the heat transfer per kg of steam. The change in KE and PE may be neglected.*

Solution

We know that, SFEE on mass basis,

$$\left[\frac{C_1^2}{2} + gZ_1 + h_1\right] + q = \left[\frac{C_2^2}{2} + gZ_2 + h_2\right] + w$$

$$(0+0+h_1)+q = (0+0+h_2)+w$$

$$h_1+q = h_2+w$$

But in a boiler since no work is done $w = 0$

$$\therefore \quad q = h_2 - h_1$$

$$q = 705 - 35$$

$$\therefore \quad \mathbf{q = 670 \text{ kJ/kg}}$$

EXAMPLE 2.5 *A piston and cylinder machine contains a fluid system which passes through a complete cycle of 4-processes. During a cycle the sum of all heat transfers is – 170 kJ. The system completes 100 cycles/min. Complete the following table showing the method for each item and calculate the net rate of work output in kW.*

Process	*Q (kJ/min)*	*W (kJ/min)*	*4E (kJ/min)*
a-b	*0*	*2170*	—
b-c	*21,000*	*0*	—
c-d	*– 2100*	—	*– 36600*
d-a	—	—	—

Solution

Process a–b

$$Q = \Delta E + W$$

$$0 = \Delta E + 2170$$

$$\mathbf{\Delta E = -2170 \text{ kJ/min}}$$

Process b–c

$$Q = \Delta E + W$$

$$21000 = \Delta E + 0$$

$$\therefore \quad \mathbf{\Delta E = 21000 \text{ kJ/min}}$$

Process c–d

$$Q = \Delta E + W$$

$$-2100 = -36600 + W$$

$$\mathbf{W = 34500 \text{ kJ/min}}$$

Process d–a

It is given that, sum of all heat transfers/cycle

i.e. $$\oint Q = -170 \text{ kJ}$$

Secondly, the system completes 100 cycles/min

$\therefore$ Total heat transfer/min = – 17000 kJ/min

$$Q_{a-b} + Q_{b-c} + Q_{c-d} + Q_{d-a} = -17000 \text{ kJ/min}$$

$$0 + 21000 - 2100 + Q_{d-a} = -1700$$

$$\therefore \quad \mathbf{Q_{d-a} = -35900 \text{ kJ/min}}$$

While studying energy a property of the system, we have studied that change in energy for the process $1-a-2 = \Delta E_a$

And change in energy for the process,

$$2-b-1 = \Delta E_b(-ve)$$

$\therefore$ Net change in energy for the cycle

$$= \Delta E_a - \Delta E_b = 0$$

$\therefore$ We can write cyclic integral of any property is zero or $\oint dE = 0$

So, $\Delta E_{a-b} + \Delta E_{b-c} + \Delta E_{c-d} + \Delta E_{d-a} = 0$

$\therefore$ $-2170 + 21000 - 36600 + \Delta E_{d-a} = 0$

$\therefore$ $\mathbf{\Delta E_{d-a} = 17770\ kJ/min}$

We know that, $Q_{d-a} = \Delta E_{d-a} + W_{d-a}$

$$W_{d-a} = Q_{d-a} - \Delta E_{d-a} = -35900 - 17770$$

$$\mathbf{W_{d-a} = -53670\ kJ/min}$$

Lastly rate of work output since

$$\oint \delta Q = \oint \delta W$$

$$= -17000\ \text{kJ / min}$$

$$\oint \delta \mathbf{W} = \textbf{Net rate of work} = \mathbf{-283.33\ kW\ or\ kJ/sec}$$

EXAMPLE 2.6. *A certain cycle consists of 3 processes. The energy transfer in each process are tabulated below.*

Process	*Q (kJ/kg)*	*w (kJ/kg)*	*Δu (kJ/kg)*
1-2	50	—	20
2-3	– 30	– 40	—
3-1	—	—	– 30

If the network done/kg of fluid is + 30 kJ/kg complete the table. Also find the power developed, if the mass of fluid in the cycle is 0.1 kg and the system completes 10 such cycles per sec.

Solution

Since all the equalities are given for unit mass,

For process 1–2 $q = \Delta u + w$

$$50 = 20 + w$$

$\therefore \quad$ **w = 30 kJ / kg**

For process 2–3 $\quad q = \Delta u + w$

$$-30 = \Delta u - 40$$

$$\mathbf{\Delta u = 10\ kJ/kg}$$

For process 3–1

It is given that, net w.d./kg $= -30$ kJ / kg

$$\therefore \quad W_{1-2} + w_{2-3} + w_{3-1} = 30 \text{ kJ / kg}$$

$$30 - 40 + w_{3-1} = 30$$

$$30 - 40 + w_{3-1} - 30 = 0$$

$$\therefore \quad \mathbf{W_{3\text{-}1} = 40\ kJ/kg}$$

$$\therefore \quad q = \Delta u + w$$

$$q = -30 + 40$$

$$\mathbf{q = 10\ kJ/kg}$$

It is given that W.D./kg = 30 kJ/kg and mass of the fluid flowing = 0.1 kg/cycle and system completes 10 cycle per sec.

$$\therefore \quad \dot{m} = 0.1 \text{ kg / cycle} \times 10 \text{ cycle / sec}$$

$$\therefore \quad \dot{m} = 1 \text{ kg / sec}$$

$\therefore$ We know that w.d. / kg $= 30$ kJ / kg

$\therefore$ Rate of w.d. or power $= 30 \text{ kJ / kg} \times 1 \text{ kg / sec} = 30 \text{ kJ / sec} = 30 \text{ kW}$

Remember: $\quad \dfrac{\delta \mathbf{w}}{\mathbf{dt}} = \dot{\mathbf{m}} \dfrac{\delta \mathbf{w}}{\mathbf{dm}}$

Unit: $\quad \dfrac{\text{kJ}}{\text{sec}} = \dfrac{\text{kg}}{\text{sec}} \times \dfrac{\text{kJ}}{\text{kg}}$

$$\frac{\text{kJ}}{\text{sec}} = \frac{\text{kJ}}{\text{sec}}$$

EXAMPLE 2.7 *The I.E. of a certain substance is given by the following equation,*

$$u = 3.56\ Pv + 84,$$

where u is in kJ/kg, P is in kPa, v is in m^3*/kg.*

A system composed of 3 kg of this substance expands from an initial pressure of 500 kPa and a volume of 0.22 m^3 *to a final pressure of 100 kPa in a process in which pressure and volume are related by* $PV^{1.2} = C$.

(i) *If the expansion is quasi-static, find Q, ΔU and W for the process.*

(ii) *In another process the same system expands according to the same pressure-volume relation as in case (i) and from the same initial state and same final state as in case (i), but the heat transfer for this case is 30 kJ. Find the work transfer for this process.*

(iii) *Explain the difference in work transfer in case (i) and (ii).*

Data : u is in kJ/kg

P is in kPa $\quad v = m^3/kg$

$$m = 3\ kg$$

$$P_1 = 500\ kPa$$

$$V_1 = 0.22\ m^3$$

$$P_2 = 100\ kPa$$

$$PV^{1.2} = C$$

$$Q = ? \quad \Delta U = ? \quad W = ?$$

Solution

It is givmen that $\quad u = 3.56\ Pv + 84$

Change in I.E. for unit mass

$$\Delta u = u_2 - u_1$$

$$= 3.56\ P_2V_2 + 84 - 3.56P_1V_1 - 84$$

$$\Delta u = 3.56(P_2V_2 - P_1V_1)$$

$\therefore$ Total change in IE $= \Delta U = 3.56(P_2V_1 - P_1V_1)$

Note: V is the total volume for 3 kg mass and it is an extensive property.

$\therefore$ It depends on mass and P is an intensive property.

Since in Eq. (1) V_2 is not given,

$\therefore$ from $\quad P_1 V_1^{1.2} = P_2 V_2^{1.2}$

$$\therefore \quad V_2 = V_1\left(\frac{P_1}{P_2}\right)^{\frac{1}{1.2}} = 0.22\left(\frac{500}{100}\right)^{\frac{1}{1.2}}$$

$$V_2 = 0.22 \times 3.83 = 0.845\ \text{m}^3$$

Now from (1) $\quad \Delta U = 3.56[100 \times 0.845 - 500 \times 0.22]\ \text{kJ}$

$$= 356[1 \times 0.845 - 5 \times 0.22]\ \text{kJ}$$

$$\mathbf{\Delta U = -91\ kJ}$$

For a quasi-static process, W.D. $= \int PdV$

Since it is a polytropic process,

$$W_{1-2} = \frac{P_2V_2 - P_1V_1}{1-n}$$

$$W_{1-2} = \frac{100\times0.845 - 500\times0.222}{1-1.2}$$

$$\mathbf{W_{1-2} = 127.5\ kJ}$$

$$\therefore \quad Q = \Delta U + W$$

$$Q = -91 + 127.5$$

$$\mathbf{Q = 36.5\ kJ}$$

Case (ii) Here $Q = 30\ \text{kJ}$

Since the end states are same as in case (i)

$$W = Q - \Delta U = 30 - (-91)$$

$$\mathbf{W = 121\ kJ}$$

(iii) The work in case (ii) is not equal to $\int PdV$. So the process is not quasi-static.

EXAMPLE 2.8 *The I. E. of a certain substance is given by the equation $U = 3.5\,PV + 80$. Where U is in kJ, pressure P is in kPa, and volume V is in m^3. A system composed of 5 kg of this substance expands from an initial pressure of 5 bar and volume of 0.22 m^3 to a final pressure of 1 bar in a process in which pressure and volume are related by $PV^{1.2} = C$.*

Find for quasi-static process,

(1) Heat transfer, (2)Total change in I.E. (3) Non-flow work.

Data: U *in kJ;* P *in kPa*

V in m^3; $m = 5\ kg$

$$P_1 = 5\ bar = 500\ kPa$$

$$V_1 = 0.22\ m^3\ ; \quad P_2 = 1\ bar = 100\ kPa$$

Solution

$$P_1V_1^{1.2} = P_2V_2^{1.2} = C$$

$$\therefore \quad V_2 = V_1 \times \left(\frac{P_1}{P_2}\right)^{\frac{1}{1.2}}$$

$$V_2 = 0.22 \times (5)^{1/1.2} = 0.84\ \text{m}^3$$

Now ΔU i.e. change in I.E. will be,

$$\Delta U = U_2 - U_1 = 3.5(P_2V_2 - P_1V_1)$$

$$= 3.5[100 \times 0.84 - 500 \times 0.22]$$

$$\mathbf{\Delta U = -91 \ kJ}$$

$$W_{1-2} = \frac{P_1V_1 - P_2V_2}{n-1}$$

$$= \frac{500 \times 0.22 - 100 \times 0.84}{0.2}$$

$$\mathbf{W_{1-2} = 130 \ kJ}$$

$$Q = \Delta U + W = -91 + 130$$

$$\mathbf{Q = 39 \ kJ}$$

EXAMPLE 2.9 *A fluid is confined in a cylinder by a spring loaded frictionless piston, so that the pressure in the fluid is a linear function of the volume, $P = a + bV$. The internal energy of the fluid is given by the following equation $U = 34 + 3.15\,PV$.*

Where U is in kJ, P in kPa, V in m^3. If the fluid changes from an initial state of 170 kPa, 0.03 m^3 to a final state of 400 kPa, 0.06 m^3 with no work other than that done on the piston. Find the direction and magnitude of the work and heat transfer.

Data: $P = a + bV$; $U = 34 + 3.15\,PV$

$P_1 = 170 \ kPa$; $V_1 = 0.03 \ \text{m}^3$

$P_2 = 400 \ kPa$; $V_2 = 0.06 \ \text{m}^3$

$W = ?$ $Q = ?$

Solution

The change in I.E. of the fluid during the process is given as,

$$U_2 - U_1 = 3.15(P_2V_2 - P_1V_1)$$

$$\left[\textbf{Note: } \text{Units} \frac{\text{kN}}{\text{m}^2} \times \text{m}^3 = \text{kNm} = \text{kJ} \right]$$

$$U_2 - U_1 = 3.15[400 \times 0.06 - 1.70 \times 0.03]$$

$$= 100 \times 3.15[4 \times 0.06 - 1.70 \times 0.03]$$

$$U_2 - U_1 = 315 \times 0.189$$

$$\mathbf{U_2 - U_1 = 59.5 \ kJ}$$

Now $P = a + bV$

$$170 = a + b \times 0.03$$

$$400 = a + b \times 0.06$$

From these two equations on solving we get,

$$a = -60 \ \text{kN/m}^2$$

$$b = 7667 \text{ kN/m}^5$$

We know that, W.D. during the process $W_{1-2} = \int PdV$

$$= \int_{V_1}^{V_2} (a+bV)dV$$

$$= a(V_2 - V_1) + b\frac{(V_2^2 - V_1^2)}{2}$$

$$= (V_2 - V_1)\left[a + \frac{b}{2}(V_1 + V_2)\right]$$

$$= 0.03 \text{ m}^3 \left[-60\frac{\text{kN}}{\text{m}^2} + \frac{7667}{2}\frac{\text{kN}}{\text{m}^5} \times 0.09 \text{ m}^3\right]$$

$$\mathbf{W_{1-2} = 8.55 \text{ kJ}}$$

Since W_{1-2} is positive, work is done by the system,

∴ Heat transfer involved is given by

$$Q = \Delta U + W$$

$$Q_{1-2} = U_2 - U_1 + W_{1-2} = 59.5 + 8.55$$

$$Q_{1-2} = 68.05 \text{ kJ}$$

Since Q_{1-2} is positive, heat flows into the system during the process.

EXAMPLE 2.10 *The expression for the I.E. of a certain closed system is given by U = M + N (PV) where P is the pressure, V is the volume and M and N are constants. The system now undergoes a reversible process where Q = 0. Show that the relation between P and V is given by, $PV^K = C$ where $K = \dfrac{N+1}{N}$.*

Solution

As given in the problem,

$$U = M + N(P.V) \quad \text{differentiating we get,} \qquad ...(1)$$

$$dU = O + N.P.dV + N.V.dP$$

We know that from first law for the process,

$$Q - W = \Delta U$$

$$Q = \Delta U + W$$

or $$\delta Q = dU + \delta W$$

$$\delta Q = dU + P.\,dV$$

Since $Q = 0,\ 0 = dU + P.\,dV$

Given, $dU = -P.\,dV$ ...(2)

Equating (1) and (2)

$$-P.\,dV = N.\,P.dV + N.V\,dP.$$

$$O = N.\,P.dV + N.V\,d\,P + P.\,dV$$

$$= (N+1)P.\,dV + N.V.\,dP$$

Dividing throughout by N, we get

$$O = \frac{N+1}{N}P.dV + V.dP$$

i.e. $$-V.dP = \frac{N+1}{N}P.\,dV$$

$$-VdP = K.\,P.\,dV \text{ as given } \frac{N+1}{N} = K$$

$$-\frac{d\,P}{P} = K.\frac{dV}{V}$$

Integrating, $$\ln\frac{1}{P} = \ln V^K$$

or $$\mathbf{PV^K = C}$$

where $$K = \frac{N+1}{N}$$

EXAMPLE 2.11 *Following data refers to a steady flow process.*

	At entrance	*At exit*
Enthalpy (kJ/kg)	4000	4100
Velocity (m/sec)	50	20
Height (m)	50	10

Mass flow rate $\dot{m} = 1$ *kg/sec*

Heat transfer rate = *200 kJ/sec to the system.*

Determine the power capacity of the system and state whether it is a power producing system or otherwise.

Solution

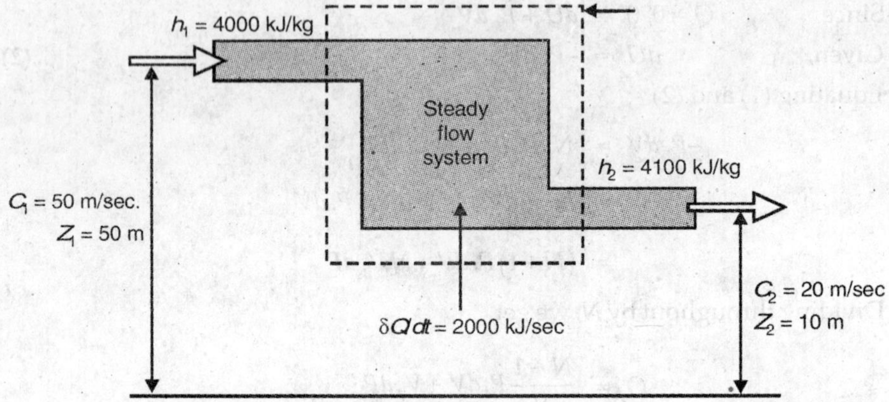

Fig. Ex. 2.11

$$\dot{m} = 1\,\text{kg}/\text{sec}$$

Since heat transfer is given in kJ/sec, we will have to use SFEE on time basis.

$$\dot{m}\left[\frac{C_1^2}{2}+gZ_1+h_1\right]+\frac{\delta Q}{dt} = \dot{m}\left[\frac{C_2^2}{2}+gZ_2+h_2\right]+\frac{\delta W}{dt}$$

Since power is to found,

$$\frac{\delta W}{dt} = \dot{m}\left[\frac{C_1^2-C_2^2}{2}+g(Z_1-Z_2)+(h_1-h_2)\right]+\frac{\delta Q}{dt}$$

$$\frac{\delta W}{dt} = \frac{1\,\text{kg}}{\text{sec}}\left[\frac{50^2-20^2}{2\times 1000}\frac{\text{kJ}}{\text{kg}}+\frac{9.81(40)}{1000}\frac{\text{kJ}}{\text{kg}}+(4000-4100)\right]+200\frac{\text{kJ}}{\text{sec}}$$

$$\frac{\delta W}{dt} = \frac{1\,\text{kg}}{\text{sec}}\left[1.05\frac{\text{kJ}}{\text{kg}}+0.3924\frac{\text{kJ}}{\text{kg}}+-100\frac{\text{kJ}}{\text{kg}}\right]+200\frac{\text{kJ}}{\text{sec}}$$

$$\frac{\delta W}{dt} = 1.05\frac{\text{kJ}}{\text{sec}}+0.3924\frac{\text{kJ}}{\text{sec}}-100\frac{\text{kJ}}{\text{sec}}+200\frac{\text{kJ}}{\text{sec}}$$

$$\frac{\delta \mathbf{W}}{\mathbf{dt}} = \mathbf{101.4484\ kJ/sec\ or\ kW}$$

Since $\frac{\delta W}{dt}$ is + ve, it is a power producing system.

EXAMPLE 2.12 *In a steady flow system 135 kJ/kg of work is done by the system. The specific volume of the fluid, pressure and velocity at the inlet are 0.37 m³/kg, 600 kPa, 16 m/sec. The inlet is 32 m above the floor level. The discharge conditions are 0.62 m³/kg, 100 kPa and 270 m/sec. The total heat loss between inlet and discharge is 9 kJ/kg. In flowing through this system, does the specific I.E. increase or decrease and by how much? The outlet is at flow level.*

Solution

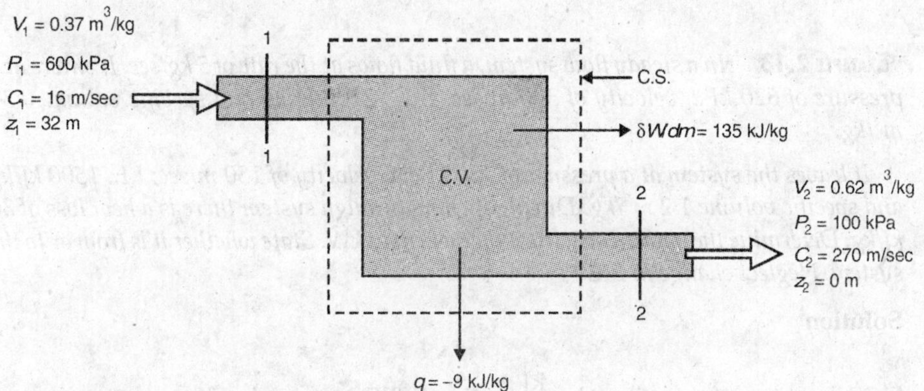

Fig. Ex. 2.12

Since work and heat transfer are given in $\frac{kJ}{kg}$

Using SFEE on mass basis,

$$\left[\frac{C_1^2}{2}+gZ_1+h_1\right]+q = \left[\frac{C_2^2}{2}+gZ_2+h_2\right]+w$$

Since pressure, specific volumes are given and change in Sp. I.E., is to be found, substitute $h_1 = u_1 + P_1v_1$ and $h_2 = u_2 + P_2v_2$ we get,

$$\left[\frac{C_1^2}{2}+gZ_1+u_1+P_1v_1\right]+q = \left[\frac{C_2^2}{2}+gZ_2+u_2+P_2v_2\right]+w$$

$$\therefore \quad u_1 - u_2 = \frac{C_2^2 - C_1^2}{2} + g(Z_2 - Z_1) + (P_2v_2 - P_1v_1) + w - q$$

$$u_1 - u_2 = \frac{270^2 - 16^2}{2 \times 1000} \frac{\text{kJ}}{\text{kg}} + \frac{9.81\,(-32)}{1000} \frac{\text{kJ}}{\text{kg}}$$

$$+ (100 \times .62 - 600 \times 0.37) \frac{\text{kJ}}{\text{kg}} + 135 \frac{\text{kJ}}{\text{kg}} - \left(-9 \frac{\text{kJ}}{\text{kg}}\right)$$

$$u_1 - u_2 = 36.45 - 0.314 - 160 + 135 + 9$$

$$u_1 - u_2 = 36.45 - 20.136 \text{ kJ/kg}$$

$$u_1 - u_2 = 20.136 \text{ kJ/kg}$$

$$\therefore \quad \mathbf{u_2 - u_1 = -20.136 \text{ kJ/kg}}$$

$\therefore$ Specific I.E. decreases by 20.236 kJ/kg

EXAMPLE 2.13. *In a steady flow system, a fluid flows at the rate of 5 kg/sec. It enters at a pressure of 620 kPa, velocity of 300 m/sec. I. E., 2100 kJ/kg and specific volume 0.37 m³/kg.*

It leaves the system at a pressure of 130 kPa, a velocity of 150 m/sec; I.E. 1500 kJ/kg and specific volume 1.2 m³/kg. During its flow through system there is a heat loss of 30 kJ/kg. Determine the power capacity of the system in kW. State whether it is from or to the system. Neglect change in P.E.

Solution

Since heat transfer is given in $\frac{\text{kJ}}{\text{kg}}$, using SFEE on mass basis.

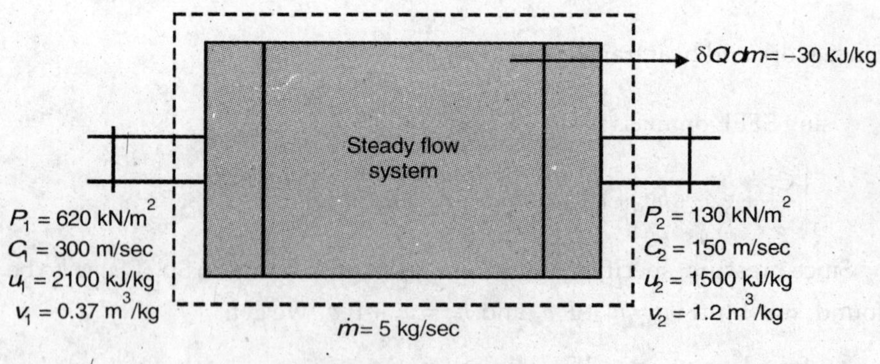

Fig. Ex. 2.13

Note: Since P, u, v are given, writing the equation in terms of these quantities.

$$\left[\frac{C_1^2}{2} + gZ_1 + u_1 + P_1 v_1\right] + \frac{\delta Q}{dm} = \left[\frac{C_2^2}{2} + gZ_2 + u_2 + P_2 v_2\right] + \frac{\delta W}{dm}$$

Since power capacity is to be found out,

$$\frac{\delta W}{dm} = \left[\frac{C_1^2 - C_2^2}{2} + g(Z_1 - Z_2) + (u_1 - u_2) + (P_1 v_1 - P_2 v_2)\right] + \frac{\delta Q}{dm}$$

Neglecting ΔPE as given in the problem,

$$\frac{\delta W}{dm} = \left[\frac{C_1^2 - C_2^2}{2} + (u_1 - u_2) + (P_1 v_1 - P_2 v_2)\right] + \frac{\delta Q}{dm}$$

$$\frac{\delta W}{dm} = \frac{300^2 - 150^2}{2 \times 1000}\frac{kJ}{kg} + (2100 - 1500) + (620 \times 0.37 - 130 \times 1.2) - 30\frac{kJ}{kg}$$

$$\frac{\delta W}{dm} = \left[\frac{67500}{2000} + 600 + 73\right]\frac{kJ}{kg} - 30\frac{kJ}{kg}$$

$$= 33.75 + 600 + 73 - 30$$

$$\mathbf{\frac{\delta W}{dm} = 676.75\ kJ/kg}$$

Since power is required to be expressed in kW,

Using $\frac{\delta W}{dt} = \dot{m}\frac{\delta W}{dm}$

Note: Mass flow rate = 5 kg/sec is given

$\therefore$ Power $= 676.75\frac{kJ}{kg} \times 5\frac{kg}{sec}$

Power = 3383.75 kW $= \mathbf{\frac{\delta W}{dt}}$

EXAMPLE 2.14 *The following data is obtained from tests on an air compressor.*

	At Inlet	*At Exit*
Pressure (kPa)	100	500
Sp. volume (m^3/kg)	0.6	0.15
Sp. Internal Energy (kJ/kg)	50	125
Velocity of air (m/sec)	8	4

Rate of flow of air is 5 kg/sec.

Heat rejected to cooling water = 45 kW

Find the power required to drive the compressor in kW.

Solution

Since heat is rejected is given in kJ / sec., using SFEE on time basis,

$$\dot{m}\left[\frac{C_1^2}{2}+gZ_1+h_1\right]+\frac{\delta Q}{dt} = \dot{m}\left[\frac{C_2^2}{2}+gZ_2+h_2\right]+\frac{\delta W}{dt}$$

$$\therefore \quad \frac{\delta Q}{dt}-\frac{\delta W}{dt} = \dot{m}\left[\frac{C_2^2-C_1^2}{2}+g(Z_2-Z_1)+(P_2v_2-P_1v_1)+(u_2-u_1)\right]$$

$$-45-\frac{\delta W}{dt} = 5\left[\frac{4^2-8^2}{2\times 1000}+0+(500\times 0.15-100\times 0.6)+(125-50)\right]$$

$$-45-\frac{\delta W}{dt} = 5\left[\frac{16-64}{2000}+(75-60)+75\right]$$

$$-45-\frac{\delta W}{dt} = 5[-0.024+15+75]$$

$$\Rightarrow \quad \mathbf{\frac{\delta W}{dm} = -494.88\ kW}$$

$\therefore$ Power required to drive the compressor = 494.88 kW.

EXAMPLE 2.15 *Steam enters a steam turbine with a velocity of 40 m/sec, enthalpy 2500 kJ/kg and leaves with a velocity of 90 m/sec and enthalpy 2030 kJ/kg. Heat losses from the turbine insulation to the surroundings are 250 kJ/min, and the steam flow rate is 5000 kg/hr. Neglect the change of P.E. Find the power developed by the turbine in kW.*

Solution

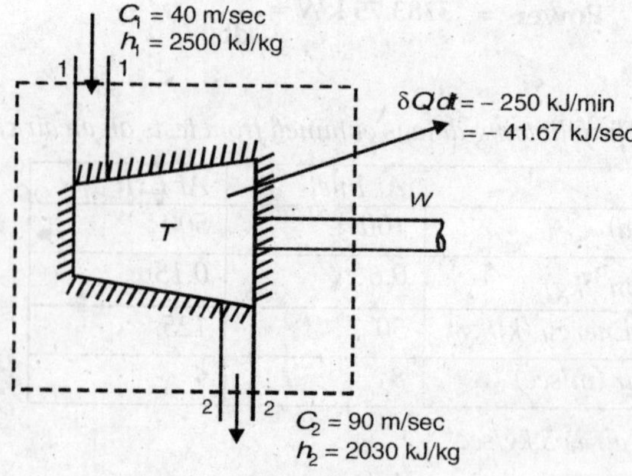

Fig. Ex. 2.15

$$\text{Steam flow rate} = 5000 \text{ kg/hr} = \frac{5000}{3600} \text{ kg/sec} = 1.389 \text{ kg/sec}$$

$$\dot{m}\left[\frac{C_1^2}{2} + gZ_1 + h_1\right] + \frac{\delta Q}{dt} = \dot{m}\left[\frac{C_2^2}{2} + gZ_2 + h_2\right] + \frac{\delta W}{dt}$$

$$\dot{m}\left[\frac{C_1^2 - C_2^2}{2} + (h_1 - h_2)\right] + \frac{\delta Q}{dt} = \frac{\delta W}{dt}$$

$$1.389 \frac{\text{kg}}{\text{sec}}\left[\frac{40^2 - 90^2}{2 \times 1000} \frac{\text{kJ}}{\text{kg}} + (2500 - 2030)\frac{\text{kJ}}{\text{kg}}\right] - 4.167 \frac{\text{kJ}}{\text{sec}} = \frac{\delta W}{dt}$$

$$1.389\left[-3.25 \frac{\text{kJ}}{\text{kg}} + (470)\frac{\text{kJ}}{\text{kg}}\right] - 4.167 \frac{\text{kJ}}{\text{sec}} = \frac{\delta W}{dt}$$

$$1.389 \frac{\text{kg}}{\text{sec}}\left[466.75 \frac{\text{kJ}}{\text{kg}}\right] - 4.167 \frac{\text{kJ}}{\text{sec}} = \frac{\delta W}{dt}$$

$$\mathbf{\frac{\delta W}{dt} = 644.149 \text{ kW}}$$

EXAMPLE 2.16 *Steam enters a steam turbine with a velocity of 16 m/sec. and specific enthalpy 2990 kJ/kg. The steam leaves the turbine with a velocity of 37 m/sec and specific enthalpy of 2530 kJ/kg. The heat lost to the surroudings is 25 kJ/kg. The steam flow rate is 3,60,000 kg/hr. Determine the work output from the turbine in kW.*

Solution

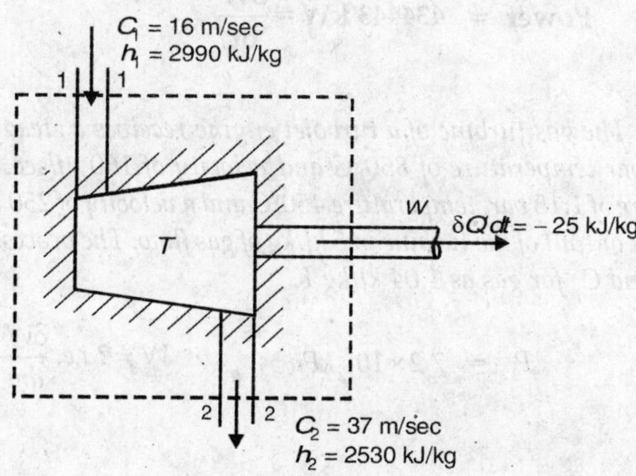

Fig. Ex. 2.16

Since heat lost is given in kJ / kg, using SFEE on mass basis,

$$\left[\frac{C_1^2}{2}+gZ_1+h_1\right]+\frac{\delta Q}{dm} = \left[\frac{C_2^2}{2}+gZ_2+h_2\right]+\frac{\delta W}{dm}$$

Since nothing is mentioned about elevation, neglecting ΔPE.

$$\therefore \quad \frac{C_1^2}{2}+h_1+\frac{\delta Q}{dm} = \frac{C_2^2}{2}+h_2+\frac{\delta W}{dm}$$

$$\therefore \quad \frac{\delta W}{dm} = \frac{(C_1^2-C_2^2)}{2}+(h_1-h_2)+\frac{\delta Q}{dm}$$

$$\frac{\delta W}{dm} = \frac{16^2-37^2}{2\times1000}\,\frac{\text{kJ}}{\text{kg}}+(2990-2530)\,\frac{\text{kJ}}{\text{kg}}-25\,\frac{\text{kJ}}{\text{kg}}$$

$$\frac{\delta W}{dm} = \frac{-1113}{2000}+460-25$$

$$\mathbf{\frac{\delta W}{dm} = 434.443\,\frac{kJ}{kg}}$$

To express power in kW, using

$$\frac{\delta W}{dt} = \dot{m}\frac{\delta W}{dm}$$

It is given that mass flow rate = 360000 kg/hr = 100 kg/sec.

$$\therefore \quad \text{Power} = 434.443\,\frac{\text{kJ}}{\text{kg}}\times100\,\frac{\text{kg}}{\text{sec}}$$

$$\mathbf{Power = 434443\ kW = \frac{\delta W}{dt}}$$

EXAMPLE 2.17 *The gas turbine of a turbojet engine receives a steady flow of gases at pressure of 7.2 bar temperature of 850°C and velocity of 160 m/sec. It discharges the gases at a pressure of 1.15 bar, temperature 450ºC and a velocity of 250 m/sec. Determine the external work output of the turbine in 5 kJ/kg of gas flow. The process may be assumed to be adiabatic and C_p for gas as 1.04 kJ/kg K.*

Data: $P_1 = 7.2\times10^2$ kPa $\qquad W = ?$ i.e. $\frac{\delta W}{dm} = ?$

Solution

$$Cp_{\text{gases}} = 1.04\ \text{kJ / kg-K}$$

$T_1 = 850°C \qquad C_1 = 160 \text{ m/sec}$

Insulation (Adiabtic Turbine) $\therefore\ Q = 0$

$$P_2 = 1.15\times10^2 \text{ kPa}$$

$$T_2 = 450° \text{ C}$$

$$C_2 = 250 \text{ m/sec}$$

We know that steady Flow Energy equation on mass basis is given by,

$$\left[\frac{C_1^2}{2}+gZ_1+h_1\right]+\frac{\delta Q}{dm} = \left[\frac{C_2^2}{2}+gZ_2+h_2\right]+\frac{\delta W}{dm} \qquad ...(1)$$

Since Adiabtic Expansion process $\frac{\delta Q}{dm}=0$

Since they have not mentioned anything about elevation $gZ_1 = 0, gZ_2 = 0$.

Then Eq. (1) becomes,

$$\left[\frac{C_1^2}{2}+h_1\right] = \left[\frac{C_2^2}{2}+h_2\right]+\frac{\delta W}{dm}$$

i.e.
$$\frac{(C_1^2-C_2^2)}{2}+(h_1-h_2) = \frac{\delta W}{dm} \qquad ...(2)$$

Since
$$h = C_P T$$

$\therefore$
$$\frac{\delta W}{dm} = (h_1-h_2)+\frac{(C_1^2-C_2^2)}{2}$$

$$= C_P(850-450)+\frac{160^2-250^2}{2\times1000}$$

$$= 1.04(850-450)+\frac{160^2-250^2}{2\times1000} \quad 416+(-18.45)$$

$$\frac{\delta \mathbf{W}}{\mathbf{dm}} = \mathbf{397.55\ kJ/kg}$$

EXAMPLE 2.18 *The power output of an adiabatic steam turbine is 5 MW, and the inlet and exit conditions are as under,*

	Inlet	*A Exit*
Pressure	*2 MPa*	*0.15 bar*
Temperature (°C)	*400*	*Dryness 0.9*
Velocity (m/sec)	*50*	*180*
Elevation m	*10*	*6*

Determine the work done per unit of mass of the steam flowing through turbine and mass flow rate of the steam.

Solution

We know that, since work done per unit of mass of steam is to be found, using SFEE on mass basis.

$$\left[\frac{C_1^2}{2}+gZ_1+h_1\right]+q = \left[\frac{C_2^2}{2}+gZ_2+h_2\right]+w$$

or
$$w = (h_1-h_2)+\frac{(C_1^2-C_2^2)}{2}+g(Z_1-Z_2)$$

Since $h_1 = 3248.7\text{ kJ/kg}$

[**Note:** From steam tables coresponding to and 20 bar (2 MPa)].

$h_2 = h_f + x \cdot h_{fg}$. Since te steam is wet at 0.15 bar from steam tables

and
$$h_2 = 226+0.9\times 2373.2 = 2361.88\text{ kJ/kg}$$

$$w = (3248.7-2361.88)+\frac{50^2-180^2}{2000}+\frac{4\times 9.81}{1000}$$

$$w = 871.91\text{ kJ/kg} = \frac{\delta W}{dm}$$

Also, we know that, $\frac{\delta W}{dt} = \dot{m}\frac{\delta W}{dm}$

$$5000\frac{\text{kJ}}{\text{sec}} = \dot{m}\times 871.91\text{ kJ/kg}$$

∴ $\dot{m}$ = **5.734 kg/sec**

Note: Problems which are to be solving by using steam tables will not be asked, since steam chapter is not included.

EXAMPLE 2.19 *A turbine operating on air has inlet conditions as10 bar, 750 K and 200 m/s. While exit conditions are 1.25 bar and 40 m/sec. The mass flow rate of air is 1000 kg/hr. The flow of air is assumed to be reversible adiabatic (isentropic). Calculate,*

(i) The temperature of air at exit.

(ii) The power output of turbine.

Assume C_p = 1.053 kJ/kg K and k = 1.375 adiabatic index.

Solution

$$\dot{m} = 1000 \text{ kg / hr} = 0.2777 \text{ kg / sec}$$

To find T_2, since the expansion process is adiabatic,

We know that, $$\frac{T_2}{T_1} = \left(\frac{P_2}{P_1}\right)^{\frac{k-1}{k}}$$

$$\Rightarrow \quad T_2 = T_1 \times \left(\frac{P_2}{P_1}\right)^{\frac{k-1}{k}} = 750 \times \left(\frac{1.25 \times 100}{10 \times 100}\right)^{\frac{1.375-1}{1.375}}$$

$$\mathbf{T_2 = 425.367 \text{ K}}$$

We also know that, the SFEE is,

$$\dot{m}\left[\frac{C_1^2}{2} + gZ_1 + h_1\right] + \frac{\delta Q}{dt} = \dot{m}\left[\frac{C_2^2}{2} + gZ_2 + h_2\right] + \frac{\delta W}{dt}$$

Neglecting the ΔPE, and $\frac{\delta Q}{dt} = 0$ since the turbine is adiabatic.

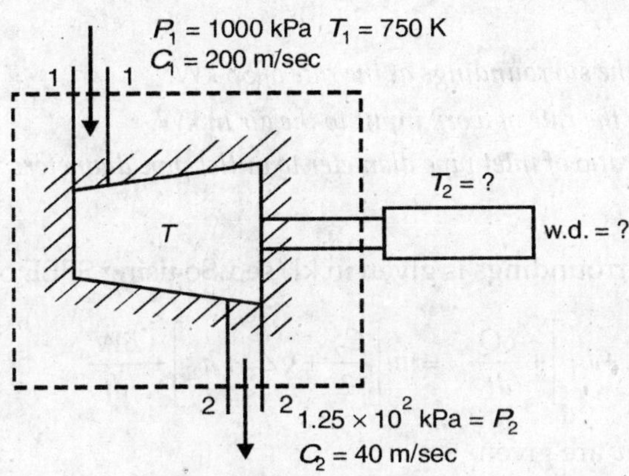

Fig. Ex. 2.19

$$\dot{m}\left[\frac{C_1^2}{2} + h_1\right] = \dot{m}\left[\frac{C_2^2}{2} + h_2\right] + \frac{\delta W}{dt}$$

$$-\frac{\delta W}{dt} = \dot{m}\left[\frac{(C_1^2 - C_2^2)}{2} + h_2 - h_1\right] \quad ...(1)$$

$$h_2 - h_1 = \Delta h = m \times C_p \,.\, \Delta T$$

$$= 1 \times 1.053 \times (425.367 - 750) \quad [\because m = 1 \text{ for unit mass}]$$

$$(h_2 - h_1) = -341.83855 \text{ kJ} \qquad \text{...(2)}$$

Substituting (2) in (1) we get,

$$-\frac{\delta W}{dt} = 0.277 \frac{\text{kg}}{\text{sec}}\left[\frac{40^2 - 200^2}{2 \times 1000} + (-341.838)\right]$$

$$-\frac{\delta W}{dt} = 0.277[-19.2 - 341.838] = 0.277[-361.038]$$

$$= -100.007 \text{ kW}$$

$$\frac{\delta \mathbf{W}}{\mathbf{dt}} = \mathbf{+100.007 \text{ kW}.}$$

EXAMPLE 2.20 *Air flows steadily at the rate of 0.5 kg/sec, through the air compressor, entering at 7 m/sec velocity, 100 kPa pressure, 0.95 m³/kg volume and leaving at 5 m/sec, 700 kPa, 0.19 m³/kg. The I.E. of the air leaving is 90 kJ/kg greater than that of air entering.*

Heat rejected to the surroundings at the rate of 58 kW.

(a) Calculate the rate of work input to the air in kW.

(b) Find the ratio of inlet pipe diameter to outlet pipe diameter.

Solution

Heat lost to the surroundings is given in kJ / sec. So using SFEE on time basis.

$$\dot{m}\left[\frac{C_1^2}{2} + gZ_1 + h_1\right] + \frac{\delta Q}{dt} = \dot{m}\left[\frac{C_2^2}{2} + gZ_2 + h_2\right] + \frac{\delta W}{dt}$$

Since, P, u, and v are given.

$$\dot{m}\left[\frac{C_1^2}{2} + gZ_1 + u_1 + P_1 v_1\right] + \frac{\delta Q}{dt} = \dot{m}\left[\frac{C_2^2}{2} + gZ_2 + u_2 + P_2 v_2\right] + \frac{\delta W}{dt}$$

Since power is to be found,

$$\frac{\delta W}{dt} = \dot{m}\left[\frac{C_2^2 - C_1^2}{2} + g(Z_2 - Z_1) + (P_2 v_2 - P_1 v_1) + (u_2 - u_1)\right] + \frac{\delta Q}{dt}$$

$$= 0.5 \frac{\text{kg}}{\text{sec}} \left[\frac{5^2 - 7^2}{2 \times 1000} + 0 + (700 \times 0.19 - 100 \times 0.95) \frac{\text{kJ}}{\text{kg}} + 90 \frac{\text{kJ}}{\text{kg}} \right] - 58 \frac{\text{kJ}}{\text{sec}}$$

Air Compressor

$\delta Q dt = -58$ kJ/sec

1, 1, 2, 2

$C_1 = 7$ m/sec, $P_1 = 100$ kPa, $v_1 = 0.95$ m^3/kg

$C_2 = 5$ m/sec, $P_2 = 700$ kPa, $v_2 = 0.19$ m^3/kg, $u_2 = u_1 + 90$

Fig. Ex. 2.20

$$= 0.5 \frac{\text{kg}}{\text{sec}} \left[(-0.012 + 38 + 90) \frac{\text{kJ}}{\text{kg}} \right] - 58 \frac{\text{kJ}}{\text{sec}}$$

$$= 0.5 \frac{\text{kg}}{\text{sec}} \left[127.988 \frac{\text{kJ}}{\text{kg}} \right] - 58 \frac{\text{kJ}}{\text{sec}}$$

$$\frac{\delta W}{dt} = -63.994 \frac{\text{kJ}}{\text{sec}} - 58 \frac{\text{kJ}}{\text{sec}}$$

$$\mathbf{\frac{\delta W}{dt} = -121.994 \text{ kW}}$$

(b) From continuity equation,

$$\dot{m} = \frac{A_1 C_1}{v_1} = \frac{A_2 C_2}{v_2}$$

$$\therefore \quad \frac{A_1}{A_2} = \frac{v_1}{v_2} \cdot \frac{C_2}{C_1} = \frac{0.95}{0.19} \times \frac{5}{7}$$

$$\mathbf{\frac{A_1}{A_2} = 3.57}$$

Also we know that, $\dfrac{A_1}{A_2} = \dfrac{\frac{\pi}{4} \cdot d_1^2}{\frac{\pi}{4} \cdot d_2^2} = 3.57$

$$\therefore \quad \frac{d_1}{d_2} = \sqrt{3.57}$$

$$\frac{d_1}{d_2} = 1.89$$

EXAMPLE 2.21 *At the inlet to a certain nozzle, the enthalpy of the fluid passing is 3000 kJ/kg and the velocity is 60 m/sec. At the discharge end the enthalpy is 2757 kJ/kg. The nozzle is horizontal. The heat loss during the flow is negligible.*

(1) Find the velocity at the exit.

(2) If the inlet area is 0.1 m² and specific volume is 0.187 m³/kg, find the mass flow rate.

(3) If the specific volume at the outlet is 0.498 m³/kg.

Find the area at the exit of the nozzle.

Solution

Since the nozzle is horizontal, $\Delta PE = 0$

Since the heat loss during the flow is negligible, $q = 0$

Since no work is done in the nozle, $w = 0$

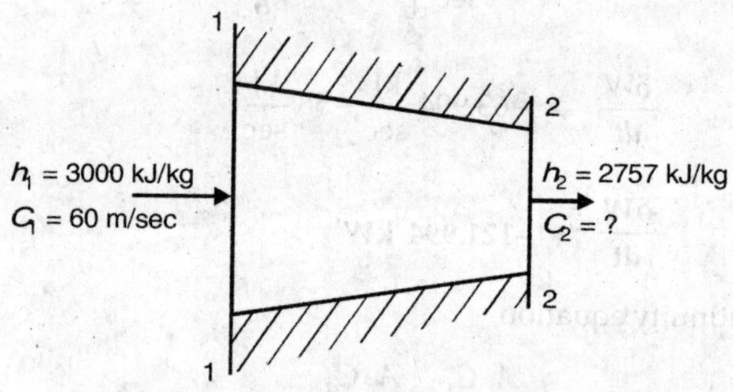

Fig. Ex. 2.21

We know that SFEE on mass basis is,

$$\left[\frac{C_1^2}{2} + gZ_1 + h_1\right] + q = \left[\frac{C_2^2}{2} + gZ_2 + h_2\right] + w \qquad \text{...(1)}$$

or

$$q - w = \frac{C_2^2 - C_1^2}{2} + g(Z_2 - Z_1) + h_2 - h_1$$

On substituting the given characteristics, Eq. (1) reduces to,

$$h_1 + \frac{C_1^2}{2} = h_2 + \frac{C_2^2}{2}$$

$$3000\frac{\text{kJ}}{\text{kg}} + \frac{60^2}{2 \times 1000} = 2757\frac{\text{kJ}}{\text{kg}} + \frac{C_2^2}{2 \times 1000}$$

$$3000 + 1.8 = 2757 + \frac{C_2^2}{2000}$$

$$244.8 = \frac{C_2^2}{2000}$$

$$\therefore \quad C_2^2 = 489600$$

$$\therefore \quad \mathbf{C_2 = 699.71\ m/sec}$$

(ii) Mass flow rate $\dot{m} = \frac{A_1 C_1}{v_1} = \frac{A_2 C_2}{v_2}$

$$= \frac{0.1\ m^2 \times 60\ m/sec}{0.187}$$

$$\mathbf{\dot{m} = 32.08\ kg/sec}$$

(iii) Area at the exit. $\dot{m} = \frac{A_1 C_1}{v_1} = \frac{A_2 C_2}{v_2}$

$$A_2 = \frac{m\,v_2}{C_2} = \frac{32.08 \times 0.498}{699.71}$$

$$\therefore \quad \mathbf{A_2 = 0.0228\ m^3}$$

EXAMPLE 2.22 *A nozzle is a device used to increase velocity of steam flowing through it. At inlet the specific enthalpy of steam is 3000 kJ/kg. Specific volume 0.187 m^3/kg and area 0.1 m^2 and at outlet sp. enthalpy 2762 kJ/kg, specific volume 0.498 m^3/kg. There is no loss of heat, nozzle is horizontal and stream enters with the velocity of 60 m/sec.*

Find: (i) Velocity of steam of inlet, (ii) Mass flow rate of steam in kg/hr, (iii) Exit area of nozzle.

Solution

For nozzle since no work is done $w = 0$.

We know that SFEE on mass basis is,

$$\left[\frac{C_1^2}{2} + gz_1 + h_1\right] + q = \left[\frac{C_2^2}{2} + gz_2 + h_2\right] + w$$

On substituting characteristic, we get,

$$h_1 + \frac{C_1^2}{2} = h_2 + \frac{C_2^2}{2}$$

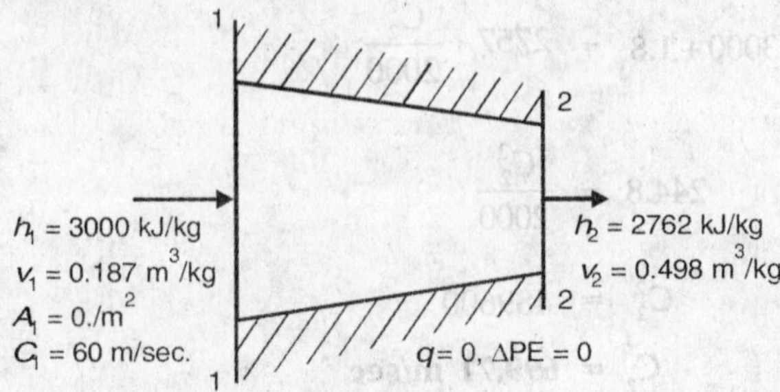

Fig. Ex. 2.22

$$3000+\frac{60^2}{2\times1000} = 2762+\frac{C_2^2}{2\times100}$$

$$\Rightarrow \quad \mathbf{C_2 = 692.53\ m/sec}$$

(ii) Mass flow rate $\dot{m} = \frac{A_1C_1}{v_1} = \frac{A_2C_2}{v_2}$

$$= \frac{0.1\times60}{0.187}$$

$$\mathbf{\dot{m} = 32.856\ kg/sec}$$

(iii) Area at the exit, $\dot{m} = \frac{A_1C_1}{v_1} = \frac{A_2C_2}{v_2}$

$$A_2 = \frac{m\,.\,v_2}{C_2} = \frac{32.856\times0.4948}{692.53}$$

$$\mathbf{A_2 = 0.0231\ m^2}$$

EXAMPLE 2.23 *Air enters a nozzle with a velocity of 40 m/sec. The decrease in the enthalpy in the nozzle is 180000 J/kg. Determine the exit velocity. Assume nozzle to be adiabatic.*

Solution

We know that, SFEE on mass basis is,

$$\left[\frac{C_1^2}{2}+gZ_1+h_1\right]+q = \left[\frac{C_2^2}{2}+gZ_2+h_2\right]+w \quad \text{...(1)}$$

Since for a nozzle characteristics are,

$$\Delta PE = 0, w = 0, q = 0 \text{ since adiabatic}$$

Equation (1) reduces to,

$$\frac{C_1^2}{2} + h_1 = \frac{C_2^2}{2} + h_2$$

$$\frac{C_1^2}{2} + (h_1 - h_2) = \frac{C_2^2}{2}$$

$$\frac{(40)^2}{2 \times 1000} + 180000 = \frac{C_2^2}{2 \times 1000}$$

$$\therefore \quad \mathbf{C_2 = 601.33 \ m/sec}$$

EXAMPLE 2.24 *Air is compressed continuously by an air compressor according to the law $PV^{1.3} = C$. The pressure at the inlet is 1 bar, and at the outlet is 5 bar. The volume of the air changes from 3 m³/kg to 0.8 m³/kg and the velocity changes 25 m/sec at inlet to 130 m/sec at the delivery. The delivery connection is 12 m above the inlet. What is the shaft power of the compressor? Is it a power absorbing or power producing system?*

Solution

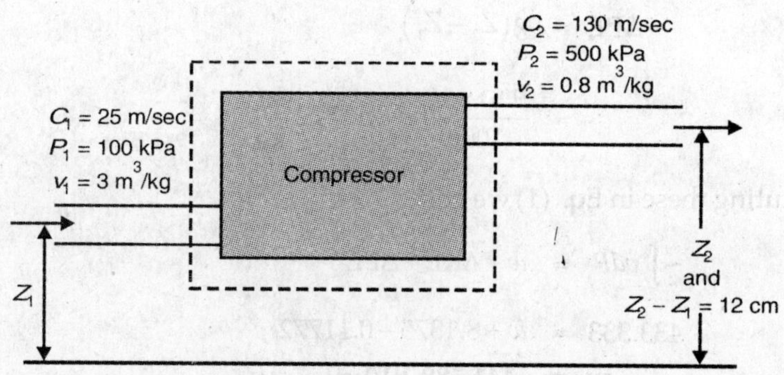

Fig. Ex. 2.24

Note. If the steady flow process or system follows. $PV^u = C$ then follow this method to solve the problem.

i.e. use $$-\int v\,dP = w + \Delta KE + \Delta PE \quad \text{...(1)}$$

We know that,

	Isothermal ($Pv = C$) (For unit mass)	Polytropic ($Pv^n = C$)
(i) W.D. in closed system $= \int P dv$	$P_1 v_1 \log_e \frac{v_2}{v_1}$	$\frac{P_2 v_2 - P_1 v_1}{1-n}$
(ii) W.D. for open system $= -\int v\, dP$	Steady flow system, $P_1 v_1 \log_e \frac{v_2}{v_1}$ or $\frac{P_1}{P_2}$	$\frac{n}{n-1} \times P_2 v_2 - P_1 v_1$

For Polytropic Process,

$$-\int_1^2 v dP = \frac{n}{n-1} \times P_1 v_1 - P_2 v_2$$

$$= \frac{1.3}{1.3-1} \times (100 \times 3 - 500 \times 0.8) \frac{\text{kJ}}{\text{kg}}$$

$$= -433.333 \text{ kJ/kg}$$

and, $$\Delta KE = \frac{1}{2}\left(\frac{C_2^2 - C_1^2}{1000}\right) = \frac{130^2 - 25^2}{2 \times 1000}$$

and $$\Delta KE = 8.1375 \frac{\text{kJ}}{\text{kg}}$$

$$\Delta PE = g(Z_2 - Z_1)$$

$$= \frac{9.81 \times 12}{1000} = 0.11772 \frac{\text{kJ}}{\text{kg}}$$

Substituting these in Eq. (1) we get,

$$\therefore \quad -\int v dP = w + \Delta KE + \Delta PE$$

$$-433.333 = w + 8.1375 + 0.11772$$

$$\therefore \quad \mathbf{w = -441.588 \text{ kJ/kg}}$$

$$= \text{Shaft work}$$

Since w is –ve work is done on the system.

∴ It is a power absorbing system.

EXAMPLE 2.25 *A fluid at the rate of 10 kg/sec is compressed adiabatically from 5 bar to 50 bar in a steady flow process. Calculate the power required assuming that the specific volume of fluid being as 0.001 m³/kg which remains almost constant.*

Solution

Since the compression process follows an adiabatic process, we have to use,

$$\text{W.D.} = -\int v dP = v.(P_2 - P_1)$$

$$\therefore \quad \text{Power required} = \dot{m} \times v \times (P_2 - P_1)$$

$$= 10 \text{ kg/sec} \times 0.001 \text{ m}^3/\text{kg} \, (50-5) \times 100$$

$$\textbf{Power required} = \textbf{45 kW}$$

EXAMPLE 2.26 *A blower handles 2 kg/sec of air at 20ºC and consumes a power of 30 kW. The inlet and outlet velocities are 100 in/sec and 150 m/sec respectively. Estimate the exit air temperature assuming adiabatic conditions.*

Take R *for air* $= 0.287$ *kJ/kg-K*

$$\gamma = \frac{C_p}{C_v} = 1.4$$

and $C_p = 1.005$ *kJ/kg-K*

Solution

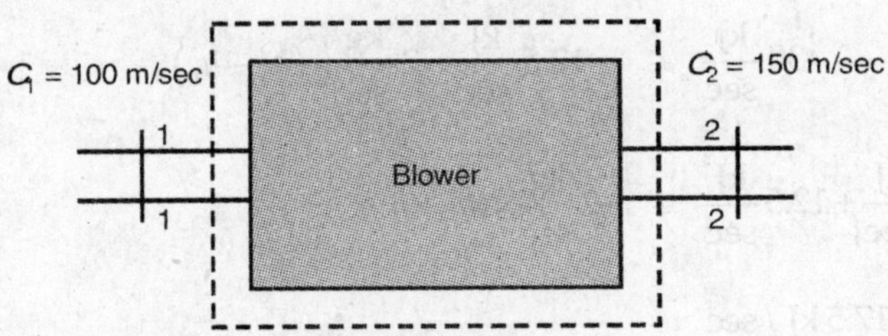

Fig. Ex. 2.26

$$\dot{m} = 2 \text{ kg/sec}$$

$$\frac{\delta W}{dt} = -30 \text{ kW} = -30 \text{ kJ/sec}$$

$$t_2 = ? \quad t_1 = 20°\text{C}$$

Since $\frac{\delta W}{dt}$ is given using the SFEE on time basis.

$$\dot{m}\left[\frac{C_1^2}{2} + gZ_1 + h_1\right] + \frac{\delta Q}{dt} = \dot{m}\left[\frac{C_2^2}{2} + gZ_2 + h_2\right] + \frac{\delta W}{dt}$$

$$\frac{\delta W}{dt} = -\dot{m}\left[\frac{C_2^2 - C_1^2}{2} + g(Z_2 - Z_1) + (h_2 - h_1)\right] + \frac{\delta Q}{dt}$$

Since adiabatic conditions are given, $\frac{\delta Q}{dt} = 0$

Neglecting the change in PE, we have,

$$\frac{\delta W}{dt} = -\dot{m}\left[\frac{C_2^2 - C_1^2}{2} + (h_2 - h_1)\right] \quad \text{...(1)}$$

$$\frac{\delta W}{dt} = -\dot{m}\left[\frac{C_2^2 - C_1^2}{2} + (h_2 - h_1)\right]$$

$$-30\,\frac{\text{kJ}}{\text{sec}} = -2\,\frac{\text{kg}}{\text{sec}}\left[\frac{150^2 - 100^2}{2 \times 1000} + (h_2 - h_1)\right]$$

$$-30\,\frac{\text{kJ}}{\text{sec}} = -2\,\frac{\text{kg}}{\text{sec}}\left[\frac{12500}{2000} + (h_2 - h_1)\right]$$

$$-30\,\frac{\text{kJ}}{\text{sec}} = -2\,\frac{\text{kg}}{\text{sec}}\left[6.25\,\frac{\text{kJ}}{\text{kg}} + (h_2 - h_1)\right]$$

$$-30\,\frac{\text{kJ}}{\text{sec}} = -12.5\,\frac{\text{kJ}}{\text{sec}} - 2\,\frac{\text{kg}}{\text{sec}} \times (h_2 - h_1)$$

$$-30\,\frac{\text{kJ}}{\text{sec}} + 12.5\,\frac{\text{kJ}}{\text{sec}} = -2\,\frac{\text{kg}}{\text{sec}} \times (h_2 - h_1)$$

$$\frac{-17.5\ \text{kJ/sec}}{2\ \text{kg/sec}} = (h_2 - h_1)$$

$$\mathbf{8.75\,\frac{kJ}{kg} = (h_2 - h_1)}$$

We also know from the definition of C_P,

$$\left(\frac{dh}{dT}\right)_P = C_P$$

or

$$dh = C_P \, .\, dT \qquad (m = 1 \text{ for unit mass})$$

or Change enthalpy $\Delta h = C_P \, \Delta T$

$$h_2 - h_1 = C_P (t_2 - t_1)$$

$$8.75 = 1.005\,(t_2 - 20)$$

$$\mathbf{t_2 = 28.75^{\circ}C}$$

EXAMPLE 2.27 *In a water cooled compressor 0.6 kg of air is compressed/sec. Power required to run the compressor is 40 kW. Heat lost to the cooling water is 30% of input and 10% of input is lost in bearings and other frictional effects. Air enters the compressor of 1 bar and 30ºC. If the changes in PE and KE are neglected, estimate the exit air temperature.*

Take $C_p = 1.005$ kJ/kg-K.

Data : $\dot{m} = 0.6$ kg / sec. $\qquad \dfrac{\delta W}{dt} = 40$ kW

$P_1 = 1 \text{ bar} = 10^5 \text{ n/m}^2 \qquad T_1 = 30^{\circ}C$

$\Delta PE = 0,\ \Delta KE = 0. \qquad T_2 = ?$

Solution

We know that, $\quad \Delta H = m\,C_P\,(T_2 - T_1)$

$$= 0.6 \times 1.005\,(T_2 - 30)$$

$$\Delta H = 0.603\,(T_2 - 30) \qquad \text{...(1)}$$

Net heat losses, $\quad = 40$ % of input

$$= \frac{40}{100} \times 40\,\text{kW}$$

$$\frac{\delta Q}{dt} = 0.4 \times 40 = 16 \text{ kW}.$$

SFEE is given by,

$$\dot{m}\left[\frac{C_1^2}{2} + gZ_1 + h_1\right] + \frac{\delta Q}{dt} = \dot{m}\left[\frac{C_2^2}{2} + gZ_2 + h_2\right] + \frac{\delta W}{dt}$$

Since $\quad \Delta KE = 0$ and $\Delta PE = 0$

$$\dot{m}\,h_1 + \frac{\delta Q}{dt} = \dot{m}\,h_2 + \frac{\delta W}{dt}$$

$$\therefore \quad \frac{\delta W}{dt} - \frac{\delta Q}{dt} = \dot{m}\,(h_1 - h_2) = \Delta H \qquad \text{...(2)}$$

From (1) and (2)

$$40 - 16 = 0.603\,(T_2 - 30)$$

$$\mathbf{T_2 = 70^{\circ}\ C}$$

EXAMPLE 2.28 *A room is fitted with 2 fans each consuming 0.2 kW power. There are three lamps in the room which are provided to heat the room each consuming 200 W. Ventilation air enters the room with the enthalpy of 85 kJ/kg and leaves the room with the enthalpy of 60 kJ/kg. The rate of air flow is 100 kg/hr. There are five persons in the room and heat generated by each person is 600 kJ/hr. Determine the rate at which the heat is to be removed by a room cooler, so that steady state is maintained in the room.*

Solution

Air flow rate = 100 kg/hr

Heat generated by each person = 600 kJ/hr

= 0.1666 kJ/sec

∴ For 5 persons = 0.8333 kJ/sec

Since work input is given in kW using SFEE on time basis.

$$\dot{m}\left[\frac{C_1^2}{2} + gZ_1 + h_1\right] + \frac{\delta Q}{dt} = \dot{m}\left[\frac{C_1^2}{2} + gZ_2 + h_2\right] + \frac{\delta W}{dt}$$

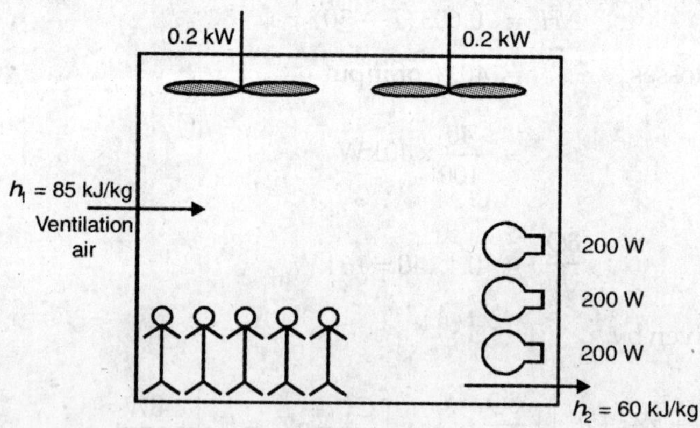

Fig. Ex. 2.28

Neglecting ΔKE and ΔPE.

$$\dot{m}\,[h_1 - h_2] + \frac{\delta Q}{dt} = \frac{\delta W}{dt}$$

$$\text{i.e. } \dot{m}\,[h_1 - h_2] + \left[\begin{pmatrix}\text{heat added}\\ \text{due to lamps}\end{pmatrix} + \begin{pmatrix}\text{heat generated}\\ \text{by 5 persons}\end{pmatrix} - \begin{pmatrix}\text{heat to be removed}\\ \text{by the cooler}\end{pmatrix}\right]$$

$$= \text{Work input to the fans (which should be } -\text{ve)}$$

$$\therefore \frac{100}{3600}\frac{\text{kJ}}{\text{sec}}\left[85\frac{\text{kJ}}{\text{kg}}-60\frac{\text{kJ}}{\text{kg}}\right]+\left[0.6\frac{\text{kJ}}{\text{sec}}+0.833\frac{\text{kJ}}{\text{sec}}-\dot{X}\right] = -0.4\frac{\text{kJ}}{\text{sec}}$$

$$0.6944+[0.6+0.833-X] = -0.4 \quad 0.6944+0.6+0.833+0.4$$

$$\therefore \qquad \mathbf{X = 2.574 \ kJ/sec.}$$ **Heat to be removed by Cooler**

EXAMPLE 2.29 *In a steady flow device, the inlet and outlet conditions are given below. Determine the heat loss/gain by the system.*

Property	*Inlet*	*Outlet*
Pressure (bar)	*10*	*0.15*
Specific volume (m^3/kg)	*0.206*	*8.93*
Specific enthalpy (kJ/kg)	*2827*	*2341*
Velocity (m/s)	*20*	*120*
Elevation m	*3.2*	*0.5*

The fluid flow rate through the device is 2.1 kg/s. The work output of the device is 750 kW.

Data:

$$Q = ?$$

$$\frac{\delta W}{dt} = +750 \text{ kW}$$

Solution

Since work output is given in $\frac{\delta W}{dt}$, using SFEE on time basis,

$$\dot{m}\left[\frac{C_1^2}{2}+gZ_1+h_1\right]+\frac{\delta Q}{dt} = \dot{m}\left[\frac{C_2^2}{2}+gZ_2+h_2\right]+\frac{\delta W}{dt}$$

$$= \dot{m}\left[\frac{C_2^2-C_1^2}{2}+g(Z_2-Z_1)+(h_2-h_1)\right]+\frac{\delta W}{dt}$$

$$= 2.1\frac{\text{kg}}{\text{sec}}\left[\frac{120^2-20^2}{2\times 1000}+\frac{9.81(0.5-3.2)}{1000}+(2341-2827)\right]+750$$

$$= 2.1[7+(-0.026487)+(-486)]+750$$

$$= 2.1[7-0.026487-486]+750$$

$$= 2.1[-479.026]+750$$

$$= -1005.9556+750$$

$$\frac{\delta Q}{dt} = \mathbf{-255.95562\ kW}$$

Note: –ve sign implies that heat will flow out of system.

EXAMPLE 2.30 *Air at a temperature of 15°C passes through a heat exchanger at a velocity of 30 m/c where its temperature is raised to 800°C. It then enters a turbine with the same velocity of 30 m/s and expands until the temperature falls to 650°C. On leaving the turbine, the air is taken at a velocity of 60 m/s to a nozzle where it expands until the temperature has fallen to 500°C. If the air flow rate is 2 kg/s calculate,*

(a) Rate of heat transfer to the air in the heat exchanger.

(b) Power output from the turbine, assuming no heat loss, and

(c) Velocity of air at exit from nozzle, assuming no heat loss, take the specific enthalpy of air as

$h = C_p T$, where $C_p = 1.005$ kJ/kg K, and T = Temperature

Solution

For heat exchanger:

$$q_{1-2} + \dot{m}\left(h_1 + \frac{C_1^2}{2} + gz_1\right) = w_{1-2} + \dot{m}\left(h_2 + \frac{C_2^2}{2} + gz_2\right).$$

Then $$q_{1-2} = \dot{m}(h_2 - h_1) \qquad (\because C_1 = C_2)$$

$$= \dot{m}\,C_P\,(T_2 - T_1)$$

$$q_{1-2} = 2\times1.005\times(800-15)$$

$$\mathbf{q_{1-2} = 1577.85\ kJ/s}$$

For the turbine:

$$q_{2-3} + \dot{m}\left(h_2 + \frac{C_2^2}{2} + gz_2\right) = w_{2-3} + \dot{m}\left(h_3 + \frac{C_3^2}{2} + gz_3\right)$$

$$\therefore \quad w_{2-3} = \dot{m}\left[(h_2 - h_3) + \frac{C_2^2 - C_3^2}{2}\right] \qquad (\because q_{2-3} = 0)$$

$$= \dot{m}\left[C_P\,(T_2 - T_3) + \frac{C_2^2 - C_3^2}{2}\right]$$

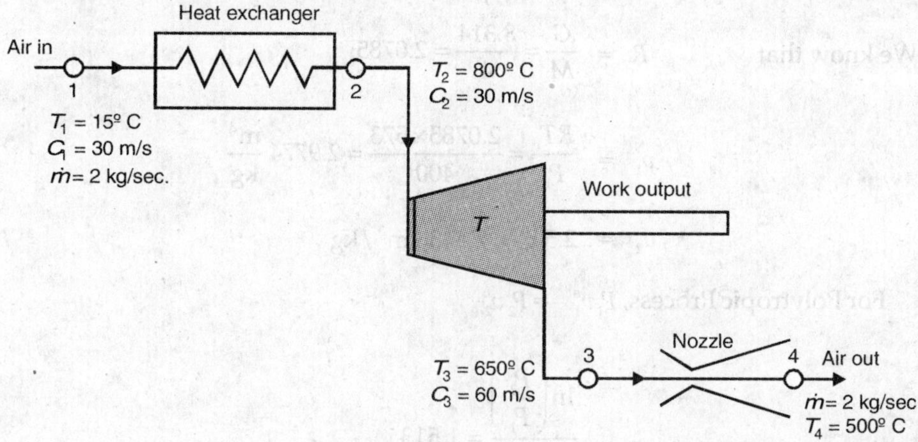

Fig. Ex. 2.30

$$= 2\left[1.005(800-650)+\left(\frac{30^2-60^2}{2\times1000}\right)\right]$$

$$\mathbf{W_{2-3} = 298.8\ kW}$$

For nozzle:

$$q_{3-4}+\dot{m}\left[h_3+\frac{C_3^2}{2}+gz_3\right] = w_{3-4}+\dot{m}\left[h_4+\frac{C_4^2}{2}+gz_4\right]$$

$$C_4^2-C_3^2 = 2\times1000\times C_P(T_3-T_4)$$

$$(\because q_{3-4}=0 \text{ and } w_{3-4}=0)$$

$$\therefore \quad C_4^2-C_3^2 = 2\times1000\times C_P(T_3-T_4)=2\times1000\times1.005(650-500)$$

$$= 301500$$

$$\therefore \quad \mathbf{C_4 = 552.36\ m/s}$$

EXAMPLE 2.31 *Helium gas is expanded polytropically in a turbine, from 4 bar, 300°C to 1 bar such that final volume is 2.5 times the volume at inlet. Velocity of gas at exit is 50 m/s. What is the mass flow rate of gas required to produce 1 MW turbine output? How much is the heat transfer during the process ? Also determine exit area of turbine.*

Assume specific heat of helium = 5.193 kJ/kg K at constant pressure.

Solution

We know that $$R = \frac{G}{M} = \frac{8.314}{4} = 2.0785$$

$$v_1 = \frac{RT_1}{P_1} = \frac{2.0785 \times 573}{400} = 2.9774\,\frac{\text{m}^3}{\text{kg}}$$

$$v_2 = 2.5\,v_1 = 7.4436\ \text{m}^3/\text{kg}$$

For Polytropic Process, $P_1 v_1^n = P_2 v_2^n$

$$\therefore \quad n = \frac{\ln\left(\dfrac{P_2}{P_1}\right)}{\ln\left(\dfrac{v_1}{v_2}\right)} = 1.513$$

and $$\frac{T_2}{T_1} = \left(\frac{P_2}{P_1}\right)^{\frac{n-1}{n}}$$

$$\therefore \quad T_2 = 358.11\ \text{K}$$

$$\text{Specific work} = -\int_1^2 v\,dP = \frac{n}{n-1} RT_1 \left[1 - \frac{P_2 v_2}{P_1 v_1}\right]$$

$$= \frac{1.513}{0.513} \times 2.0785 \times 573 \times \left[1 - \frac{1 \times 2.5}{4}\right]$$

$$= 1317.2\ \text{kJ/kg}$$

$$\Delta\text{KE} = \frac{1}{2}C_2^2 = 1250\,\frac{\text{J}}{\text{kg}} \ \text{or} \ 1.25\ \text{kJ/kg}$$

$$q - w_s = \Delta h + \Delta\text{KE}$$

and $$-\int v\,dP = w_s + \Delta\text{KE} \quad \text{as} \quad \Delta\text{PE} = 0$$

$$\therefore \quad W_s = 1317.2 - 1.25 = 1315.95\ \text{kJ/kg}$$

and $q = 1315.95 + C_P(T_2 - T_1) + 1.25 = 1315.95 + 5.193(358.11 - 573) + 1.25$

$$q = 201.28 \text{ kJ/kg}$$

Then $$\dot{m} = \frac{100}{w_s} \text{ kJ/s} = 0.7599 \text{ kg/s}$$

$\therefore$ $$\text{Exit area } A_2 = \frac{\dot{m}\, v_2}{C_2} = 0.1131 \text{ m}^2$$

THEORY QUESTIONS

1. Write a note on 'Joule's Experiment'?
2. State and explain the first law of thermodynamics for a closed system undergoing a cyclic change.
3. Derive an equation for energy transfer for a closed system.
4. Explain the First Law of Thermodynamics and hence show their cyclic integral of δQ and cyclic integral of δW are equal for a closed system.
5. State and explain the first law of thermodynamics for a closed system undergoing a process.
6. Prove that 'Internal Energy' is a property of a system.
7. Write a note on different forms of stored energies.
8. Define the term 'Enthalpy' (H).
9. Write a note on'Internal Energy' (U).
10. Distinguish between heat and internal energy.
11. Define specific heat at constant pressure and at constant volume for a perfect gas.
12. Define Adiabatic Index (g).
13. What is Perpetual Motion Machine of first kind (PMM-1) ?
14. Define a flow process and explain the terms Control Volume and control surface.
15. Explain the term flow work and discuss how it differs from the work done by a closed system.
16. Write a note on Flow work or Flow energy.
17. How is 'Steady flow system characterized'?
18. What are the conditions for a steady flow process?
19. State the conditions of First Law of Thermodynamics for a Steady Flow Processes.

20. Explain the essentials of non-flow and flow systems. Give suitable examples of each. What are advantages of steady flow system over non-flow system.
21. Derive the Steady Flow Energy Equation (SFEE).
22. Apply first law of thermodynamics to steady flow system and derive the equation of energy.
23. Derive Steady Flow Energy Equation on mass basis.
24. Explain the significance of $-\int v\,dP$.
25. Explain the significance of $\int Pdv$ in case of steady flow process and non-flow process.
26. Prove that, $-\int vdP = Q - \Delta H$.
27. Prove that, $\int Pdv = Q - \Delta U$.
28. Obtain the energy equations for the following open systems:
    (a) Boiler (b) Turbine
    (c) Compressor (d) Throttling process
    (e) Nozzles (f) Condenser
29. Derive Continuity equation.

PROBLEMS FOR PRACTICE

1. A centrifugal air compressor delivers 15 kg of air / min. Air enters at 1 bar, 10 m / sec, 0.5 m^3 / kg. It is discharged at 7 bar, 80 m / sec and 0.15 m^3 / kg. During the process, its enthalpy increases by 160 kJ / kg and the air rejects 720 kJ / min of heat to the surroundings. Find,
   (i) Power required to run the compressor.
   (ii) Ratio of inlet to outlet pipe diameter. Neglect ΔPE.
2. For a steady flow system, following particulars are noted,

	Inlet	Exit
1. Pressure (bar)	1	5
2. Vol (m^3/kg)	2	0.58
3. Velocity (m/sec)	20	30
4. Elevation (m)	0	2

If the mass flow rate of the working fluid is 1 kg / sec., find the power of the system. Is it a power absorbing or power producing system?
Assume that the steady flow system follows $Pv^{1.3} = C$.

3. One kg of working substance undergoes a reversible constant pressure process at 1.2 bar during which its volume changes from 1 m^3 to 1.8 m^3 and temperature changes from 50°C to 37°C. The sp. heat of a substance at constant pressure is given by $C_p = \left(1.1 + \frac{40}{t+30}\right)$ kg/kg°C where t is in °C.

Find out

(a) Heat supplied (b) Work done

(c) Change in I.E. (d) Change in enthalpy

4. A fluid is contained in a cylinder piston arrangement. The piston is spring loaded and frictionless such that the pressure of the fluid varies with its volume according to the law $P = a + bV$. The internal energy of the fluid is given by $V = 30 + 3.2\,PV$ where U is in kJ. P in kPa, V in m^3. The fluid undergoes a change of state from 150 kPa and 0.025 m^3 to 450 kPa and 0.05 m^3. Assuming no work other than that on the piston determine the direction and magnitude of work and heat transfer.

5. What at the rate of 10 kg/sec is compressed adiabatically from 5 bar to 50 bar in a steady flow process. Calculate the power required assuming that sp. volume of water being as 0.001 m^3/kg which remains almost constant.

6. Steam enters a turbine steadily at 10 mPa and 550ºC with a velocity of 60 m/s and leaves at 25 kPa with a dryness fraction of 0.95. A heat loss of 30 kJ/kg occurs during the process. The inlet area of the turbine is 150 cm^2 and exit area is 1400 cm^2. Determine : (1) mass flow rate of steam. (2) exit velocity (3) power output.

7. A closed system receives 200 kg of heat at constant volume. The system then undergoes a constant pressure process during which 150 kg of heat is rejected and 50 kJ of work is done on the system. Calculate how much of work is required to be done if the system is to be restored to its original state by means of an adiabatic process. If the initial internal energy is 50 kJ, determine the values of internal energy at the end of the constant volume and constant pressure processes.

8. In a reversible steady flow process 225 kW of heat is rejected per kg of the fluid passing through the system. During the process, pressure and volume change from 6 bar and 0.075 m^3/kg to 2 bar and 0.2 m^3/kg. Assuming that flow process follows the law PV^n = constant and ignoring changes in K.E. and P.E., determine the change in enthalpy of the fluid during the process.

9. Air flow steadily at the rate of 0.4 kg/sec through an air compressor, entering at 6 m/sec with pressure of 1 bar and sp. volume of 0.85 m^3/kg and leaving at 4.5 m/sec with a pressure of 6.9 bar and sp. volume of 0.16 m^3/kg. The sp. internal energy of the air leaving is 88 kJ/kg greater than that of the air entering, cooling water in the jacket surrounding the cylinder absorbs heat from the air at the rate of 59 kg/sec. Calculate the power required to drive the compressor and the inlet and outlet pipe areas.

10. In the turbine of a gas turbine unit the gases flow through the turbine at 17 kg/sec and the power developed by the turbine is 14 MW. The sp. enthalpies of the gases at inlet and outlet are 1200 kJ/kg and 360 kJ/kg respectively. Calculate the rate at which heat is rejected from the turbine. Find also the area of the inlet pipe given that sp. volume of the gases at inlet is 0.5 m^3/kg.

11. A steam turbine operates under steady flow conditions, receiving steam and leaving the steam at the following rate.

	Inlet	**Outlet**
Pressure	15 bar	130 kPa
I.E. (kJ/kg)	2500	1500
Velocity (m/sec)	300	200
Elevation (m)	3	0
Volume (m^3)	0.15	1.2 m^3 /kg

12. A closed system undergoes a cycle 1–2–3–1. If $Q_{12} = 30$, $Q_{23} = 10$, $W_{12} = 5$, $W_{31} = 25$ and $\Delta E_{31} = 15$, determine W_{23} and ΔE_{23}.

13. The following data is available for a test on air compressor for an air flow rate of 6 kg/sec.

	Inlet	**Outlet**
Pressure	95 kPa	7.5 bar
Sp. Internal Energy (kJ/kg)	40	150
Sp. Volume (m^3/kg)	0.95	0.162
Velocity of air (m/sec)	6	1

Estimate: (1) Power required to drive the compressor in kW. (2) Ratio of inlet pipe diameter to outlet pipe diameter.

14. A non-flow system undergoes a frictionless process according to the law $P = \frac{4.5}{v} + 2$ where P is in bar, V is in m^3/kg. During this process volume changes from 0.12 m^3/kg to 0.04 m^3/kg and temperature increases by 133ºC. The change in I.E. of fluid is given by $du = C_v \, dT$ where $C_V = 0.71$ kJ/kg K and dT is temperature change. Find out (1) Heat transfer, (2) Change in enthalpy. Assume fluid mass of 5 kg.

❑❑❑

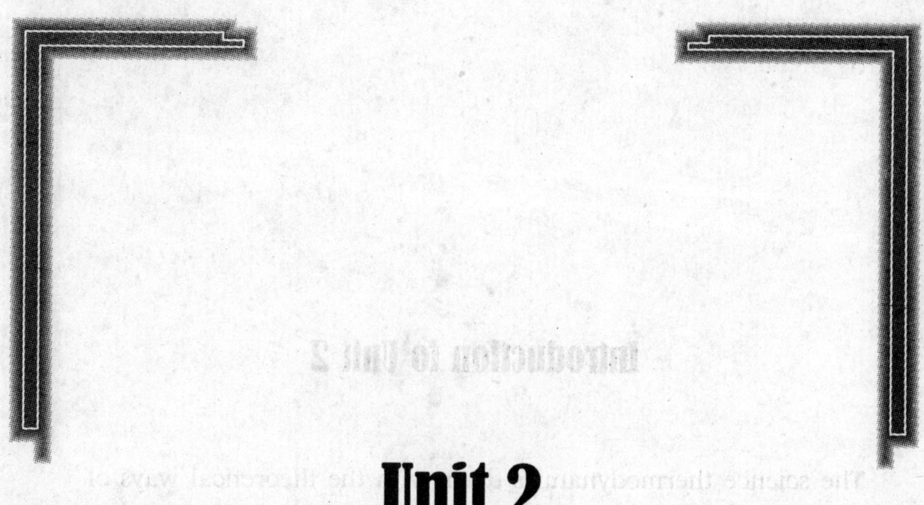

Unit 2

Power producing / Absorbing Devices

Syllabus

Power Producing Devices — Boiler and steam turbines, Reciprocating I. C. engines, Gas turbines, Hydraulic turbines, Compressed air motor (theoretical study using schematic diagrams only).

Power Absorbing Devices — Reciprocating pumps and compressors, centrifugal pumps, rotary compressors, blowers. Study of household refrigerators and window air conditioners using schematic diagrams (elementary treatment only).

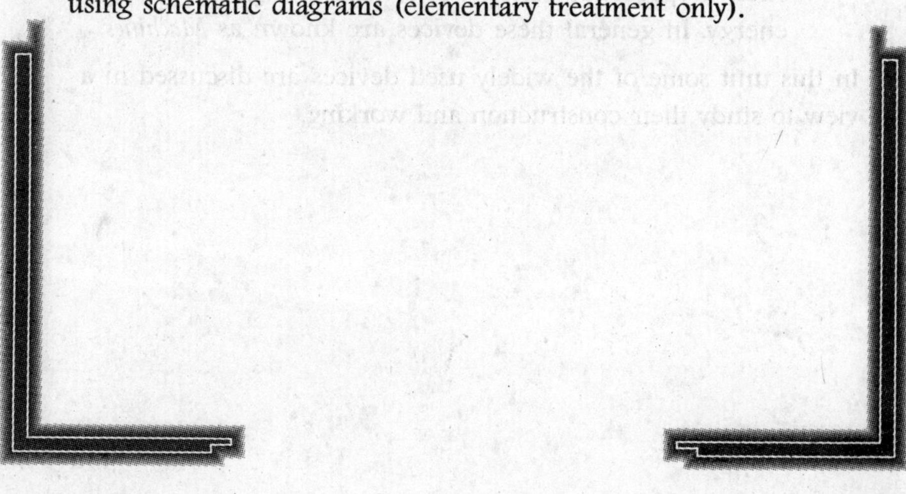

Introduction to Unit 2

The science thermodynamics deals with the theoretical ways of conversion of one form of energy in to the other. But, in order to make such energy conversions useful for human being, they must be associated with application based devices. This unit introduces such devices in which energy conversion is used to achieve certain useful task.

In a broad sense, such devices can be classified in to two types based on the direction in which energy is transformed from one form to the other. They are

1. **Power producing devices** — In these devices, the different forms of energies such as chemical, thermal, hydraulic etc. is converted into mechanical or shaft power. In general, these devices are known as *Prime Movers*.
2. **Power absorbing devices** — In this type of devices, the mechanical energy is transformed in the different forms such as pressure energy, potential energy or thermal energy. In general these devices are known as *Machines*.

In this unit some of the widely used devices are discussed in a view to study their construction and working.

CHAPTER

3 Power Producing Devices

3.1 INTRODUCTION

Upto now we have considered a fluid which under normal conditions remains in a gaseous state. Such substances are oxygen, nitrogen, air etc. Now we consider fluids, which under normal working conditions may be either in a liquid state or in a gaseous state. Such substances are known as vapours. Examples of vapours employed in engineering practice, are steam, ammonia, freon, mercury etc. The behaviour of vapours is typically represented by steam and water.

With vapours, the simple laws of perfect gas does not even approximately apply and therefore sometimes the definition of vapour is given as: "*A vapour is any substance in the gaseous state which does not even approximately follow the general gas laws.*" The reason for this departure of the performance of vapours from the general gas law is that the vapours from the general gas law is that the vapours when heated or cooled experience some desegregation work while the general gas laws are based on a perfect gas, which would experience no desegregation work when heated or cooled. It should be remembered that all known actual gases do experience a small amount of desegregation work. So we can use the formula derived for perfect gas for vapour calculations. For this purpose we use values which have been determined by experiments and listed in tables. Highly superheated vapours approximately follow the perfect gas laws. *Vapour obeys its own laws.*

The equipments which produce power are known as *Power Producing Devices or Prime–movers.*

3.2 FORMS OF MATTER

There are three forms of matter, viz. (i) Solid, (ii) Liquid and (iii) Gaseous.

In order to explain the changes in form of matter, a hypothetical experiment may be described as follows.

A substance in the solid state is placed in a cylinder which is fitted with a frictionless right fitting piston being loaded by a weight as shown in Fig. 3.1. Heat is supplied to the cylinder. The substance may change in volume during the heating process but the pressure is kept constant by the piston and weights on it. The change in temperature of the substance will be recorded by a thermometer.

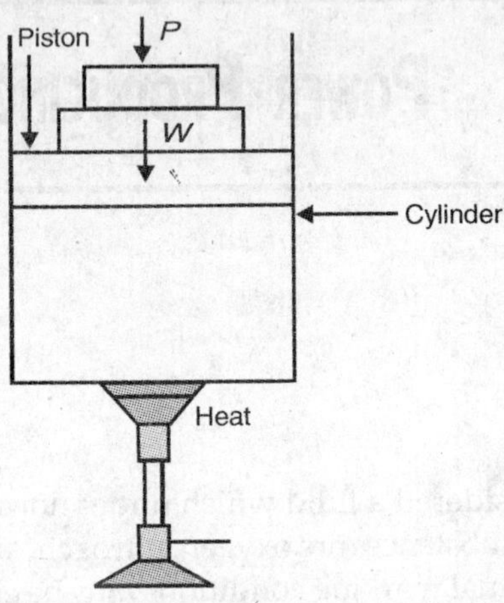

Fig. 3.1

At first, when the heat is added, the temperature of the solid rises and a certain value of temperature will be reached after which a considerable amount of heat is added, without a change in temperature. Through a sight window provided in the cylinder wall it can be seen that the solid is melting. When the solid has been completely transformed into liquid, further addition of heat again causes a rise in temperature. This rise in temperature continues till a point is reached where further addition of heat does not change the temperature of the substance but on observation we see that the liquid is boiling. When the liquid has been completely transformed into vapour, the temperature again rises when the heat is added. The relation between the heat added, entropy change and the temperature of the substance is shown in Fig. 3.2.

1–2 Shows the stage during which the rise in temperature of a substance in solid state takes place.

2–3 Shows the stage during which the substance is being transformed from a solid state to a liquid state without rise in temperature.

3–4 Shows the stage during which rise in temperature of the substance in liquid state takes place.

4–5 Shows the stage during which the substance is being transformed from a liquid state to a vapour state without rise in temperature.

5–6 Shows the stage, during which rise in temperature of the substance may take place in a vapour state.

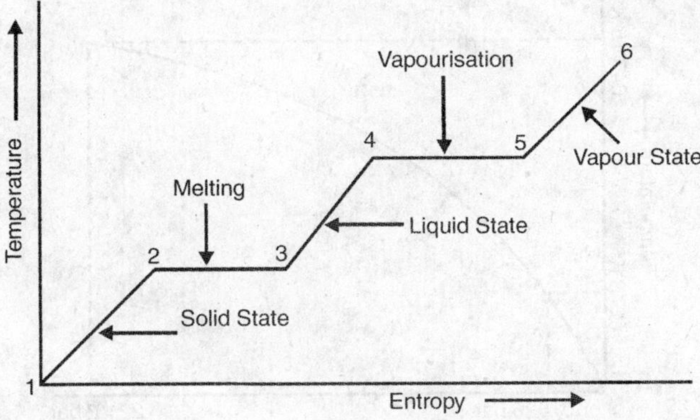

Fig. 3.2

If the hot vapour is cooled the curve will be exactly retraced with all changes taking place in reverse manner. If the weight *W* is increased and the experiment is repeated the same general behaviour is obtained but the temperature will be higher than that in the first case. If the weight is reduced still the behaviour will be the same but temperature will be lower than that in first case.

The above hypothetical experiment brings out certain important facts.

- When a solid melts the temperature remains constant and is always the same for a given substance at a given pressure. When a liquid freezes the temperature remains constant. *This temperature is called the freezing point. The quantity of heat removed or added during the change of state is constant and is known as latent heat of fusion. The latent heat of fusion depends upon pressure.*
- When the liquid boils the temperature remains constant and is always the same for a given substance at a given pressure. If the pressure is increased, the temperature at which the change of state takes place also increases. Similarly when the vapour condenses the temperature also remains constant. *This temperature is called the boiling point or saturated temperature of the liquid. The definite quantity of heat removed or added during the change of state is called the Latent heat of vaporisation. The latent heat of vaporisation depends upon pressure.*

The freezing point, the latent heat of fusion, the boiling point and the latent heat of vaporisation vary widely with different substances.

When the pressure is very low, an interesting phenomenon is noticed. A solid when heated directly transforms into a vapour without passing through an intermediate liquid state. Such a phenomenon is known as sublimation. A familiar example is the conversion of solid CO_2 (dry ice) directly into vapour.

Figure 3.3 shows a temperature pressure curve. In practice, steam is generated in a device called steam boiler.

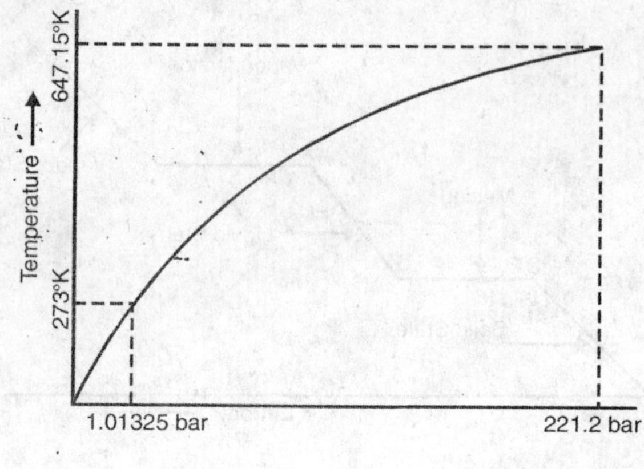

Fig. 3.3

3.3 STEAM BOILERS

The function of a steam boiler is to transfer heat produced by burning of fuel to water and thus to produce steam. The construction and appearance of a boiler depend upon the arrangements made for burning the fuel and effecting the transfer of heat.

A boiler may be used for supplying

(i) Steam to a steam engine or a turbine in steam power plant.

(ii) Steam for industrial process work

(iii) Steam for heating installations e.g. drying of gound nuts in oil mills.

3.4 CLASSIFICATION OF BOILERS

1. According to the tube contents

(a) Fire Tube Boilers. In this case flue gasses flow through the tubes, which are surrounded by the water which is to be evaporated. (Fig. 3.4)

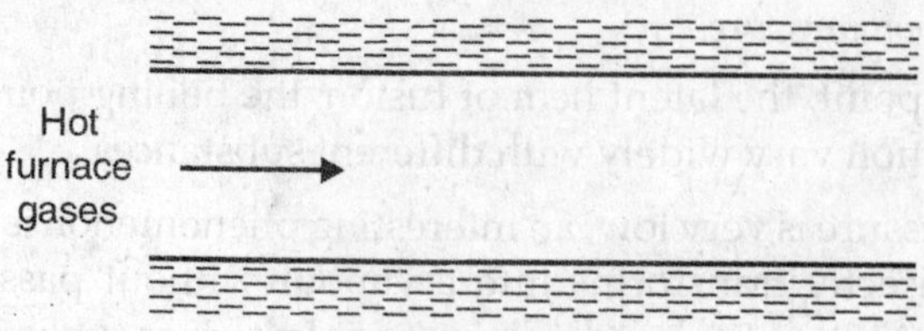

Fig. 3.4

Example: Lancashire, Cornish, Cochran, Locomotive, Scotch marine boiler, etc.

(*b*) *Water Tube Boilers.* In this case water flows through the tubes and hot gases heat the tubes from outside. (Fig. 3.5)

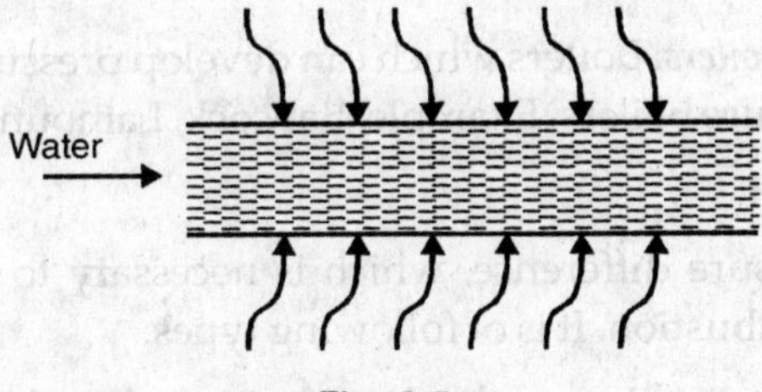

Fig. 3.5

Example: Babcock Wilcox, Stirling etc.

2. By the Method of Firing

(*a*) *Internally Fired.* In this case the furnace is located inside the boiler shell as in case of Lancashire boiler.

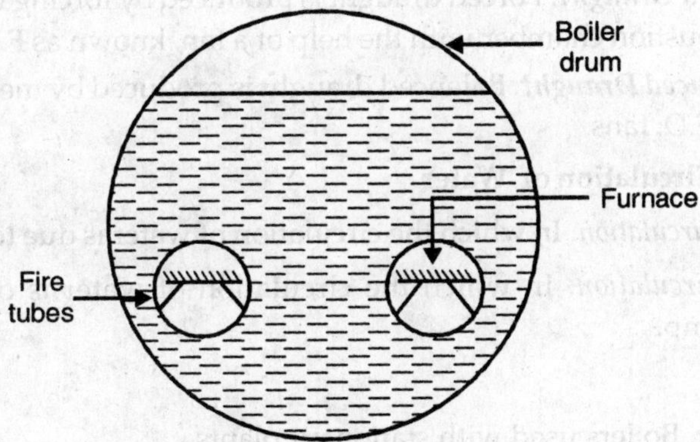

Fig. 3.6. Internal fired.

(*b*) *Externally Fired.* The furnace is located outside the boiler shell as in case of Babcock Wilcox boiler.

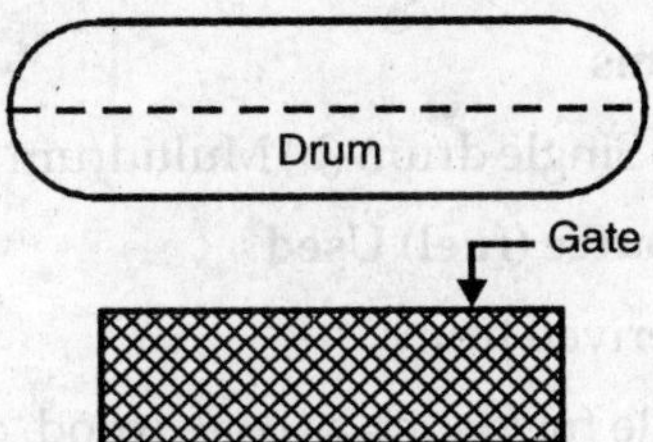

Fig. 3.7. External fired.

3. According to Pressure of Steam

(a) *Low Pressure boilers.* Boilers which can develop pressures below 80 bar are known as low pressure boilers. Example, Lancashire, Cochran, Cornish etc.

(b) *High Pressure boilers.* Boilers which can develop pressures above 80 bar are known as high pressure boilers. Example, Babcock, Lamount, Velox, Benson etc.

4. Nature of Draught

Draught is the pressure difference, which is necessary to draw the sufficient quantity of it for combustion. It is of following types:

(a) *Natural draught.* In this case the draught is produced by mean of chimney only. The amount of draught directly depends upon the height of chimney.

(b) *Artificial draught.* Artificial draught is produced by means of fans.

(i) ***Induced Draught.*** Induced Draught is produced by sucking the flue gases from chamber. It is obtained by providing a fan between the boiler and the chimney. This fan is known as I.D. Fan.

(ii) ***Forced Draught.*** Forced draught is produced by forcing the air into the combustion chamber with the help of a fan, known as F.D. fan.

(iii) ***Balanced Draught.*** Balanced draught is produced by means of I.D. fan and F.D. fans.

5. Method of Circulation of Water

(a) *Natural circulation.* In which the circulation of water is due to gravity.

(b) *Forced circulation.* In which the circulation of water is obtained by a centrifugal pump.

6. By the Use

(a) *Land type.* Boilers used with stationary plants.

(b) *Marine type*

(c) *Locomotive boilers.*

7. By the Design of Flue Gas Passages

The flue gas may follow a single pass, return pass or multipass.

8. By the Number of Drums

There are two types, (a) Single drum (b) Multidrum boilers.

9. According to Energy Source (fuel) Used

The heat energy may be derived from,

(a) Combustion of fossile fuels such as coal, wood, oil or natural gas etc.

(b) Electric or Nuclear energy.

(c) Hot waste gases of other chemical reactions.

10. According to the Material of Construction of the Boiler Shell

(a) *Steel Boilers*. Power boilers are generally fabricated out of steel plates.

(b) *C.I. Boilers*. Generally low pressure boilers are sometimes built out of castings.

3.5 TYPES OF BOILERS

3.5.1 Lancashire Boiler

It is fire tube, stationary, horizontal straight tube, internally fired, natural circulation boiler. Normal working pressure of this boiler is about 15 bar and steam generation capacity is upto $8T$ /hr.

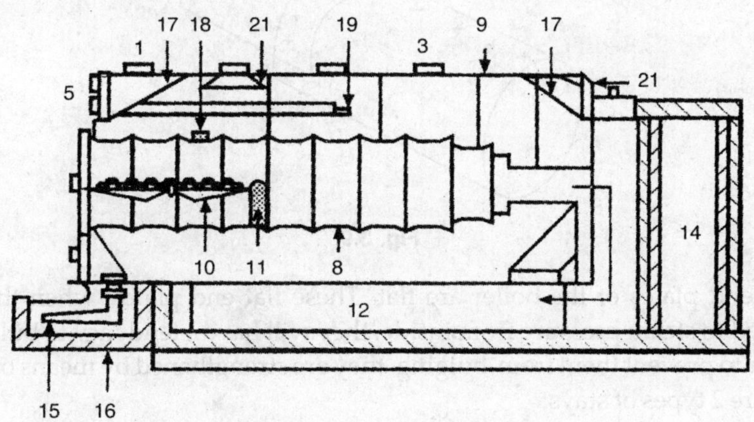

Legend:

1. Dead weight safety valve.
2. High steam low water alarm.
3. Manhole.
4. Steam stop valve and Antipriming Device.
5. Pressure gauge.
6. Water level indicator.
7. Feed check valve and perforated feed pipe.
8. Fire tubes.
9. Boiler shell.
10. Fire bars.
11. Brick work Bridge.
12. Bottom flue.
13. Side flues.
14. Main flue.
15. Blow-off cock.
16. Blow-off pit.
17. Gusset stays.
18. Fusible plug.
19. Scum tray.
20. Scum valve.
21. End plate.

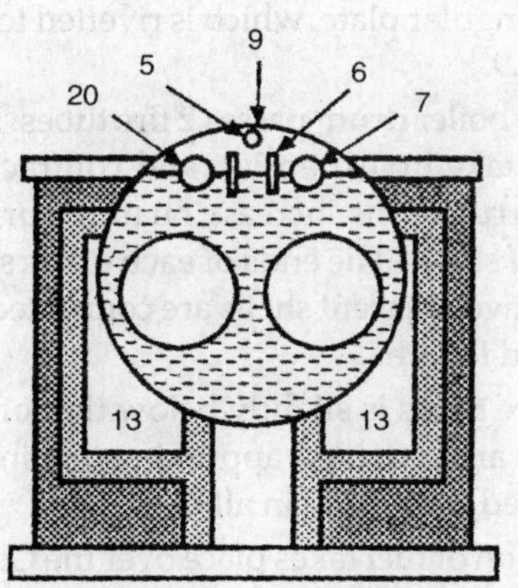

Fig. 3.8. Lancashire boiler.

Constructional Details

The main parts of the boiler are as follows.

1. Boiler Drum. This boiler consists of a very large boiler drum. The size of the boiler drum is approximately 7–9 mts in length and 2–3 mts in diameter. This is a very old type of boiler, in this boiler, the meeting edges of the plate are connected by means of rivetted joints. Steel plates are rolled to form shell and the meetings edges of the shell are connected by means of rivetted joints. The alternate shells are made smaller in a diameter, so that the longer shells of required length can be obtained by inserting one shell into the other and rivetting over the circumference.

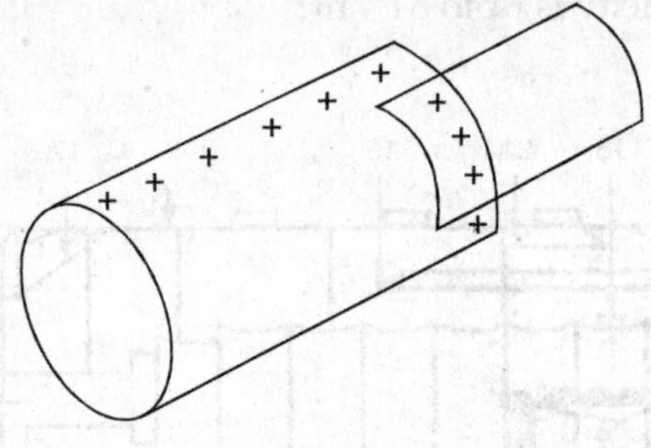

Fig. 3.9

The end plates of the boiler are flat. These flat end plates, when they are subjected to steam pressure from inside, they will have a tendency to bulge out. In order to prevent them from bulging, they are strengthened by means of stays. There are 2 types of stays.

(i) *Longitudinal stay*. It consist of round bar which extends from one end plate to the other and are connected to the end plates by means of lock nuts.

(ii) *Diagonal Gusset stay*. It consists of a triangular plate, which is rivetted to the shell plate and end plate as shown in Fig. 3.9.

2. Fire Tubes or Flue Tubes. Through the boiler drum passes 2 fire tubes. Part of each of fire tubes is corrugated, so as to take up expansion and contraction when the boiler gets hot and cold. Also corrugations increase heating surface area. These tubes are also made of number of shells. The ends of each of the shell are flanged outwards and the flanges of the two adjacent shells are connected by means of rivets, so as to get shells of required length.

It is to be noted that, the centre line of fire tubes is slightly below that of the centre line of boiler shell. Because fire tubes are provided approximately in the centre of water, so that heat can be transferred uniformly in all directions.

3. Grate, Brickwork Bridge etc. Combustion of fuel takes place over the Grate. Top surface of the grate is made up of fire-bars. Below the grate Ash-pit is provided.

When the combustion of fuel takes place, hot gases are generated and the Brickwork bridge deflects the gases to move upwards.

4. Manhole. It gives access to the inside of the boiler drum. It will be sufficiently large enough for a man to pass through it, for cleaning and inspection purpose.

5. Mountings. The boiler is fitted with various mountings. Mountings are necessary for the safe and efficient operation of the boiler.

The term mountings refers to the items such as,

(i) *Safety valves*

(a) Dead weight safety valve.

(b) Spring loaded safety valve.

(c) High steam low water alarm.

When the pressure of steam in the boiler drum increases beyond safe value, then these valves will open and allow the surplus steam to escape to the atmosphere, till the pressure of steam in the boiler drum comes to normal working pressure.

(ii) *Pressure gauge*. It indicates pressure inside the boiler drum.

(iii) *Feed check valve*. Water enters the boiler drum through feed check valve. On the inner side of the feed check valve, perforated feed pipe is fitted for feeding the water uniformly in the boiler drum. Below the feed pipe scum tray is fitted which separates scum or froth and is removed through the scum valve.

(iv) *Water level indicator*. It indicates the water level in the drum.

(v) *Blow-off valve*. It blows out impurities collected due to gravity.

(vi) *Steam stop valve*. Through this steam will be taken out for use.

(vii) *Antipriming device*. It is always desirable that the steam should leave the boiler as dry as possible, for that purpose an antipriming device or steam collecting pipe is provided in the steam space. It consists of a pipe and rectangular openings are provided in its bottom surface. It is fitted by means of baffles from inside. When the wet steam (steam and water particles) enters the pipe, the water particles being heavier than steam, after striking the baffles fall back to water space and steam almost dry will be taken out through the steam stop valve.

6. Accessories. Accessories are used for improving the efficiency of boiler plant. Generally Super heater and Economiser are used as accessories.

7. Path of Flue Gases. When the combustion of fuel takes place, hot gases are generated and the brick work bridge deflects these gases to move upwards.

During their first pass, these gases flow from the front end of the boiler to the rear end through central tubes. During their I-pass, flues gases heat the water centre outwards. After their I-pass, they pass through the superheater, then they enter the common bottom flue at the rear end. Now their II-pass starts, during this gases flow from rear end to front end through the common bottom flue. At the end of II-pass, the gases are bifurcated into two side flues, when the flue gases enter side flues, their III-pass starts end during their III-pass, they flow from the front end to rear end through side flues and during this pass they heat the water from sides.

After their third pass they meet in main common flue, from where they pass through the economiser and then they are exhausted through the chimney.

3.5.2 Cornish Boiler

The only difference between Cornish boiler and Lancashire boiler is that, in case of Cornish boiler there is only one fire tube, whereas in case of Lanchashire boiler there are two fire tubes, rest all the arrangements are same.

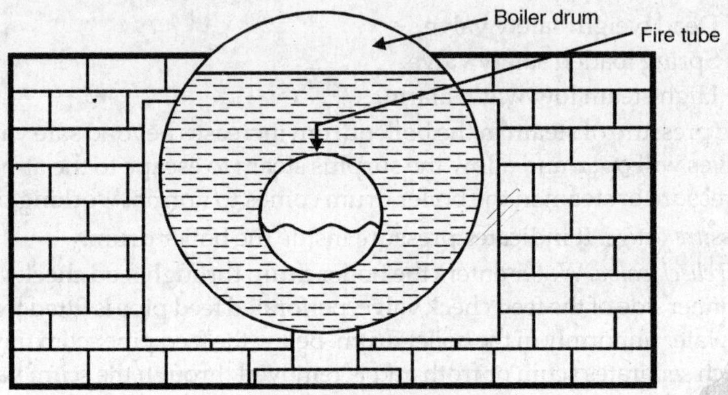

Fig. 3.10. Cornish boiler.

3.5.3 Locomotive Boiler

It is a Fire tube, Horizontal, Multitubular, Internally fired, Natural circulation boiler. As shown in Fig. 3.11 it consist of a large cylindrical shell. This shell is fitted

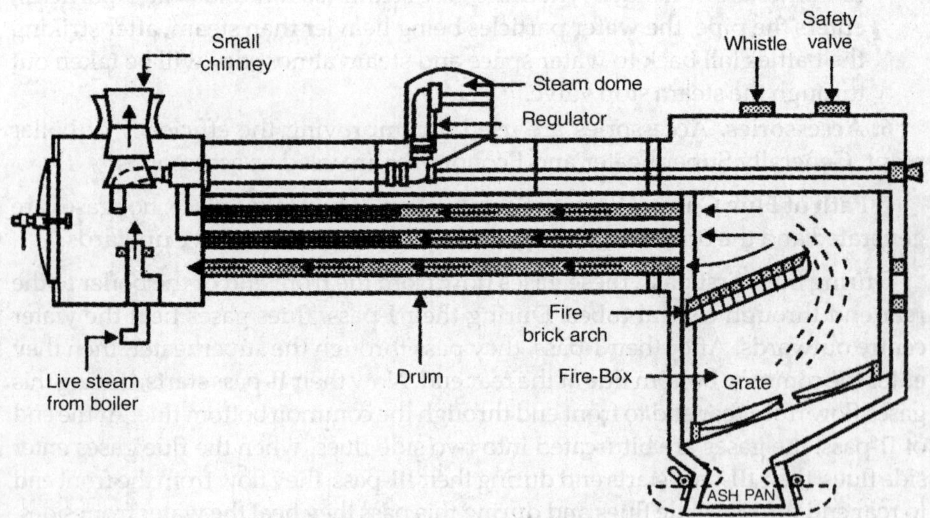

Fig. 3.11. Locomotive boiler.

to a rectangular fire box at one end and smoke box at other end. In between the smoke box and fire box a number of tubes are fitted.

When the combustion of fuel (coal) takes place over the grate, hot gases are generated. These hot gases rising from the grate are deflected by the Fire brick arch, so that they come in contact with the complete heating area of the fire box. The dempers as shown, control the flow of air for combustion. These hot gases then flow through the fire tubes to the smoke box, when they flow through the tubes, water will be converted into steam and accumulates in the steam dome and gases afterwards are exhausted to the atmosphere through the small chimney.

The regulator can be operated from the cabin by turning the lever. In order to have superheated steam, the steam is passed through the superheater tubes. These superheater tubes are of smaller diameter and are laid in the large diameter fire tubes as shown. Then this superheated steam is supplied to the engine. This boiler is fitted with various mountings as shown in Fig. 3.11.

3.5.4 Cochran Boiler

It is a Fire tube, Stationary, Multitubular, Internally fired, Vertical boiler.

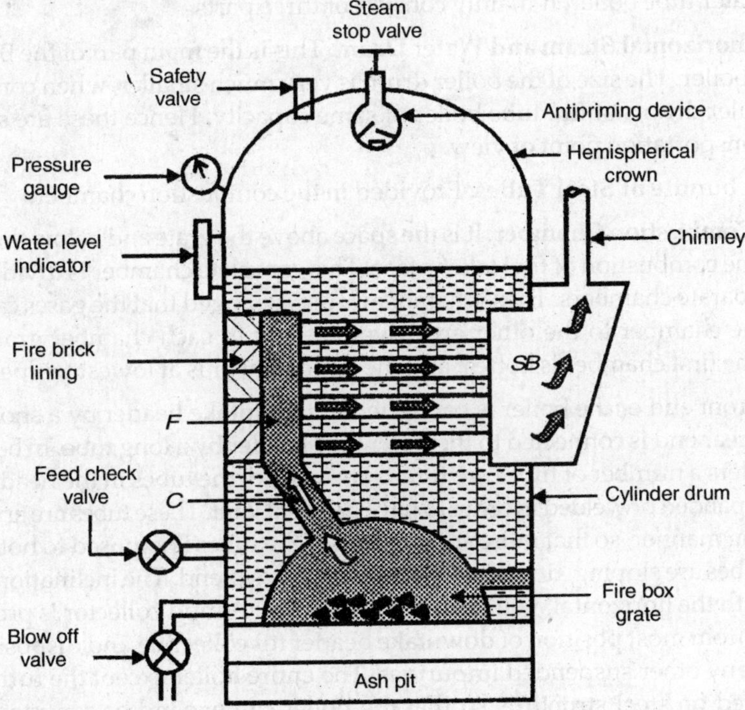

Fig. 3.12. Cochran boiler.

Cochran boiler is made in many sizes capable of producing steam upto 3T / hour and working pressures upto 17 bar. This boiler gives Thermal efficiency (ηth) of 70% with coal firing and about 75% with oil firing.

The boiler consists of a vertical cylindrical shell, with its crown having hemispherical shape. Such a shape of the crown plate, gives sufficient strength to withstand the bulging effect.

The furnace or fire box is also crown shaped or dome shaped, which deflects the unburnt fuel back to the grate. Combustion of fuel takes place over the grate. Hot gases generated in the fire box, enter into the chamber *C*, though a short flue pipe *F* and strikes the boiler shell plate or Back plate which is lined by fire bricks. The back plate directs the gases into the smoke or fire tubes. The gases after passing through the horizontal tubes enter the smoke box *SB* and from where they are exhausted through the chimney.

The hot gases passing through the horizontal tubes, give their heat to the water and in doing so, convert water into steam, which gets accumulated in the steam space.

3.5.5 Babcock Wilcox Boiler

It is a water tube boiler. It mainly consists of three parts.

(i) A horizontal Steam and Water Drum. This is the main part of the Babcock Wilcox boiler. The size of the boiler drum is very much smaller, when compared with boiler drums of fire tube boiler of same capacity. Hence these are simpler from transportation point of view.

(ii) A bundle of Steel Tubes. Provided in the combustion chamber.

(iii) Combustion Chamber. It is the space above the grate and below the drum where the combustion of fuel takes place. This complete chamber is divided into three separate chambers. These chambers are so arranged that the gases can flow from one chamber to the other and give out heat in each chamber gradually. Hence the first chamber is hottest and the last chamber is at lowest temperature.

The front end of the boiler is connected to the uptake header by a short tube and the rear end is connected to the downtake header by a long tube. In between the headers a number of tubes are fitted. The ends of the tubes in the headers are either expanded or welded in order to have gas tight joint. These tubes are arranged in zig-zag manner, so that more surface area of the tubes is exposed to hot gases. These tubes are sloping, downwards towards the rear end. The inclination of the tubes with the horizontal, will be in between 5–15°. A mud collector is provided at the bottom most position of downtake header for collecting and disposing the mud or any other suspended impurities. The entire boiler except the furnace in suspended on steel structure, so that the boiler can expand or contract freely without producing the cracks into the brick work, which encloses the boiler and furnace. This brickwork is lined from inside by means of fire bricks. Doors provided give access for cleaning, inspection and repair purpose. Dampers provided are used for regulating draught.

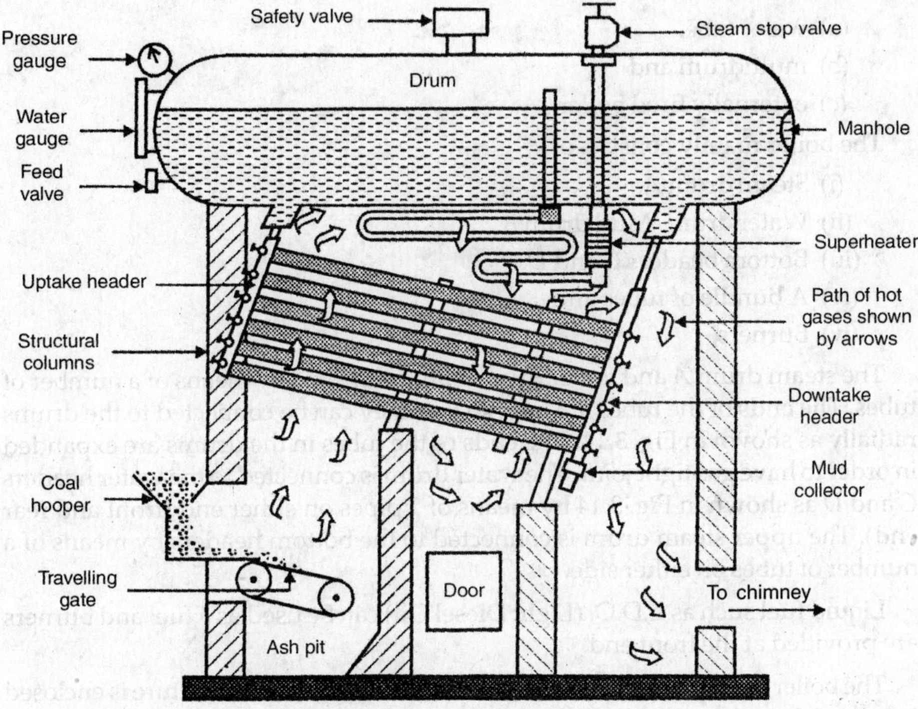

Fig. 3.13. Babcock Wilcox boiler.

This boiler is fitted with various mountings viz. Safety valve, Steam stop valve, Pressure gauge, Water level indicator, Fusible plug, Feed check valve, Blow-off valve etc. This boiler is also fitted with superheater as shown in Fig. 3.13.

Working. Water will be pumped by means of a feed pump and water enters the drum though feed check valve. Water will be filled upto the prespecified level, so that headers and tubes are flooded. We know that chamber above the grate is at hottest temperature, middle chamber is at medium hot temperature and last chamber is at lowest temperature i.e. temperature goes on decreasing towards the rear end. Due to this difference in temperature, difference in densities of water is created and water starts rising from downtake to uptake header. As the water rises some of the water gets converted into steam. Then the mixture of water and steam enters the durm. The steam gets accumulated in the steam space and water which is not converted into steam, flows back to rear end and the cycle repeats.

The steam so produced is in contact with water so it is wet. In order to get superheated steam, it is made to pass through the superheater, afterwards it is used for power generation purpose.

3.5.6 Sterling Bent Tube Boiler

It is a,

(a) water tube,
(b) multidrum and
(c) externally fired boiler.

The boiler mainly consists of,

(i) Steam drum *A*.
(ii) Water drum / Mud drum *B*.
(iii) Bottom beaders *C* and *D*.
(iv) A bundle of tubes and
(iv) Burners.

The steam drum *A* and water drum *B* are connected by means of a number of tubes. The ends of the tubes are bent so that they can be connected to the drums radially as shown in Fig. 3.14. The ends of the tubes in the drums are expanded in order to have gas tight joint. The water drum is connected to the water headers *C* and *D* as shown in Fig. 3.14 by means of 2 pipes on either end (front and rear end). The upper steam drum is connected to the bottom headers by means of a number of tubes on either side.

Liquid fuel such as L.D.O. (Light Diesel Oil) can be used as a fuel and burners are provided at the front end.

The boiler is suspended over the steel structure. The steel structure is enclosed in red brickwork and which is lined from inside by means of fire bricks.

E—are the doors which give access to the inside of the combustion chamber for inspection and repair purpose.

F—are the baffles which divide the combustion chamber into 3 separate chambers (I, II and III). These 3-chambers are so connected that the hot gases can flow from the I-chamber to III-chamber and they are made to pass through economiser and finally they are exhausted through chimney. The water in the tubes will absorb heat in each chamber, so the temperature of flue gases will go on decreasing towards the rear end.

The boiler is fitted with various mountings viz. :

(a) Spring loaded safety valves. (b) Pressure gauges.
(c) Thermometers. (d) Water level indicators.
(e) Feed check valve. (f) Steam stop valve.
(g) Blow off valve.

Also superheater and economiser are connected in the boiler circuit as accessories.

Water will be pumped by means of a centrifugal multi-stage feed pump. Water first enter the bottom header of the economiser and when it rises to through the

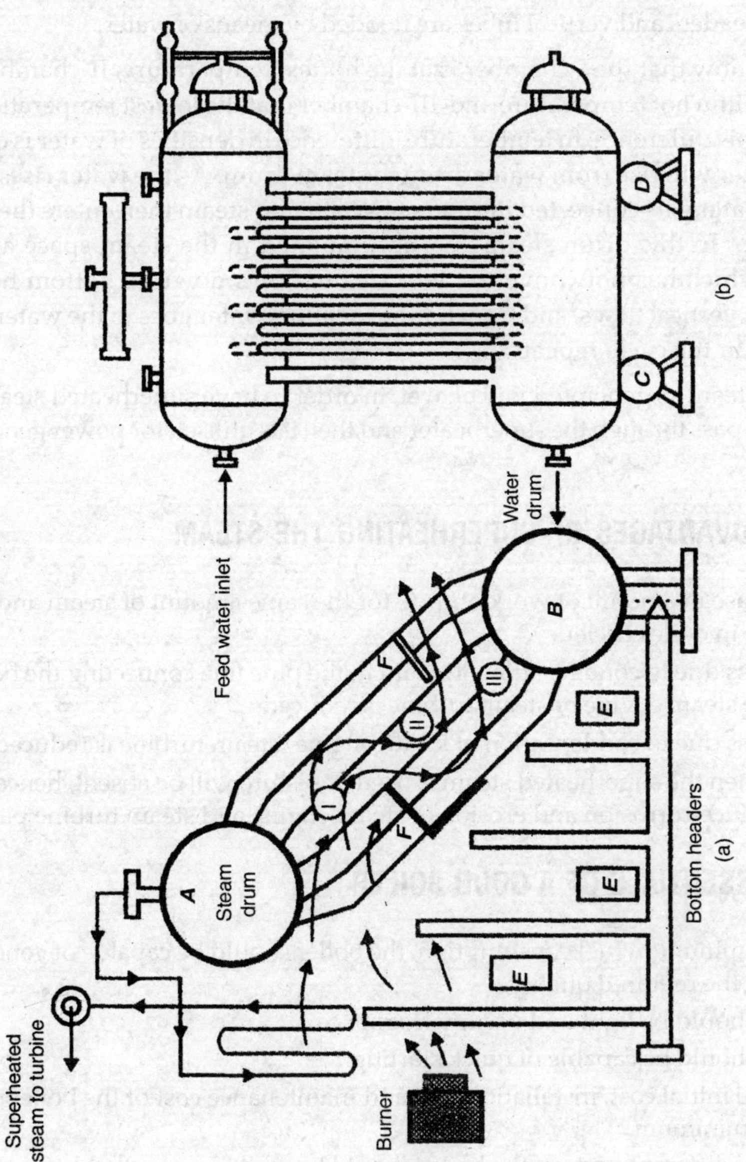

Fig. 3.14. Stirling boiler.

tubes, it recovers some of the heat of the flue gases and becomes hot. The hot water will then be collected into the upper header of the economiser and from where it goes to the upper drum.

Water is filled in the drum upto a pre-specified level, so that tubes, water drum, headers and vertical tubes are flooded by means of water.

We know that the I-chamber is at the hottest temperature, II-chamber is at the medium hot temperature and III-chamber is at the lowest temperature. So, due to this difference, in temperature, difference in densities of water is created and water will rise from water drum to steam drum. As the water rises, some of the water gets converted into steam. Water and steam then enters the steam drum *A*. In this drum steam gets accumulated in the steam space and the water which has not converted into steam, flows down to bottom headers through vertical tubes, and from bottom headers water goes to the water drum and again the cycle repeats.

The steam so generated will be wet, in order to have superheated steam, it is made to pass through the superheater and then it is utilised for power generation purpose.

3.6 ADVANTAGES OF SUPERHEATING THE STEAM

1. Increase in amount of work output, for the same amount of steam and hence increase in cycle efficiency.
2. Loss due to condensation of steam in the pipe line connecting the boiler to the steam engine or steam turbine is reduced.
3. Loss due to condensation of steam engine/steam turbine is reduced.
4. When the superheated steam is used, moisture will be absent, hence it will reduce corrosion and erosion of steam engine and steam turbine parts.

3.7 ESSENTIALS OF A GOOD BOILER

1. With minimum fuel consumption, the boiler should be capable of generating steam at the required quality.
2. It should be light and occupy less space.
3. It should be capable of quick starting.
4. The initial cost, installation cost and maintenance cost of the boiler should be minimum.
5. The different parts of the boiler should be easily approachable for repairs.
6. The boiler should have minimum number of joints and those too should be away from direct flames.
7. There should be no deposition of mud and other deposits on the inner surface of the boiler drum.

8. Boiler should conform to safety regulations, laid down in Boiler act [Indian Boiler Regulations (IBR) Act].

3.8 COMPARISON BETWEEN WATER TUBE AND FIRE TUBE BOILERS

Water Tube Boiler	Fire Tube Boiler
1. Water tube boilers can generate steam upto a pressure of 225 bar.	1. These generate steam only upto a pressure of 20–25 bar.
2. The rate of steam generation is more.	2. The rate of steam generation is less.
3. It has small shell i.e. area required by this type of boiler is less.	3. It has large shell and the area required by this boiler is more.
4. Construction is very much complex, hence more skill is required for operation.	4. Construction is relatively simple and hence less skill is required.
5. Steam generation capacity is more and hence large power plants use these boilers.	5. It is not suitable for power plants, but is used in Dairy, Laundry and some process industries.
6. Water passes through the tubes and hot gases surrounds them.	6. Hot gases pass through the tubes and water surrounds them.
7. Examples: Badcock Wilcox boiler, Stirling boiler.	7. Examples: Lancahire boiler, Locomotive boiler.

3.9 STEAM TURBINES

Introduction. The steam turbine is a prime mover in which the potential energy of the steam is transformed into Kinetic energy, and later in its turn is transformed into the mechanical energy of rotation of the turbine shaft. The turbine shaft, directly or with the help of a reduction gearing, is connected with the driver mechanism. Depending on the type of the driver mechanism a steam turbine may be utilized in most diverse fields of industry, for power generation and for transport.

Steam turbine is a rotary machine and consists of curved blades or vanes mounted on a wheel which is fixed to the shaft supported in bearings. This assembly is housed in a casing. Figure 3.15 shows these different parts.

3.9.1 Classification of Steam Turbines

There are several ways in which the steam turbines may be classified. The most important and common division being with respect to the action of the steam, as

(i) Impulse turbine

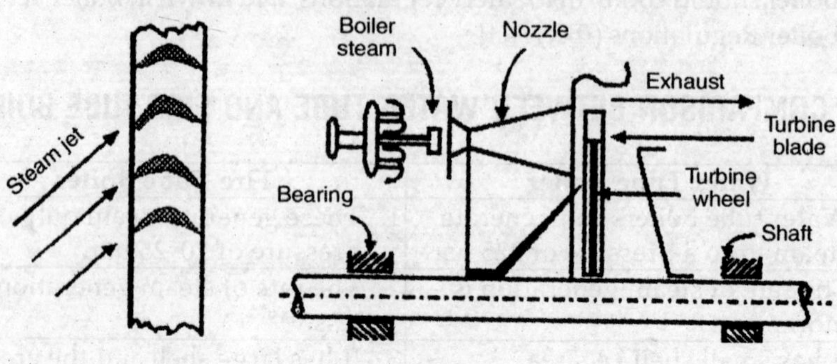

Fig. 3.15

(ii) Reaction turbine

(iii) Combination of impulse and reaction turbine

Other classifications are:

(i) According to the number of pressure stages: (a) Single stage and (b) Multi stage

(ii) According to the direction of flow of steam:

(a) Axial turbines in which steam flows in the direction of axis of the shaft.

(b) Radial turbines in which steam flows in a direction perpendicular to the axis of the turbine.

(iii) According to the number of cylinders:

(a) Single cylinder and (b) Multi cylinder turbines.

(iv) According to the methods of governing:

(a) Throttle governed turbine

(b) Nozzle governed turbine

(c) By-pass governed turbine

(v) According to the heat drop processes:

(a) Condensing turbines

(b) Back pressure turbines

(c) Extraction type turbines

(d) Topping turbines

(e) Mixed pressured turbines

(vi) According to steam conditions at inlet to turbine

(a) Low pressure turbines, using steam at 1.2 to 2 bar.

(b) Medium pressure turbines using steam upto 40 bar.

(c) High pressure turbines using steam above 40 bar.

(d) Very high pressures using steam at 170 bar and higher temperatures of 550°C and higher.

(e) Turbines with supercritical pressures using steam at 225 bar and above.

(vii) According to their usage in industry:

(a) Stationary turbines and constant speed used for driving alternators is in power plants.

(b) Stationary turbines with variable speed meant for driving turbo-blowers, air circulators, pumps etc.

(c) Non-stationary turbines with variable speed used for steamers, ships and railway locomotives.

3.9.2 Advantages of Steam Turbine

The following are the principal advantages of steam turbine:

1. The power generation in a steam-turbine is at a uniform rate and therefore the necessity to use flywheel is not felt.
2. Much higher speeds are possible.
3. High power generation is possible as in case of power station.
4. With the absence of reciprocating parts, the balancing problem is minimised.
5. No internal lubrication as there are no rubbing parts in the steam turbines.
6. It can use high vacuum very advantageously.
7. Considerable overloads can be carried at the expense of slight reduction in overall efficiency.

3.9.3 Working of Common Types of Turbines

The common types of steam turbines are

1. Simple impulse turbine
2. Reaction turbines

The main difference between these turbines lies in the way in which the steam is expanded while it flows through them. In the former type steam expands in the stationary blades or parts called nozzles and its pressure does not alter, change when it moves over the moving blades while in the reaction turbines steam expands continuously as it passes over the blades and thus there is gradual fall in the pressure during expansion.

In impulse turbine, the pressure and total heat (enthalpy) drop and this heat drop produces a jet of steam through nozzle. This jet of passes over the curved blades producing force on the blades. This force is acting tangential to be wheel producing torque and hence the mechanical work is produced.

The main or principal example of this turbine is the well known De-Laval turbine.

In reaction turbine, there is a gradual pressure drop and takes place continuously over the fixed and moving blades. The function of the fixed blades is the same as nozzles which direct the steam at a required angle. Through these fixed blades, steam expands and we get jet of steam. This steam jet passes over the moving blades and here also pressure drop and velocity increase take place. This kinetic energy is absorbed in moving blades and the wheel rotates producing work or power.

Devices such as internal combustion engines, steam engines, water/steam/ gas turbines etc. which produce power are called power producing devices or simply prime movers.

3.9.4 Steam Engines (Double Acting)

Steam Engines are not used these days but it really ruled the early industrial era and dominated the scene for very long time. It is said that the **steam engine owes less to science than science owes to it.** As such its brief description from historical point of view only is given here. Its main components are:

(a) Main frame. It supports and hold the moving parts of the engine.

(b) Cylinder block. On one side of the cylinder block is cylinder head while on the other side is stuffing box that prevents leakages of steam. The stuffing box has packging of asbestos or rubber to prevent steam leakage.

(c) Piston. The piston with piston rings moves inside the cylinder.

(d) Piston rod. It is circular rod connected to piston by screwed joint. This rod comes out of the cylinder through stuffing box and gland.

(f) Cross head and guide bars. The other end of the piston is connected to a block by means of a cotter. Cross head moves to and fro in the guide bars.

(e) Connecting rod. Smaller end of the connecting rod is conneted to cross-head by means of gudgeon pin.

(g)Valve gear. Consisting of D-slide valve, eccentric rod, eccentric.

(h) Fly wheel. Its function is to maintain constant angular motion during the cycle by storing energy.

(k) Governor. It maintains the speed constant at all loads.

Steam at high pressure enters the steam chest and then into the cylinder through the admission port. High pressure steam pushes the piston, which through the connecting rod rotates the crankshaft producing the power. Steam continues to enter till the eccentric moves in the opposite direction to close the

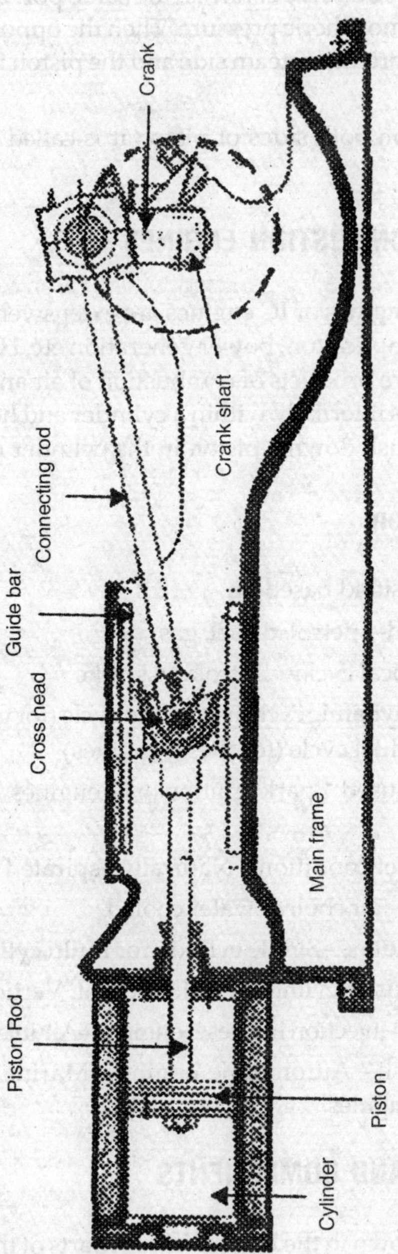

Fig. 3.16. Steam engine.

admission port. This point is known as cut off point. The steam in the cylinder expands and does work on the piston. The volume of steam increases and the pressure falls. The D-slide valve connects the same port to the exhaust side. The pressure falls to the atmospheric pressure. Then the opposite side of the piston is connected to the hgih pressure steam side and the piston is pushed in the reverse direction.

As the steam acts on both sides of piston it is called a double acting steam engine.

3.10 INTERNAL COMBUSTION ENGINES

Internal combustion engines or IC engines are extensively used in automobiles, locomotives, marine applicaiton, power generation etc. Here the working media is hot and high pressure products of combustion of air and gasoline / diesel fuel. The combustion occurs internally within a cylinder and hence the name. The hot high pressure gases push down a piston in the cylinder producing power.

3.10.1 Classification

IC engines can be classified based on

(a) Type of fuel used—petrol, diesel, gas

(b) Type of mechanical cycle—2 stroke, 4 stroke

(c) Type of thermodynamic cycle used. Otto cycle (for gasoline engine and gas engine), diesel, dual cycle (for diesel engines)

(d) Type of ignition used. Spark ignition or SI engines / Compression Ignition or CI engines.

(e) According to inlet conditions. Naturally aspirated or supercharged

(f) Type of cooling—air cooled, water cooled

(g) Number of cylinders—Single cylinder or multi cylinder

(h) Accoriding to axis of cylinders—Horizontal, Vertical or Radial

(i) According to fuel injection in Diesel engine—Air injection or solid injection.

(j) According to use—Automobile engines, Marine engines, Locomotive engines, Land engines.

3.10.2 IC ENGINE AND COMPONENTS

4-stroke SI engine is shown in the Fig. 3.17. Main parts of the internal combustion engine are as following.

(a) **Cylinder.** It is the main body of the engine. It has finely machined bore in which piston with rings closely fits and is able to reciprocate. In the cylinder

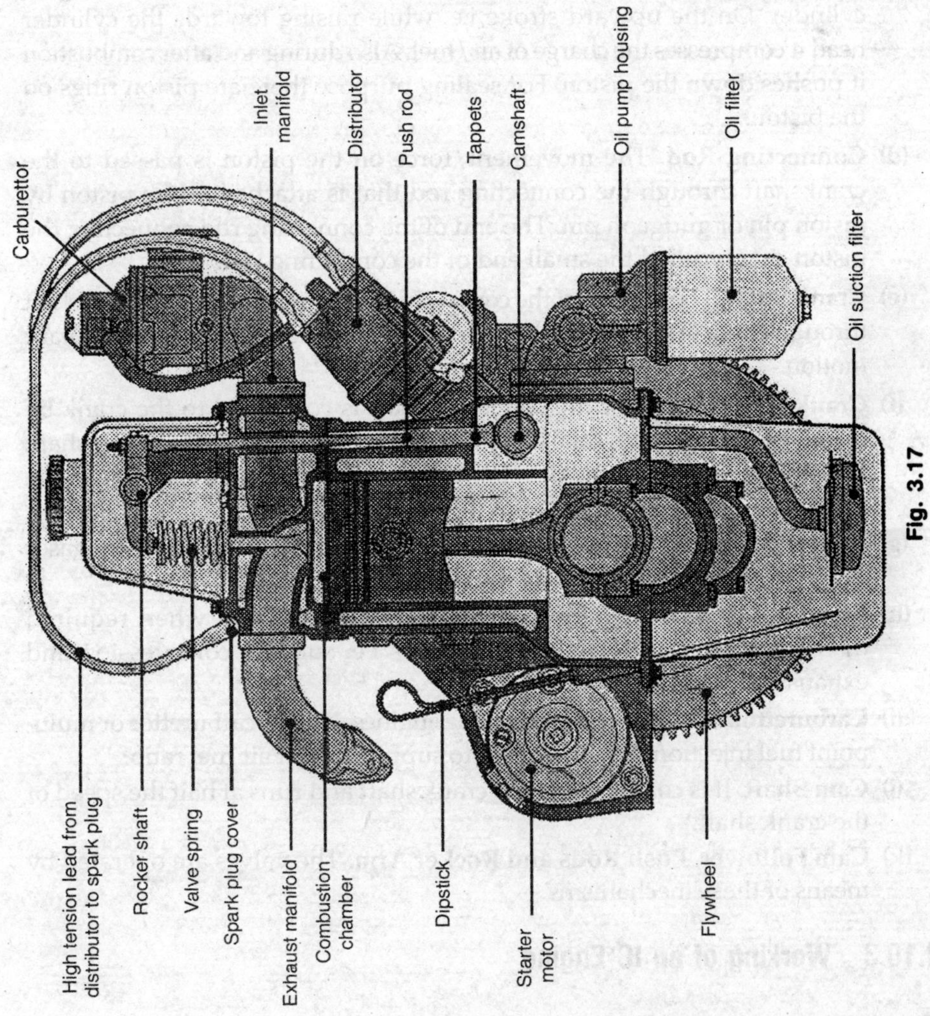
Carburettor
Inlet manifold
Distributor
Push rod
Tappets
Camshaft
Oil pump housing
Oil filter
Oil suction filter
High tension lead from distributor to spark plug
Rocker shaft
Valve spring
Spark plug cover
Exhaust manifold
Combustion chamber
Dipstick
Starter motor
Flywheel

Fig. 3.17

combustion takes place. It should be able to withstand high temperature and pressure. It is necessary to cool the engine to avoid damage.

(b) **Cylinder Head.** The cylinder head closes the top side of the cylinder. It has inlet and exhaust valves and the spark plug to initiate combustion, in case of petrol engine. In case of diesel engines it has fuel injector instead of the spark plug.

(c) **Piston.** The piston is able to reciprocate i.e. move up and down in the cylinder. On the upward stroke i.e. while raising towards the cylinder head it compresses the charge of air / fuel. Also during and after combustion it pushes down the piston. For sealing purpose there are piston rings on the piston.

(d) **Connecting Rod.** The movement / force on the piston is passed to the crankshaft through the connecting rod that is attached to the piston by piston pin or gudgeon pin. The end of the connecting rod connecting the piston side is called the small end of the connecting rod.

(e) **Crank Shaft.** The big end of the connecting rod is attached to the crankshaft through crank that converts the reciprocating motion of the piston to rotary motion.

(f) **Crank.** The big end of the connecting rod is connected to the crank by means of a crank pin while the other end is connected to the crank shaft. Cranks may be centre crank in between two bearings or over hung crank on one side of the bearing.

(g) **Inlet and Exhaust Valve.** These allow fresh charge and let out spent gases respectively at proper times in the cycle.

(h) **Flywheel.** It stores energy during power stroke and when required distributes energy during other storkes i.e. suction, compression and exhaust.

(i) **Carburettor/MPFI.** In case of gasoline engines—either carburettor or multi-point fuel injection system (MPFI) to supply correct air fuel ratio.

(j) **Cam Shaft.** It is connected to the crank shaft and runs at half the speed of the crank shaft.

(k) **Cam Follower, Push Rods and Rocker Arm.** The valves are operated by means of these mechanisms.

3.10.3 Working of an IC Engine

The working of a 4-stroke SI engine is shown Fig. 3.18. The cycle is completed in four strokes of the piston. The working has been summarized in the table given below. The position when the piston is at the topmost position is called top dead center position (TDC) / IDC inner dead centre in case of horizontal engine. During one cycle there are four strokes and the crank moves through two revolution.

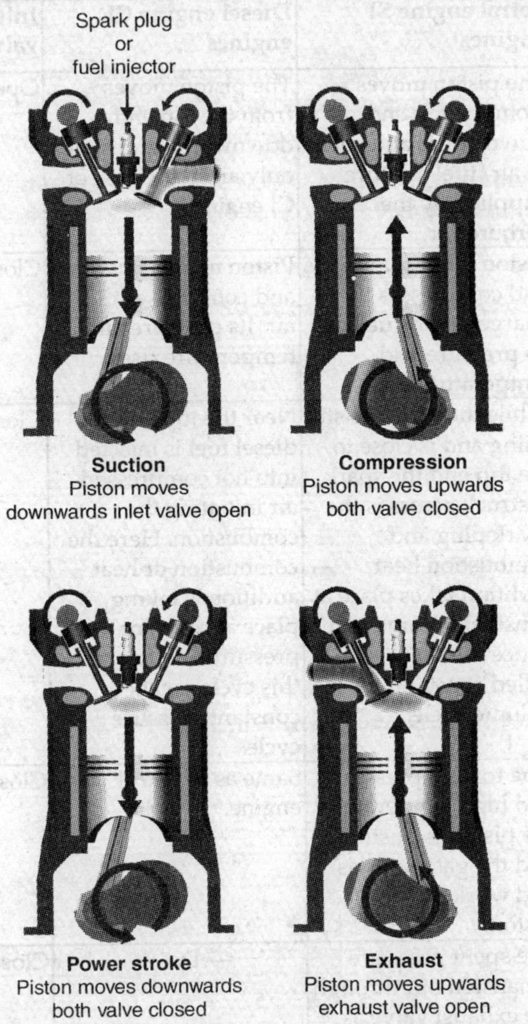

Fig. 3.18. Working of 4-stroke IC engine.

3.11 TWO STROKE SI ENGINE

In this engine the cycle is completed in two strokes of the piston or one revolution of the crankshaft (in four stroke engine crank rotates through two revolution in one cycle).

(a) Ports. This engine has no valves but there are three ports inlet, exhaust and transfer ports as shown. These ports get opened and closed due to the position of the piston.

Stroke	Petrol engine SI engines	Diesel engine CI engines	Inlet valve	Exhaust valve
First stroke suction stroke	The piston moves from the top end downwards and sucks in air/fuel mixture supplied by the carburettor.	The piston moves from the top end downwards and sucks only air in the case of Cl engine.	Open	Closed
Second stroke compression stroke	Piston moves upwards and compresses the charge of air/fuel and its pressure and temperature rise.	Piston moves upwards and compresses the air. Its pressure and temperature rises.	Closed	Closed
Ignition	While the piston is still rising and is close to the top end the spark is struck across is the spark plug and combustion heat addition takes place at constant volume and hence the cycle is called constant volume cycle.	Near the top end the diesel fuel is injected into hot compressed air initiating the combustion. Here the combustion or heat addition is taking place at constant pressure and hence this cycle is called constant pressure cycle.	Closed	Closed
Third stroke working or expansion stroke or power stroke	Due to high pressure and high temperature the piston is pushed and the gas expands and work on the piston.	Same as in petrol engine.	Closed	Closed
Fourth stroke exhaust stroke	The spent gases are exhausted out through the exhaust valve as the piston moves upwards.	— Do —	Closed	Open

(b) Piston. The piston has a deflector on its crown which prevents loss of fresh charge while allowing the burnt gases to be exhausted.

(c) Crank case. Charge is compressed in the crank case when the piston is coming down and is transferred to cylinder via. transfer port.

Working. Its working is shown in the Fig. 3.19.

(a) Starting from top dead centre position (TDC), as the piston goes downwards first the exhaust port gets uncovered. The burnt gases are going out through this port.

(b) With further downward movement of the piston the inlet port is opened. Fresh air fuel mixture is admitted into the crank case. The downward movement of the piston acts as pump and slightly compresses the fresh charge in the crank case.

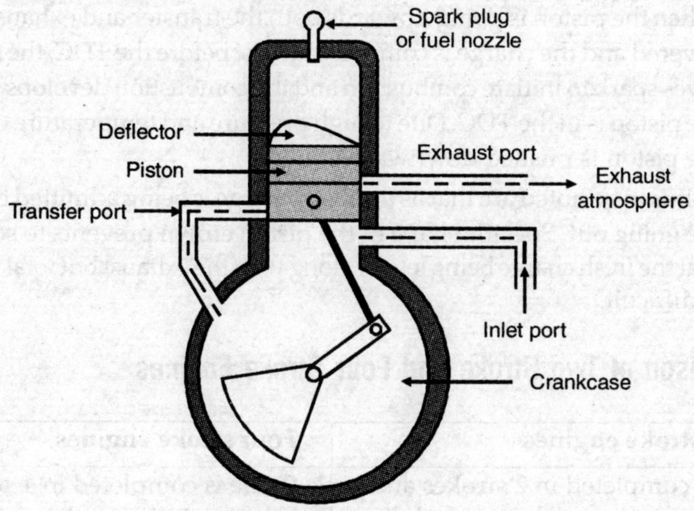

Fig. 3.19. Two stroke engine.

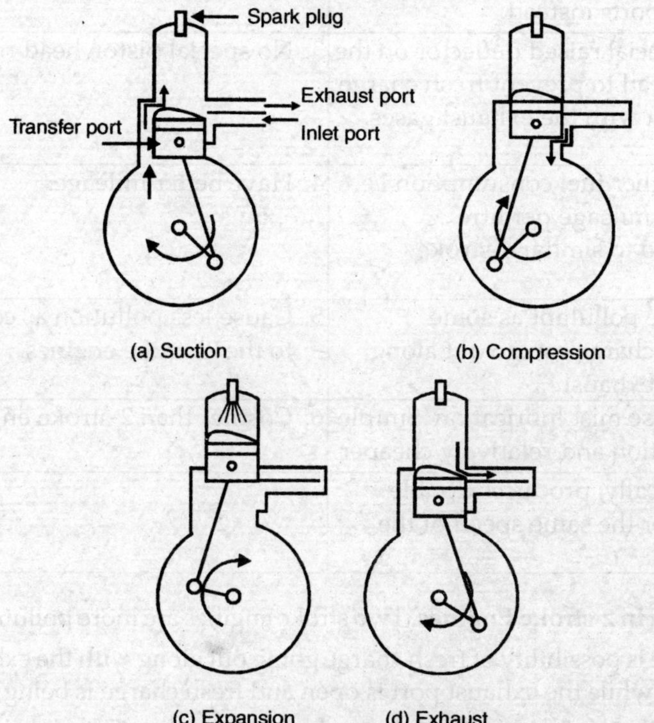

Fig. 3.20. Two stroke petrol engine.

(c) Further movement of the piston uncovers the transfer port allowing the air—fuel mixture charge to be admitted into the cylinder.

(d) When the piston is going upwards both the transfer and exhaust ports are covered and the charge is compressed. Just before the TDC, the spark plug gives spark to initiate combustion and the combustion develops fully when the piston is in the TDC. Due to high pressure and temperature of the gases the piston is pushed downwards.

(e) Points to be noted are that as the fresh charge is being admitted burnt gases are going out. Specific shape of the piston crown prevents to some extent that the fresh charge being let out along with the exhaust but total prevention is difficult.

Comparison of Two Stroke and Four Stroke Engines

Two stroke engines	Four stroke engines
1. Cycle completed in 2 strokes and one revolution of the crank shaft.	1. Cycle is completed in 4 strokes and two revolution of the crank shaft.
2. No valves, it has inlet, exhaust and transfer ports instead.	2. It has inlet and exhaust valves.
3. It has special raised deflector on the piston head to prevent fresh charge going out with the exhaust gases.	3. No special piston head required.
4. Have higher fuel consumption i.e. give less mileage per litre compared to similar 4-stroke engine.	4. Have better mileage.
5. Are more pollutant as some unbrunt charge can go out along with the exhaust.	5. Cause less pollution as compared to the 2-stroke engines.
6. Mostly use mist lubrication. Simple construction and relatively cheaper.	6. Costlier than 2-stroke engines
7. Theoretically, produces double power for the same speed of the engine.	

Pollution in 2-stroke Engines. Two stroke engines are more pollutant because

(a) There is possibility of fresh charge going out along with the exhaust gases since while the exhaust port is open and fresh charge is being admitted.

(b) Usually the 2-stroke engines use mist lubrication. The lubricant mixed with the petrol does not get fully burnt and is pollutant.

3.12 GAS TURBINE POWER PLANT

Gas turbine using natural gas or kerosene it can be used for producing electricity. The power plant consists of—

(a) Axial flow compressor — If compresses the atmospheric air.

(b) Combustion chamber — Here natural gas or fuels like kerosene, turbine fuels are burnt to produce hot gas.

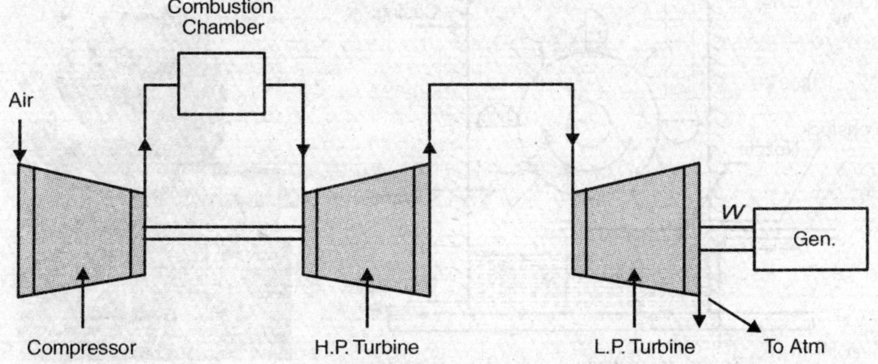

Fig. 3.21. Gas Turbine.

(c) Turbine — The gases, at high temperature and pressure run the turbine. The power developed by the turbine is used to run the air compressor and generators and pressure.

(d) We have gas turbine power plant at Uran. Here the waste heat from the gas turbines is recovered to convert water into steam. The steam is used to run steam turbine which also generate power.

3.13 HYDRAULIC TURBINES

If the turbines are driven by means of water then they are known as Hydraulic turbines. Following turbines are generally used.

(i) Pelton Turbine. It is an impulse turbine used for higher heads. As shown in Fig. 3.21 consists of turbine runner on which buckets are fixed. When the water in the form of a jet from the nozzle strikes the buckets, it produces rotary motion of the turbine shaft. This rotary motion of the turbine shaft is used to drive the generators for generating power.

(ii) Francis Turbine. It is inward flow reaction turbine. It is used for producing power under medium heads.

Figure 3.23 shows the Francis turbine runner. When the high energy water from the penstocks or pipes, enter the turbine it drives the turbine for producing power.

(iii) Kaplan Turbine. It is an axial flow reaction turbine. In this flow of water is parallel to the shaft. It is used when large quantity of water is available at smaller heads.

Figure 3.24 shows the Kaplan turbine runner. When the high energy water from the penstocks, enters the turbines, it drives the turbines for producing power.

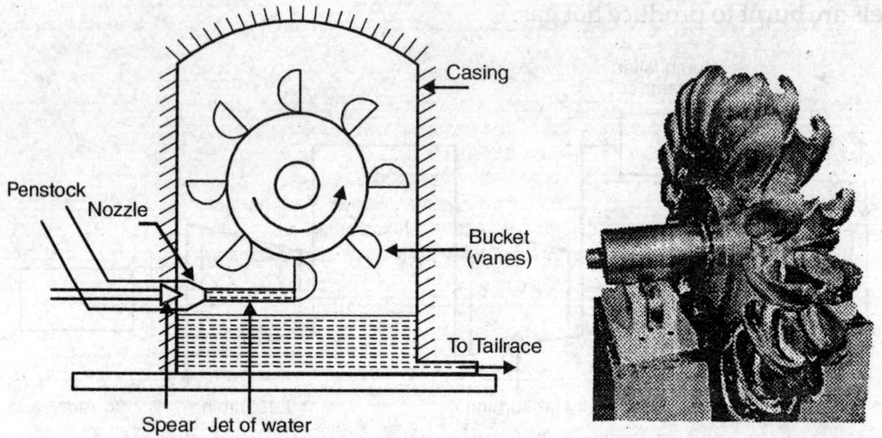

Fig. 3.22 (a) Pelton turbine.

Fig. 3.22 (b) Runner of pelton wheel.

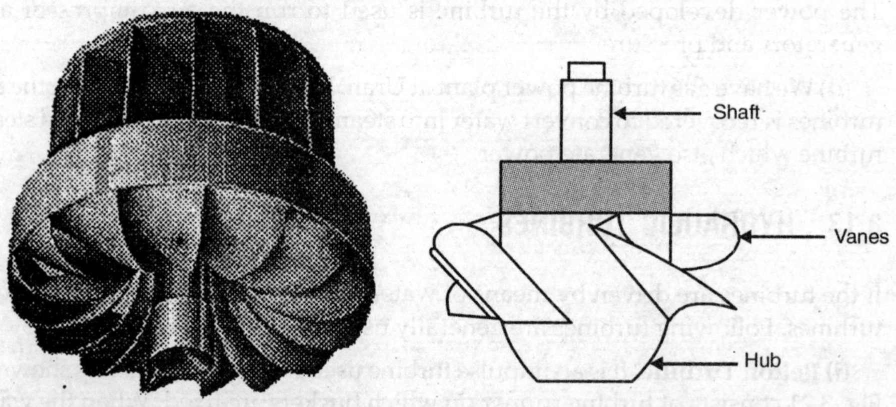

Fig. 3.23. Francis turbine runner.

Fig. 3.24. Kaplan turbine runner.

3.14 COMPRESSED AIR MOTORS

Such motors are extensively used in industry to operate variety of pneumatic tools like vibrations, drilling, hammers. riveting machines etc. In mines where safety is major consideration normally compressed air motors are used. Motors are mainly of two types—Reciprocating or Rotary expanders.

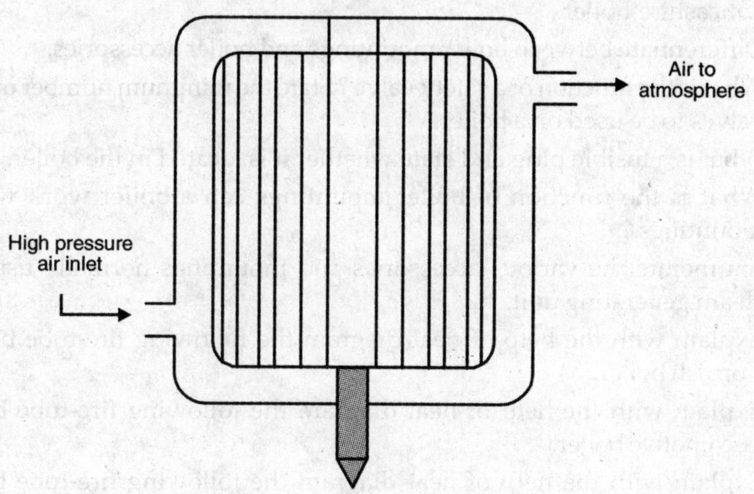

Fig. 3.25. Rotary air motor.

(a) Reciprocating. It is reverse of the reciprocating compressor. By supplying compressed air from a receiver to the cylinder the air expands from high pressure to the ambient condition. During this the piston is pushed down. The cycle gets completed after the air is exhausted and the piston returns to TDC when fresh compressed air is admitted.

(b) Rotary Type. The rotary types of compressed air turbines are small in size and valve less. The compressed air runs a turbine and power can be used at the turbine shaft as shown in Fig. 3.25.

THEORY QUESTIONS

1. Compare a two stroke cycle engine with a four stroke cycle engine.
2. Explain the working of a four stroke petrol engine by means of a sketch.
3. Explain the working of two-stroke petrol engine.
4. Which engine causes more air pollution ? A four stroke or two stroke petrol engine. Give reasons.
5. Describe the basic principles of operation and application of compressed air motor.
6. Gas Turbine.
7. How boilers are classified ?
8. Give the comparison between water tube and fire tube boilers.

9. Explain with the help of neat diagram the following : Fire-tube boilers, Lancashire boiler.

10. Differentiate between boiler mountings and boiler accessories.

11. What is the function of a safety valve ? State the minimum number of safety valves to be used on a boiler.

12. What is a fusible plug and state whether it is located in the boiler.

13. What is the function of boiler mountings can a boiler work without mountings ?

14. Enumerate the various accessories and mountings normally used in a steam generating unit.

15. Explain with the help of neat diagram the following fire-tube boilers : Cornish boiler.

16. Explain with the help of neat diagram the following fire-tube boilers, Locomotive boiler.

17. Explain with the help of neat diagram the following fire-tube boilers, Cochran boiler.

18. Give the construction and working of the following water tube boilers.

19. Babcock and Wilcox boiler.

20. Stirling boiler.

21. What advantages are obtainqd if superheated steam is used in steam prime movers.

22. List the primary requirements of boilers.

23. What are water trubines ? Explain their various types.

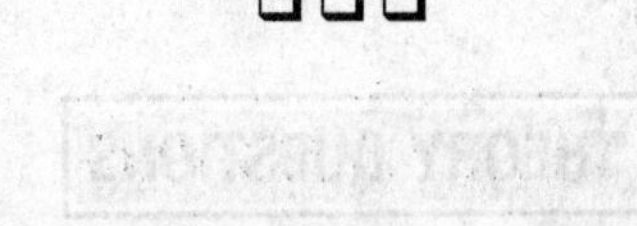

CHAPTER 4

POWER ABSORBING DEVICES

4.1 INTRODUCTION

The equipments or devices that consume power for working are called power-absorbing devices. Examples are pumps, compressors, refrigerators etc.

4.2 HYDRAULIC PUMPS

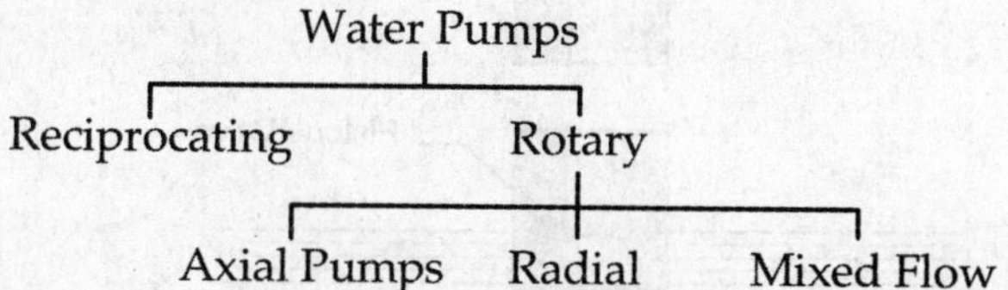

The hydraulic machines which convert the mechanical energy into hydraulic energy are called as *pumps.* The hydraulic energy is in the form of pressure energy. Two types of pumps commonly used are centrifugal and axial flow pumps. They are so named because of the general nature of the fluid flow through the impeller. The combination of centrifugal and axial flow pumps is called as *mixed flow pump* wherein part of the liquid flow in the impeller is axial and part is radial.

The hydraulic machines that convert mechanical energy into pressure energy, by means of centrifugal pump is similar in construction to the Francis turbine. But the difference is that the fluid flow is in a direction opposite to that in the turbine.

4.3 RECIPROCATING PUMPS

(a) Hand pump. Reciprocating water pumps are extenstively used in our villages where ground water level is high. It consists of a suction pipe, filter, piston/

valve assembly which is operated by hand handle or by motor. During upward motion of piston, vacuum is created which allows the water to enter the cylinder and water is transferred to the delivery side. The water is delivered intermittently due to reciprocating action. Hand pump is shown in Fig. 4.1.

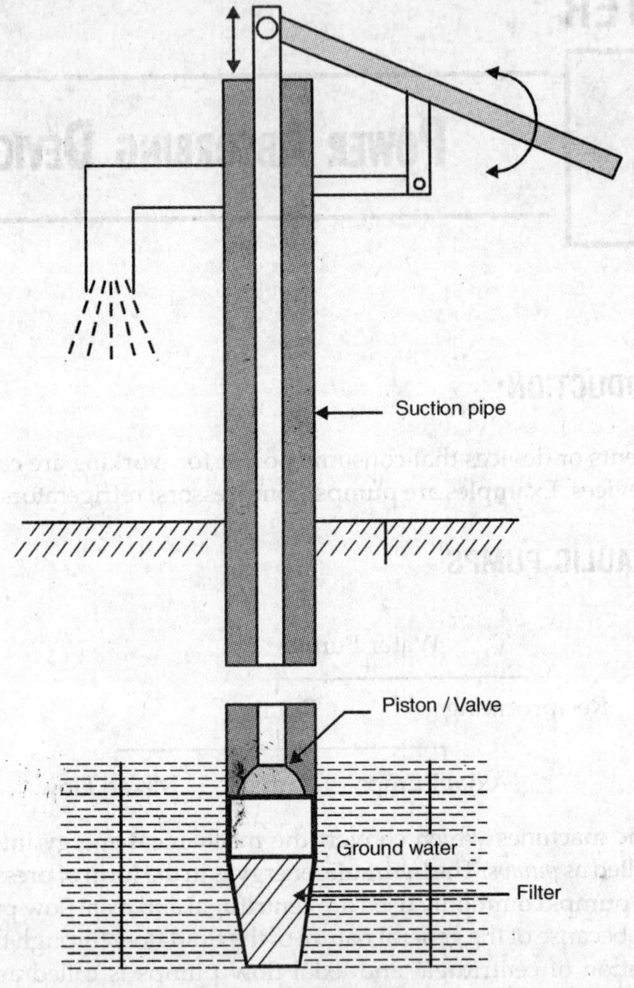

Fig. 4.1. Hand Pump.

4.4 RECIPROCATING PUMP

The parts of the reciprocating pump are shown in Fig. 4.2. When piston moves towards right, vacuum is created thus opening the suction valve and water enters the cylinder. When piston moves towards the left, piston pushes water closing the suction valve and opening the delivery valve. Water is pushed and delivered to the overhead storage tank or for any other purpose.

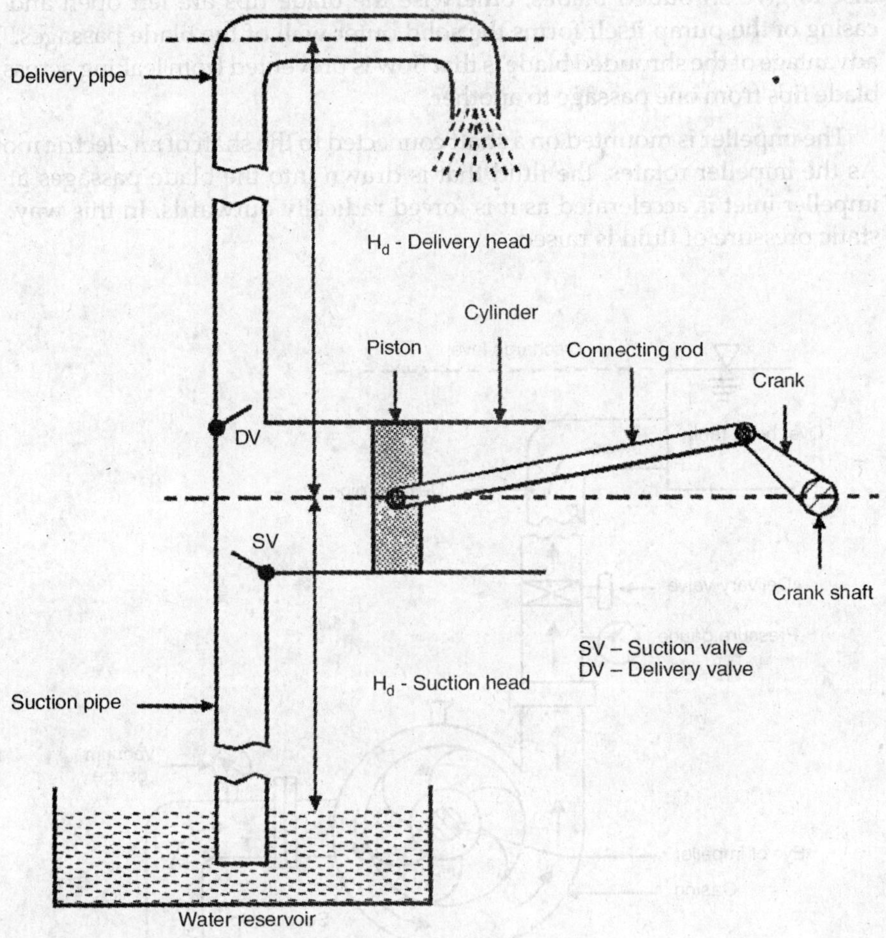

Fig. 4.2. Reciprocating Pump.

4.5 CENTRIFUGAL PUMP

Main parts of a centrifugal pump. The parts of a centrifugal pump are similar to those of a centrifugal compressor. The three important parts of a centrifugal pump are (1) Impeller, (2) Casing and (3) Suction and Delivery pipes. (Fig. 4.3)

4.5.1 Impeller

The rotating part of the centrifugal pump is called the *Impeller*. The impeller is a rotating solid disc with curved blades standing out vertically from the face of the disc.

The tips of the blades in the impeller are sometimes covered by another flat disc to give shrouded blades, otherwise the blade tips are left open and the casing of the pump itself forms the solid outer wall of the blade passages. The advantage of the shrouded blade is that flow is prevented from leaking across the blade tips from one passage to another.

The impeller is mounted on a shaft connected to the shaft of an electric motor. As the impeller rotates, the fluid that is drawn into the blade passages at the impeller inlet is accelerated as it is forced radically outwards. In this way, the static pressure of fluid is raised.

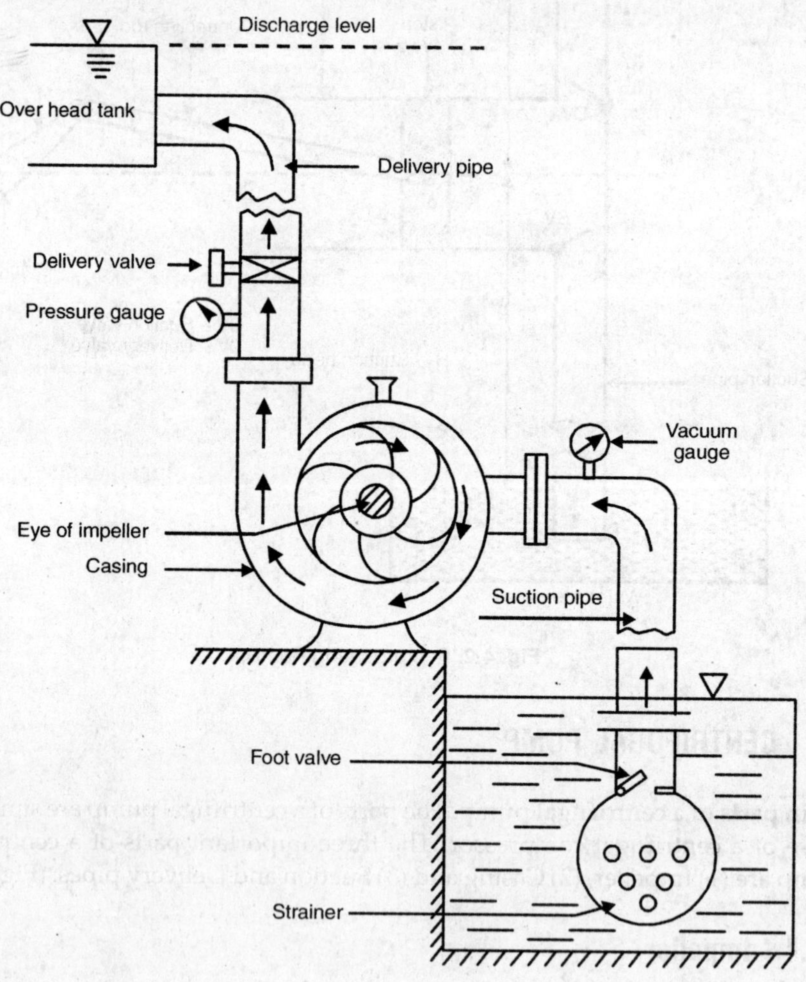

(a)

4.5.2 Casing

It is an air-tight passage surrounding the impeller which converts the K.E of water leaving the impeller into pressure energy before the water leaves the casing and enters the delivery pipe. The three commonly used casings are discussed as below.

(a) Volute casing. Casing that surrounds the impeller is of spiral type in which flow area increase gradually, The increase in area of flow, decreases the velocity of flow and thus increases the pressure of water. The efficiency of centrifugal pump having this casing is reduced due to the formation of eddies.

(b) Vortex casing [Fig 4.3 (b)]. If a circular chamber is introduced between the casing and the impeller, then that casing is known as *vortex casing*. This considerably reduces the loss of energy due to the formation of eddies. Thus, the efficiency of the pump is more than the efficiency of volute casing centrifugal pump.

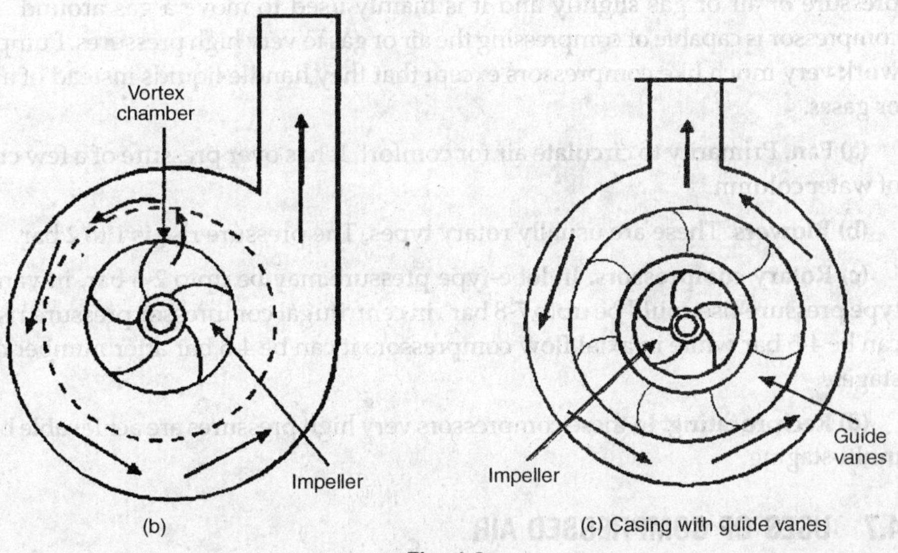

Fig. 4.3

(c) Casing with guide blades [Fig. 4.3 (c)]. In this type of casing, the impeller is shrouded by a series of guide blades mounted on a ring which is known as *diffuser*. The guide vanes are designed in such a way that the water from the impeller enters the guide vanes without shock. The area of the guide vanes increases, thus, reducing the flow velocity through guide vanes and consequently increasing the pressure of water. The water from the guide vanes then passes through the surrounding casing, which is in most cases concentric with the impeller. The diffuser is optional and may not be present in a particular design depending upon the size and cost of the pump.

4.5.4 Vortex Casing

A pipe whose one end is connected to the inlet of the pump and other end diped into the water in a sump is known as the suction pipe. A foot valves, fitted at the lower end of the suction pipe, opens only in the upward direction. A strainer to remove debris from water, is also fitted at the lower end of the suction pipe. The pipe whose one end is connected to the outlet of pump and the other end delivers the water at a required height is known as delivery pipe.

4.6 AIR COMPRESSORS

Air compressors are used to compress the atmospheric air to higher pressure. Air Compressors are extensively used in our every day lives and in industry.

Compressors as well as pumps and fans are the devices used to increase the pressure of a fluid. But they differ in the tasks they perform. A fan increases the pressure of air or gas slightly and it is mainly used to move a gas around. A compressor is capable of compressing the air or gas to very high pressures. Pumps work very much like compressors except that they handle liquids instead of air or gases.

(a) Fan. Primarily to circulate air for comfort. It has over pressure of a few cm of water column.

(b) Blowers. These are usually rotary types. The pressure rise is 1 to 2 bar.

(c) Rotary compressors. In lobe-type pressure.may be upto 2-3 bar, in vane type pressure rise could be upto 7-8 bar , in centrifugal compressor pressure rise can be 4-5 bar while in axial flow compressors it can be 4-5 bar after number of stages.

(d) Reciprocating. In these compressors very high pressures are achievable by multi-staging.

4.7 USES OF COMPRESSED AIR

Compressed air is extensively used in industry and other purposes like—

(a) Air motors (in mines where IC engines or electrical power cannot be used due to safety reasons).
(b) Inflating tyres / tubes.
(c) In spray painting.
(d) For cleaning purposes—in garages along with water for washing cars etc., sand blasting to clean castings / jobs.
(e) In blast furnaces.
(f) In gas turbines.

(g) In diesel engines.
(h) Air brakes.

4.7.1 Classification of Air Compressors

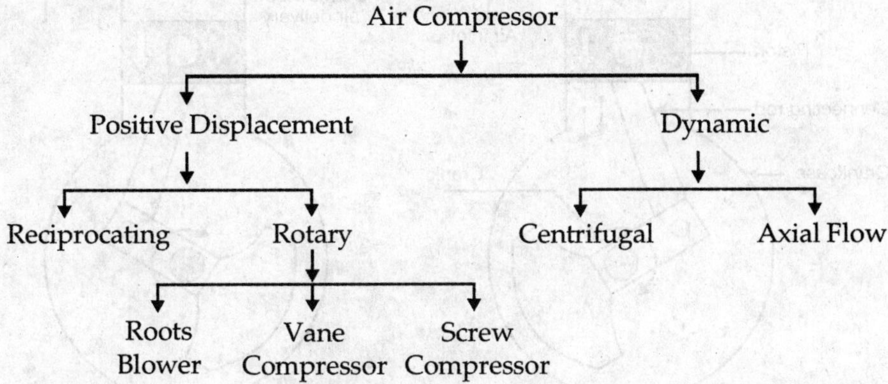

(a) Positive Displacement Type Compressor. In these compressors air between fixed boundaries is taken in and the same air after compression is delivered out. In such compressors the pressure is increased by reduction in volume of air or by back pressure. Examples are reciprocating compressors and rotary type (Roots blower / vane type and screw compressors) thus classification under this category are—

(i) Reciprocating
(ii) Rotary — Roots blower, Vane compressor and screw compressor

(b) Dynamic Compressors. Here pressure rise is achieved by imparting kinetic energy to the air. Use of diffuser further increases the pressure. Examples are centrifugal and axial flow compressors.

(i) Centrifugal compressors
(ii) Axial flow compressors

4.8 RECIPROCATING AIR COMPRESSOR

It is similar in constructions to an IC engine i.e. it has piston / cylinder arrangement, crank and valves. The crank shaft is rotated by external power to provide reciprocating movement to the pistons. The construction is shown in Fig. 4.4.

(a) The compressor can be single acting if it admits and deliver air from one side of the piston. If it admits and delivers air from both sides (i.e. when it is delivering compressed air from the top side of piston, it is admitting air from the bottom side) it is called double acting.

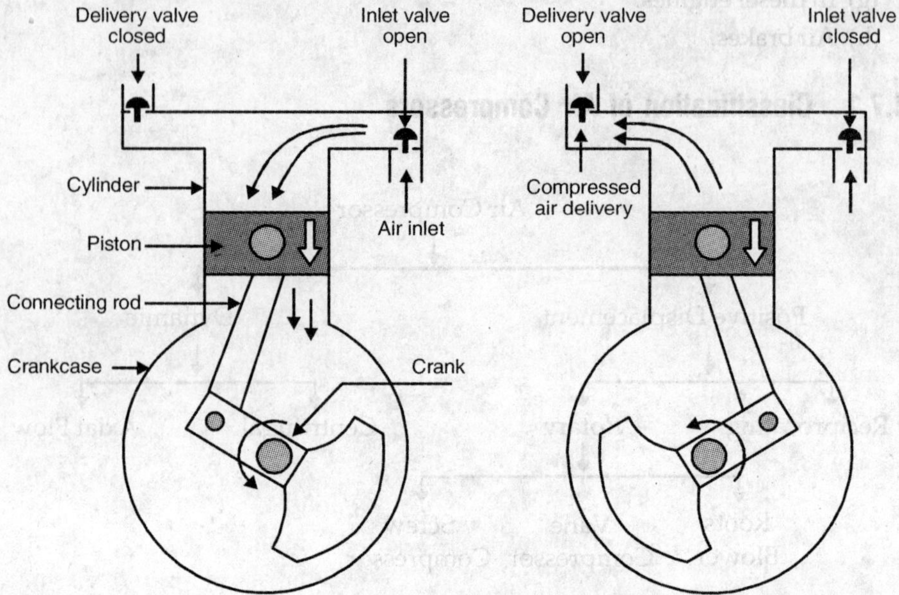

Fig. 4.4. Reciprocating air compressor.

(b) In a multi stage compressor, the compressed air of the first / previous stage is the inlet to the second / next compressor. By multistaging compression ratio can be increased considerably. Normally there is inter cooling between stages.

4.8.1 Working

The piston reciprocates in the cylinder. It has inlet and delivery valves which work on the predetermined pressure settings.

(a) The inlet valve opens when the pressure inside the cylinder becomes less than the atmospheric pressure.

(b) The delivery valve will operate when the pressure of the compressed air inside the cylinder becomes more than the reservoir / pressure.

(c) When the piston is at the top dead centre position (TDC) it starts moving downwards. The air in the clearance volume will expand, delivery valve will close. At some stage the pressure will fall below the atmospheric pressure. Then the inlet valve will open admitting the fresh air.

(d) The suction is continued till bottom dead centre position (BDC). On the upward movement of the piston when the pressure of the air is more than atmospheric pressure the inlet valve closes. Compression of air is continued during upward movement. When pressure of air increases beyond delivery valve setting, the air is delivered to reservoir.

(e) The supply of compressed air from the reciprocating compressor is intermittent and the compressed air is stroed in a reservoir.

4.9 ROTARY POSITIVE DISPLACEMENT COMPRESSORS

(a) Roots Blower. It is as shown in Fig. 4.5. It has two or three lobes but the three lobe version is very popular because of higher pressure ratio. One of the lobes is connected to the drive while the other is driven by gear. In this way, the rotors rotate in phase. The profiles of the lobes are in volute form, giving correst mating of the lobes to seal delivery side from inlet side. The sealing continues till delivery commences.

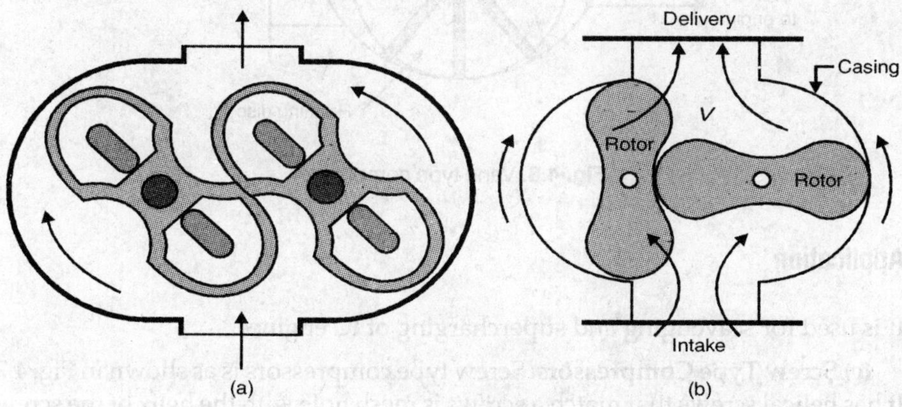

Fig. 4.5. Roots blower.

There must be some clearance between the lobes and between the casing and lobes to reduce friction and wear. This decreases the volumetric efficiency as pressure ratio increases because it forms a leakage path.

(b) Vane type Rotary Compressor. Vane type rotary compressor is as shown in Fig. 4.6. In this compressor the rotor in this eccentrically mounted in the casing. The disc has radial slots as shown and contain vanes, which are free to slide radially into the slots.

When the rotor rotates the disc, the vanes are pressed against the casing due to centrifugal force and form an air tight seal. As the disc rotates air is trapped in the pockets between the vanes and casing and is gradually supplied to the receiver, where the pressure increases by squeezing action.

Advantages

1. This compressor requires less work for given flow of air and gives more pressure ratio compared to Roots blower.
2. The floating drum type which rotates between the rotot and casing having no contact of vanes with casing. This arrangement reduces friction between vane and casing.

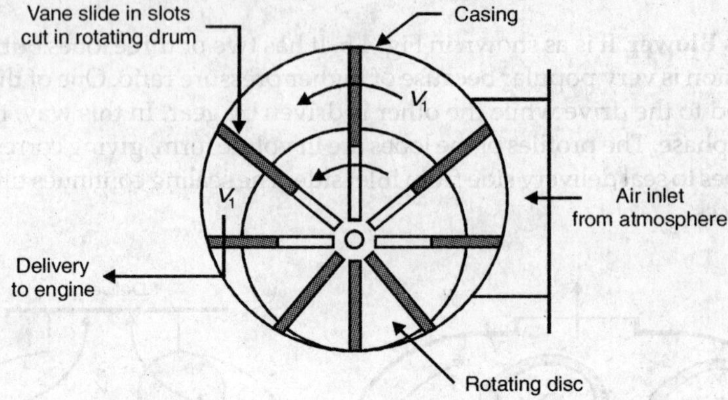

Fig. 4.6. Vane type compressor.

Application

It is used for scavenging and supercharging of IC engines.

(c) Screw Type Compressors. Screw type compressors is as shown in Fig. 4.7. It has helical screws that match a screws is mesh hole with the helix of the screw. The air is taken in from one side. The entrapped air progressively passes through the narrow passage ways formed by the locks and cover.

When the entrapped air reaches the wall it gets compressed and is taken out of the outlet valve. This process happens in a fraction of a second because the screw is rotated at a very high speed.

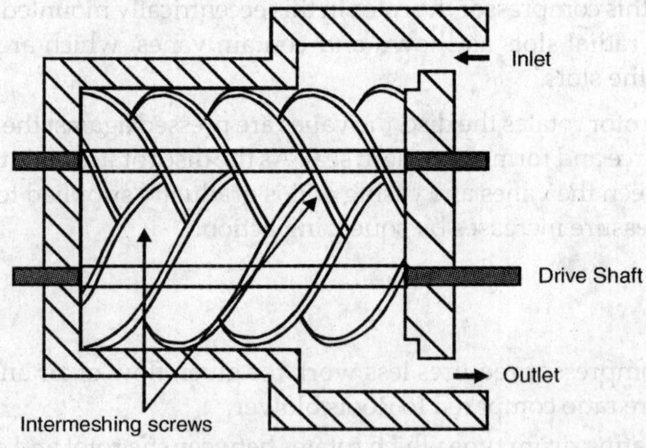

Fig. 4.7. Screw type compressor.

4.10 CENTRIFUGAL COMPRESSOR

It is extensively used in industry for supplying large quantity of air at relatively lower pressure. Its construction is as shown in Fig. 4.8. It consists of an impeller with blades. Air velocity and pressure increases after passing over the blades. High velocity then goes through a passage of increasing cross section called diffuser where the velocity gets converted into pressure. (Velocity reduces and pressure increases.)

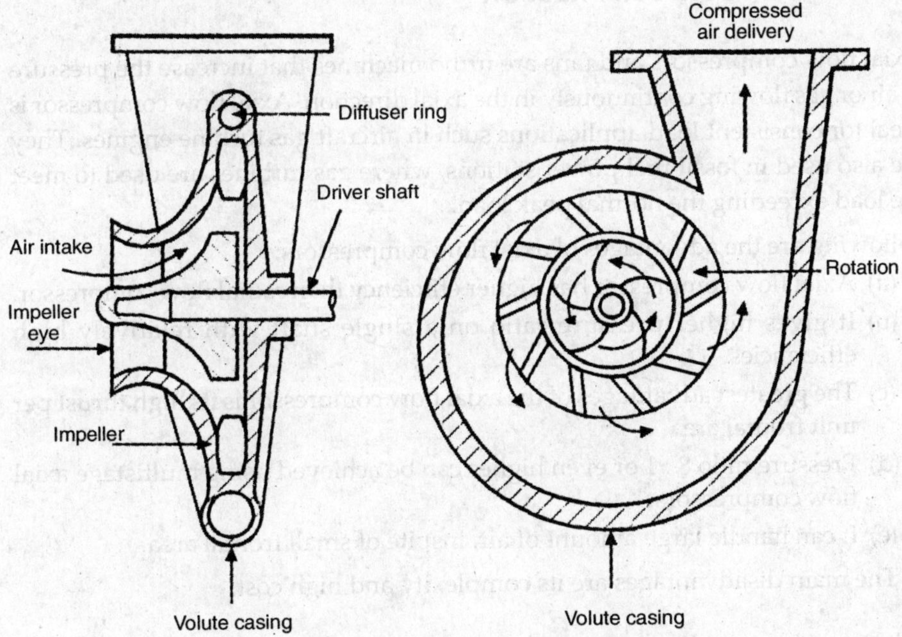

Fig. 4.8. Centrifugal compressor.

Centrifugal compressors and fans are turbo-machines employing centrifugal effects to increase pressure of the fluid. Single stage centrifugal compressors have the pressure ratio of 4 : 1. The best efficiencies are generally 3 to 4 per cent below those obtained from an axial 3 to 4 per cent below those obtained from an axial compressor for the same duty.

The advantages of centrifugal compressor over the axial flow compressor are (1) small length (2) wide range of mass flow rate of air or gas.

Figure 4.8 shows a typical centrifugal compressor. The principal components are the impeller and diffuser. When the impeller is rotating at high speed, air is drawn in through the eye of the impeller. The absolute velocity of the inflow air is axial. The magnitude and direction of the entering relative velocity depends upon the linear velocity of the impeller. The air then flows radially outwards through the impeller passages due to centrifugal force. The total mechanical energy driving the compressor is transmitted to fluid stream in the impeller where it is converted into kinetic energy, pressure and heat due to friction. The function of the diffuser is to convert the kinetic energy of air leaving the impeller into pressure. The air leaving the diffuser is collected in a spiral passage (scroll or volute) from which it is discharged to the reservoir.

4.11 AXIAL FLOW COMPRESSOR

Axial flow compressors and fans are turbo machines that increase the pressure of air or gas flowing continuously in the axial direction. Axial flow compressor is ideal for consistent load applications such in aircraft gas turbine engines. They are also used in fossil fuel power stations, where gas turbines are used to meet the load exceeding the normal peak load.

Following are the advantages of axial flow compressors:

(a) Axial flow compressor has higher efficiency than radial flow compressor.
(b) It gives higher pressure ratio on a single shaft with relatively high efficiencies.
(c) The greatest advantages of the axial flow compressor is its high thrust per unit frontal area.
(d) Pressure ratio 8 : 1 or even higher can be achieved using multistage axial flow compressors.
(e) It can handle large amount of air, inspite of small frontal area.

The main disadvantages are its complexity and high cost.

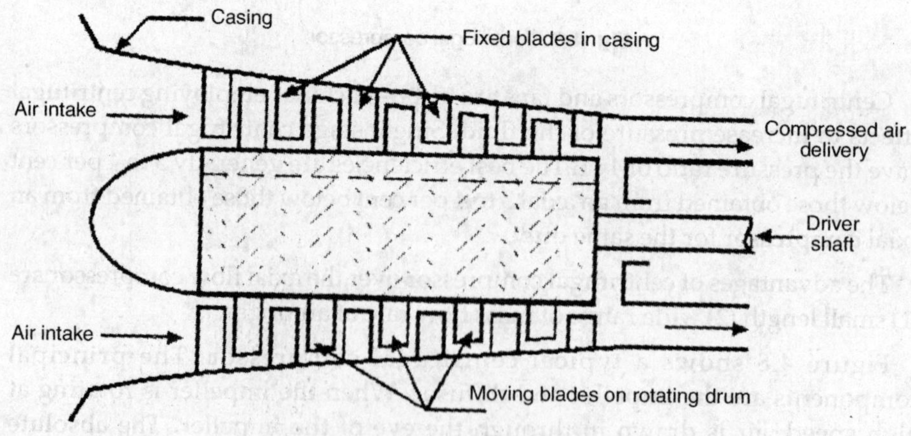

Fig. 4.9. Axial Compressor.

4.12 HOUSEHOLD REFRIGERATOR

Refrigeration is a cyclic process in which temperature less than atmospheric temperature is produced. Refrigerating cycle is the reversed power cycle. In power cycle heat is added at higher temperature while in refrigerating cycle heat is rejected at higher temperature. Similarly in power cycle heat is rejected at lower temperature while in refrigerating cycle heat is added to the working substance at the lower temperature. This is shown in Fig. 4.10.

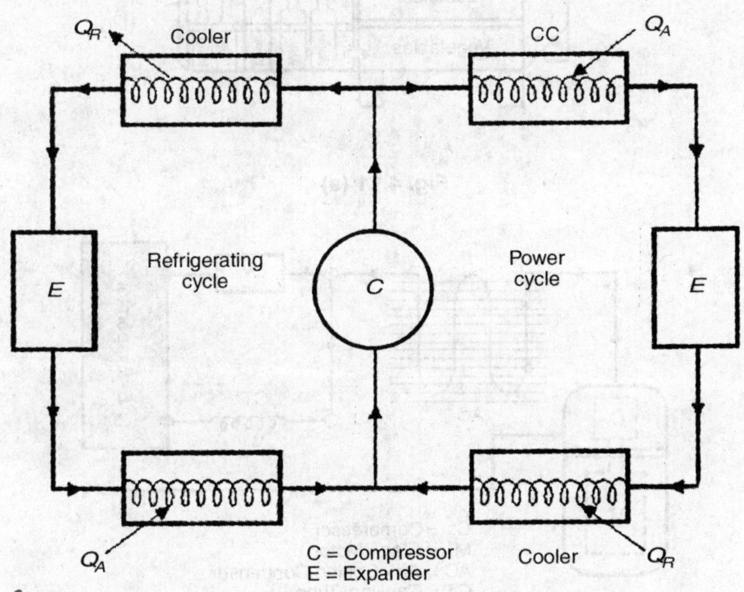

Fig. 4.10

Refrigerating device used for household purposes is called a Household Refrigerator.

A schematic household refrigerator is shown in Fig. 4.11 and is self explanatory, (a) is the cabinet of the refrigerator and Fig. 4.11 (b) shows the components used in refrigeration cycle. The figures are self explanatory.

The working substance used in the refrigerating cycle is called refrigerant. Generally freon-12(R-12) is used as the refrigerant in the household refrigerator.

To understand how lower temperature is produced, we will consider a very simple example.

Consider 1 kg of air at 1 bar pressure and 27° C (300K). This air is compressed to such a pressure such that the temperature of air is doubled i.e. 600 K. Then this air at the higher pressure is cooled so that the temperature of air is 300 K (ideally).

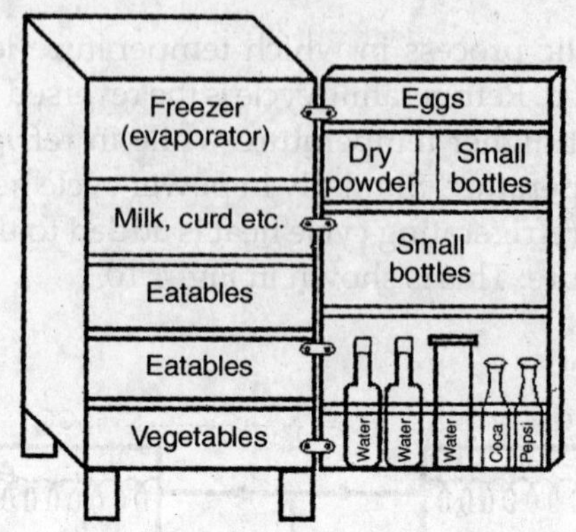

Fig. 4.11 (a)

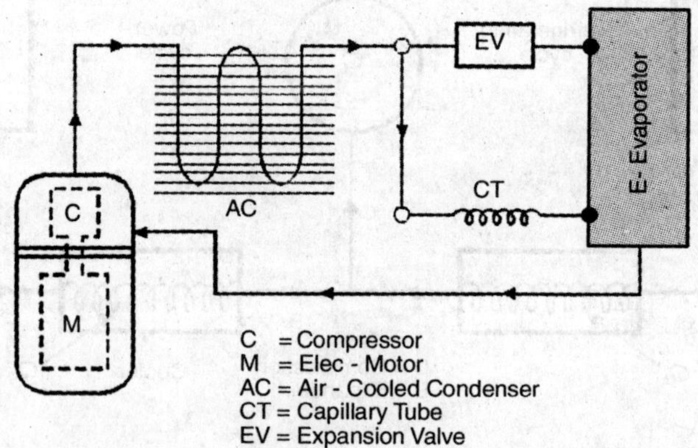

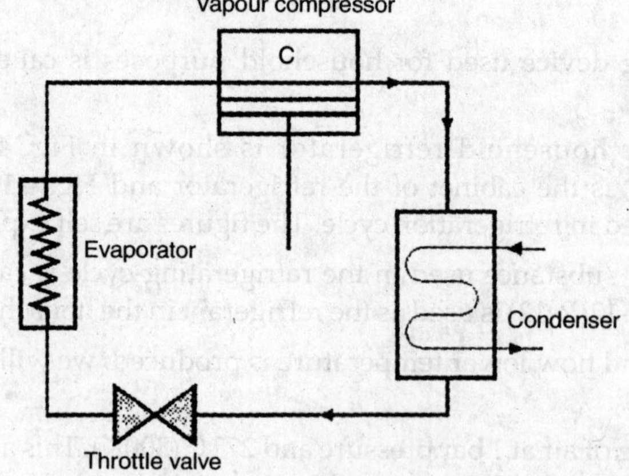

Fig. 4.11 (b)

After cooling, this air at higher pressure is expanded through the same pressure range. So that the temperature of air is halved, i.e. 300/2 = 150 K, i.e. –123°C. If this air is passed through the coil of tubes,

(a) dipped in water, water is cooled as in water cooler or

(b) over which surrounding air is passed, air is cooled and is circulated through the room as in window air conditioner.

A household refrigerator is necessity to sotre food items which otherwise would get spoiled at room temperature. The capacities of refrigerators are normally expressed in litre of internal volume and freezer volume. A schematic sketch of the household refrigerator showing various components is given below.

(a) Compressor. The compressors are hermetically sealed rotary or reciprocating type compressors with fractional power motors of say 75, 125, 180 or 350 W depending on the fridge capacity.

(b) Condenser. In earlier type of refrigerators the condenser coils used to be behind the cabinet with mesh of small sized tubes, for better cooling of the compressed refrigerant. In the new types of refrigerators condenses are consealed.

(c) Capillary tubes. Normally capillary tubes are used for the thermostatic expansion process.

(d) Freezer. The freezer compartment at the top portion where temperatures would be the lowest say 5 to 10º C. The temperatures would go on increasing in the lower portion. The evaporator coils are in the freezer portion.

(e) Thermostat. It is attached to the evaporator and operates at the temperature when cut off temperature is reached: If say it is set to –15ºC the thermostat would cut off operation of compressor when this temperature is reached and vice versa.

(f) Defrosting. While defrosting, the ice formed due to freezing of water vapours in the freezer is melted by switching off refrigerator. In modern Frost household refrigerators a small fan blows air over the evaporator and does not allow water vapour to freeze.

Working. The household refrigerator works on vapour compression cycle. The refrigerants like Freon-12 are alternately compressed and expanded where they undergo phase change. Freon-12 has boiling temp. of –30º C and operating pressure of 1 bar. The most important is to understand that the boiling points or the condensation points of the refrigerants vary at high and low pressures. The refrigerant at lower pressure has very low boiling point and absorb heat from the freezer or cabinet at lower temperature.

(a) Compression. The low pressure vapour from evaporator is compressed in the compressor. Its pressure and temperature increase to say 8 to 10 bar, temperature 60º C if it is Freon-12.

(b) Condensation. The refrigerant loses its heat in the condenser and its temperature gets reduced. It is high pressure liquid of pressure at 8 to 10 bar and temperature 30° C.

(c) Throttling. During throttling process the high pressure liquid refrigerant become low pressure vapour part liquid part vapour at very low temperature. Typically the pressure could be 1 to 2 bar and temperature –30°C.

(d) Evaporator. In evaporator the liquid portion of the refrigerant takes away heat from the cabinet and gets converted into low pressure vapour which is again compressed in the next cycle.

4.13 AIR CONDITIONING

Air conditioning means conditioning the air in a room or a building like—cinema halls, auditorium, factories for human comfort. The criteria are:

(a) *Temperature* — 21 to 26º C is comfortable temperature.

(b) *Humidity* — Humidity . range of 20 to 60% comfortable.

(c) *Purity of air* — The air should be pure from impurities and suspended particles. CO_2 not to exceed 0.15%

(d) *Motion of air* — It should not exceed 0.15 m/s.

4.13.1 One Ton of Refrigeration

Air conditioners are sold in terms of tonnage of refrigeration. It is the cooling capacity equivalent to removal of heat at such a rate as would freeze one tonne (short) of water (2000 lb = 907.187 kg) at °C into ice at °C in 24 hours. Since the latent heat of ice is, it works out that one ton of refrigeration effect is equivalent to cooling which will remove 12659 kJ/hr or 334.9 kJ/kg.

4.14 ROOM AIR CONDITIONER (WINDOW AIR CONDITIONER)

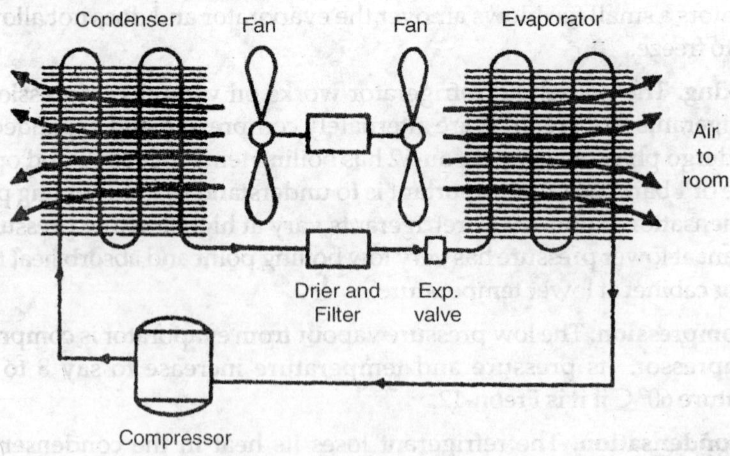

Fig. 4.12. (a) Schematic diagram of refrigeration unit for room air conditioner.

A room air conditioner is an encased assembly designed as a unit primarily for mounting in a window or through a wall. These units are made to deliver cool or warm conditioned air to the room generally, without ducts. This unit includes the main sources of refrigeration, dehumidification and means of circulating and cleaning air and also may include means of heating.

The basic function of any air conditioning plant is to provide comfort by cooling or heating, humidifying or dehumidifying, filtering or cleaning and recirculating the space air. All these functions are present in a room air conditioner or window air conditioner. It may provide ventilation by introducing outside air into the room and / or exhausting the room air to the outside. Room temperature can be controlled by providing a thermostatic setting in the window air conditioner. The conditioner may provide heating by heat pump operation, electric resistance elements or by a combination of both.

Figure 4.12 (b) shows the schematic diagram of a typical room air conditioner or window air conditioner. Working or operation of this air conditioner is as follows.

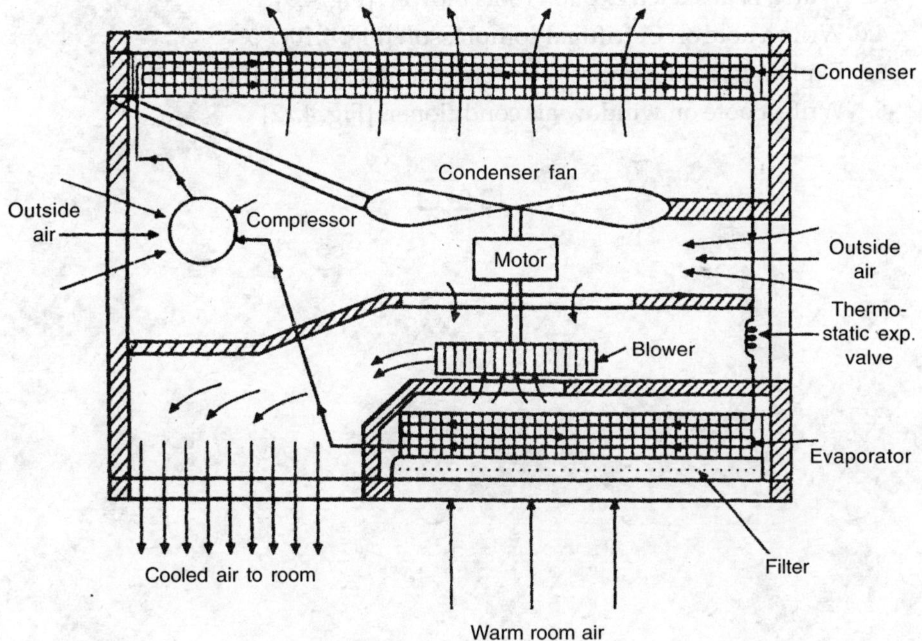

Fig. 4.12. (b) Room/window air conditioner.

Warm room air (for recirculation) passes over the cooling or evaporator coil and in the process, gives up its sensible and latent heat also (in case of dehumidification is required). This conditioned air along with the fresh air (ventilation air) is then recirculated in the room by a fan or blower.

The heat from the warm air vapourises the cold liquid refrigerant flowing through the evaporator. The vapour then carries heat to the compressor, which compresses the vapour and increase its temperature to a value higher than the temperature of the outdoor air. In the condenser the hot refrigerant vapour liquefies and gives up the heat from the room air to the outdoor air. The high pressure liquid refrigerant then passes through a restrictor thermostatic expansion valve or capillary tube—which reduces its pressure and temperature. The cold and low pressure liquid refrigerant then re-enters the evaporator to repeat this refrigeration cycle. The cooling capacities of commercially available window air conditioners range from 0.75 to 10 kW.

THEORY QUESTIONS

1. Explain Reciprocating pump with a sketch. [Fig. 4.4]
2. Explain Reciprocating compressor. [Fig. 4.8]
3. With a neat sketch explain roots blower. [Fig. 4.9]
4. Write a note on Cetrifugal compressor. [Fig. 4.10]
5. Explain vapour compression refrigeration system. [Fig. 4.12]
6. Write a note on window air conditioner. [Fig. 4.12]

❑❑❑

Unit 3

Energy Sources and Heat Transfer

Syllabus

Conventional and Non–Conventional Energy Sources — Thermal, Geothermal, Hydraulic, Nuclear, Wind Solar, Tidal, Wave, Biogas, Ocean thermal energy, biomass, fuel cells (Schematic of plant. layout).

Heat Transfer — Basic modes of heat transfer : Conduction convection and radiation.

Statement and Explanation of — Fourier's law of heat conduction, Netwon's law of cooling, Stefan—Boltzmann's law of radiation.

Emissivity and its value for practical interpretation.

Conducting and insulating materials and their properties.

Use and types of extended surfaces like find. (Descriptive treatments only).

Description and type of heat exchangers.

Introduction to Unit 3

Prime movers are devices that produce the shaft power by consuming energy in different forms such as chemical, thermal, pressure or potential. This energy is made available to the prime mover from various sources. The first part of this unit deals with different sources of energy, their merits and demerits.

Some of these sources are being used since olden days. These are known as *Conventional Energy Sources*.

While due to limited availability of such sources, the scientists are continuously working in search of new sources, these are known as *Non—conventional Energy Sources*.

In the second part of this unit, the fundamentals of a science known as Heat transfer are discussed.

The subject Heat Transfer deals with the energy in transit because of temperature difference. Flow of energy is the basic cause behind any process in the world. Hence, any process or phenomenon in any science, this subject is bound to have role in it.

We shall limit the scope of this unit to introduction of basic modes of Heat transfer namely conduction, convection & radiation and some applications of heat transfer as Pin—fin, Heat Exchanger etc.

CHAPTER 5

CONVENTIONAL AND NON-CONVENTIONAL ENERGY SOURCES

5.1 INTRODUCTION

Since ages man has been trying to harness energy to get more power which can be used to do work. Energy appears in many forms but it has ability to produce dynamic and vital effect. Energy is always associated with physical substances but it is not a substance itself. Energy manifests itself by excited state of the matter which gets capacity to do work.

5.2 SOURCES OF ENERGY

All energy eventually has common origin from the sun. Energy, especially electric energy is important for day-to-day living and the industry. Energy from various sources—like fossil fuels, coal, water etc. is converted into electrical energy and then distributed wherever needed.

5.3 IMPORTANCE OF ELECTRIC ENERGY

Per capita consumption of electricity is an index of the progress and industrialisation of a country. It is convenient to convert energy from various sources into electric energy for ease of transmission. For example, the potential energy of stored water in a hydroelectric power plant is used to run water turbines, which run electric generators that produce electricity. The electricity so generated is distributed through transmission lines up to our homes and to the factories. The electric power is reconverted into mechanical energy to run fans, compressors, and machines or for cooling/heating or lighting.

In our country demand for electricity is growing rapidly—roughly at about 8 – 10% per year. Present per capita consumption is about 500 units (kWh) as compared to approximately 36 times of this in Canada. Power generation by

Table. 5.1

(I) Energy-Basis-Conventional / Non-conventional

- Conventional
  - (1) Fossil fuels (Coal, coke)
    (2) Vegetation (Wood, alcohol)
  - Animal waste (Gobar)
  - Water or hydraulic energy
  - Nuclear energy (Fission of nuclear fuels like U^{235}, Pu^{239})
- Non-conventional
  - Solar energy
  - Wind energy
  - Ocean wave energy
  - Ocean thermal energy
  - Geothermal or Terrestrial energy
  - Tidal energy

(II) Energy-Basis- enewable / Non-renewable

- Renewable or Regenerative
  1. Vegetation
  2. Hydraulic energy
  3. Solar energy
  4. Wind energy
  5. Tidal energy
  6. Animal waste
  7. Ocean thermal energy
- Non-renewable or Non-regenerative
  1. Fossil fuels
  2. Nuclear fuels
  3. Geothermal energy

(III) Energy-Basis-Capital / Celestial energy

- Capital energy (Existing in or on earth)
  1. Fossil fuels
  2. Nuclear fuels
  3. Geothermal energy
  4. Vegetation
  5. Animal
- Incoming or celestial energy (Reaching earth from outer space)
  1. Solar energy
  2. Wind energy
  3. Tidal energy

thermal power plants is about (83%), by hydro-power stations (14%) and nuclear plants (3%).

5.4 CLASSIFICATION OF SOURCES OF ENERGY

Various sources of energy can be classified as described below:

(a) Conventional/Non-renewable. These sources of energy get consumed and are not renewable.

(b) Non-conventional/Renewable/Perennial. These sources are ever lasting and get renewed.

A broad classification of the sources of energy on various basis is given in Table 5.1 shown on previous page.

5.5 ADVANTAGES/DISADVANTAGES OF NON-CONVENTIONAL AND CONVENTIONAL ENERGY SOURCES

Advantages

1. Such sources are everlasting, renewable, are available in abundance in nature and virtually come free like solar energy, wind energy or energy from water.
2. They are nonpolluting.

Disadvantages

1. The energy is not concentrated in such sources. Energy conversion into electric energy is relatively costly.
2. The energy like solar energy or energy from wind is weather and time dependent.

ENERGY CONVERSION – NON-CONVENTIONAL/ RENEWABLE SOURCES

5.6 SOLAR ENERGY

In our country sunshine is available throughout the year and can be used for heating. The solar energy however has low energy density and therefore cannot be used directly for power generation. Solar energy received on earth rarely exceeds 1 kW/ m^2 and the total radiation over a day at best is about 7 kWh/m^2. These are low values for conversion. The solar energy can be concentrated by using parabolic mirror collectors. The solar energy can be collected by flat plate or parabolic mirrors depending on the temperature requirements.

5.6.1 Flat Plate Heaters

(a) These are used for water or space heating, and drying etc.

(b) **Parabolic mirrors.** These are used in vapour engines, process heating, cooking.

(c) **Parabolic mirror arrays (For high temperature > 200º C).** These are used for steam generation.

5.6.2 Flat Plate Solar Heaters

Flat plate water heater can give hot water up to 70–80º C. Schematic diagram is shown in Fig. 5.1. Water is heated usually by passing through tubes which are exposed to the sun's heat rays.

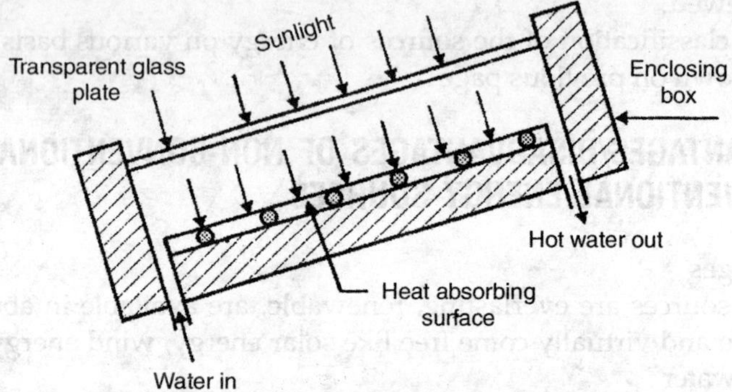

Fig. 5.1. Flat plate collector.

5.6.3 Solar Power Plant

From flat plate collector we get hot water. The temperature of water is 70–80°C maximum. This hot water is then passed through Isobutane boiler. Water gives its heat to Butane and Butane vapour is generated. This vapour will be then made to drive the butane turbine for generating the power (shown in Fig. 5.2).

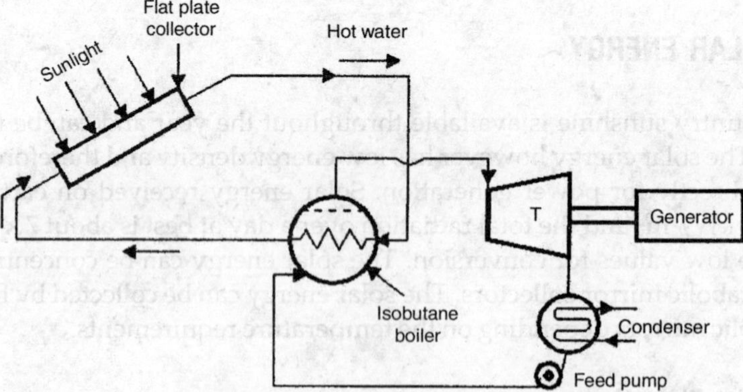

Fig. 5.2. Solar power plant.

Applications

1. Hot water can be used to vaporise Isobutene in a boiler. The vapours run a turbine that is coupled to an electric generator to produce electricity. Schematic diagram of a solar power plant using Isobutene or Freon as working substance is shown in Fig. 5.2.

2. Parabolic concentrating type mirror collectors are used to intensify the solar energy that can be used to heat water to produce steam at low pressure of 2 – 2.5 bar or to heat air. Steam and hot air used to run turbines that are coupled to electric generators.

5.7 WIND ENERGY

Wind speeds are reasonably high throughout the year. It can be used to produce electricity from wind mills. India has commissioned wind energy projects to produce 1340 MW and is presently ranked fifth in harnessing this source of energy after USA, Germany and Denmark. Maharasthra state leads in wind power development.

The wind mill consists of a propeller driven generator installed on a tower. The propeller is connected to the generator through gears. The propeller is kept in the direction of the wind and it drives a generator which produces electricity. Block diagram of the wind mill is given below.

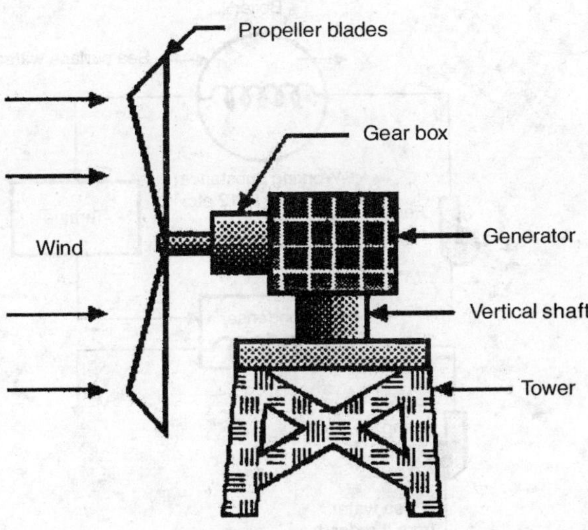

Fig. 5.3. Wind mill.

5.8 OCEAN THERMAL ENERGY

5.8.1 The Ocean Energy

The ocean can produce two types of energy, (1) Thermal energy, and (2) Mechanical energy.

(1) *Thermal energy* from the sun's heat and (2) *Mechanical energy* from the tides and waves. Oceans cover more than 70% of Earth's surface, making them the world's largest solar collectors. The sun's heat warms the water surface a lot more than the deep ocean water, and this temperature difference creates thermal energy. Due to this, in tropical climates there is large temperature difference between the surface and the lower levels. The temperature difference may be up to 20ºC within a 200–300 m depth.

5.8.2 Ocean Thermal Energy Conversion (OTEC)

Ocean thermal energy is used for many applications, including electricity generation. There are three types of electricity conversion systems: *closed-cycle, open-cycle,* and *hybrid*.

(a) Closed-cycle systems use the ocean's warm surface water to vaporise a *working fluid* like ammonia, propane or R-22 which have low boiling points. The vapour expands and runs turbine coupled to a generator to produce electricity. Vapour after expansion is condensed using cold water from depth by pumps. However a lot of energy is used to run the pump. (Fig. 5.4)

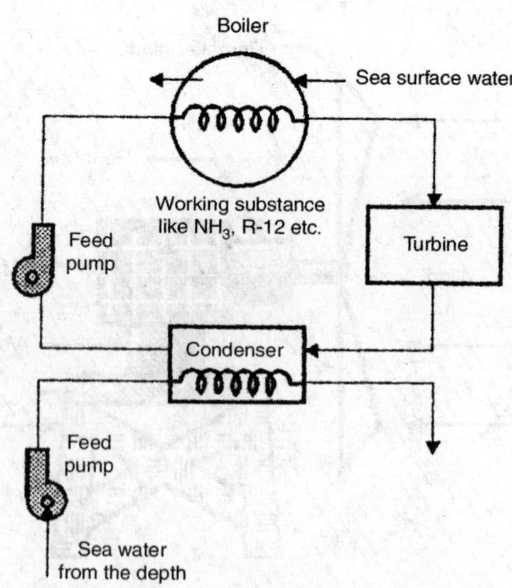

Fig. 5.4. Closed sytem ocean thermal energy plant.

(b) In open-cycle systems sea water is boiled to generate steam at low pressures. The steam in turn operates a turbine/generator. Hybrid systems combine both closed-cycle and open-cycle systems.

5.9 GEOTHERMAL ENERGY

The earth, having been originally thrown off from the sun according to some theories, still has a molten core. Evidence for this lies in the volcanic action that takes place in many regions of the earth's surface. At some places steam, geysers and hot water exudes from the earth.

The geothermal steam has been used for many years in Italy for power generation and space heating. Recently projects are in operation in Mexico, Newzealand and California. Figure 5.5 shows the schematic layout of plant. Steam is tapped from drilled wells and passes through a drum to separate out moisture. From here the dry steam enters turbines which are connected to electric generators.

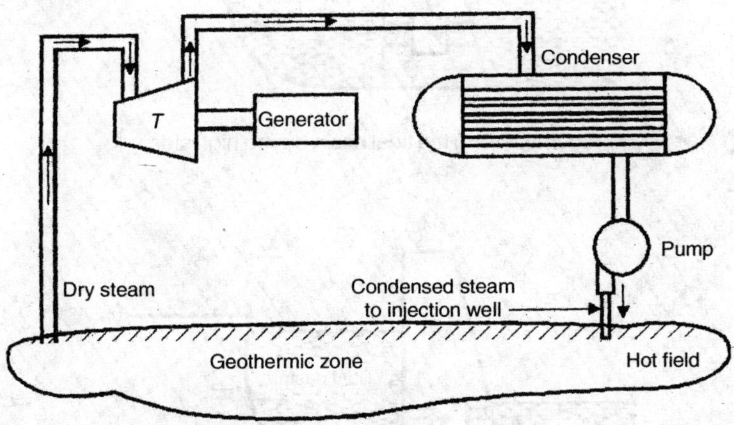

Fig. 5.5. Geothermal energy plant.

5.10 TIDAL AND WAVE ENERGY

There are two ways to tap the ocean for the mechanical work (a) using tidal energy or (b) making use of the energy in the waves. Ocean mechanical energy is quite different from ocean thermal energy. Even though the sun affects all ocean activity, tides are driven primarily by the gravitational pull of the moon, and waves are driven primarily by the winds. As a result, tides and waves are intermittent sources of energy, while ocean thermal energy is fairly constant. Also, unlike thermal energy, the electricity conversion of both tidal and wave energy usually involves mechanical devices.

A. Tidal Energy

(i) Tides are generated through combination of forces exerted by sun, moon and augmented by rotation of the earth. Local effects such as funneling, shelving and reflection increase the tides. An increase of about 5 m at low tide and high tide is needed. There are only a few places where this tide change occurs around the earth Gulf of Cambay and Kutch in our country have lot of potential of the tidal energy.

(ii) Sea water can be trapped in reservoirs behind dams on the sea shores at the time of high tides. When the tide drops, the water behind the dam is let out just like in a regular hydroelectric power plant. For tidal energy to work well we need large increases in tides.

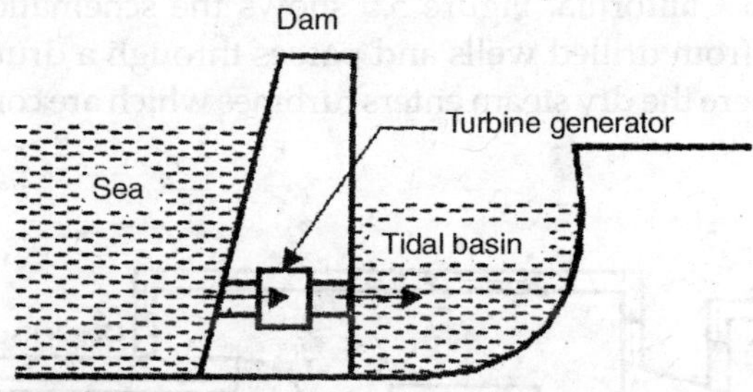

Fig. 5.6. Filling the tidal basin during high side.

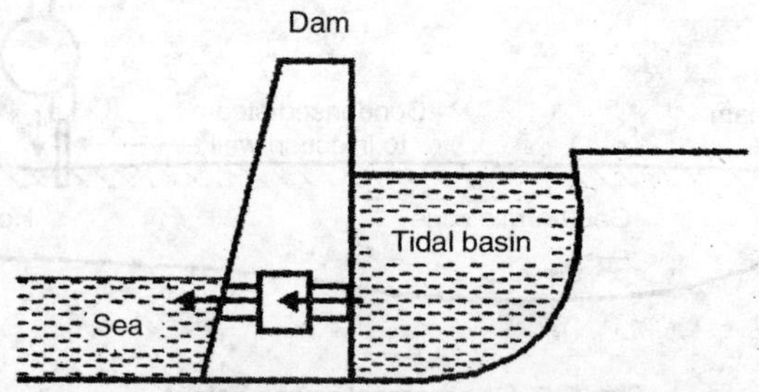

Fig. 5.7. Power generation during low tide.

As shown in Fig. 5.6 during high tide conditions water from sea flows to the tidal basin via turbine. When the water flows through the turbine, it drives the turbine, which in turn drives the generator for generating power.

Advantages	Disadvantages
1. It is pollution free	1. It can work where tides are quite
2. Its cost of running is very low	2. It is liable to corrosion
3. It does not depend on rainfall	3. Its operation is intermittent

During low tide conditions as shwon in Fig. 5.7 water from tidal basin flows to the sea. As water flows through turbine, it will rotate the turbine, which in turn drives the generator to generator power.

B. Wave Energy

(i) For wave energy conversion, there are three basic systems:

- *Channel systems* that funnel the waves into reservoirs
- *Float systems* that drive hydraulic pumps; and
- *Oscillating water column systems* that use the waves to compress air within a container.

The mechanical power created from these systems either directly activates a generator or transfers to a working fluid, water, or air, which then drives a turbine/ generator.

(ii) Ocean waves have potential energy due to height of the waves and kinetic energies due to wavelength. Electricity generation possibility is 5 to 15 kW per metre length of the waves. India with coastline of some 6000 km can exploit this source of energy.

(iii) In Fig. 5.8 the wave rises into a chamber. The rising water forces the air out of the chamber. The moving air spins a turbine which can turn a generator. When the wave goes down, air flows through the turbine and back into the chamber through doors that are normally closed. Most wave-energy systems are very small. But, they can be used to power a warning buoy in seas or a small light house.

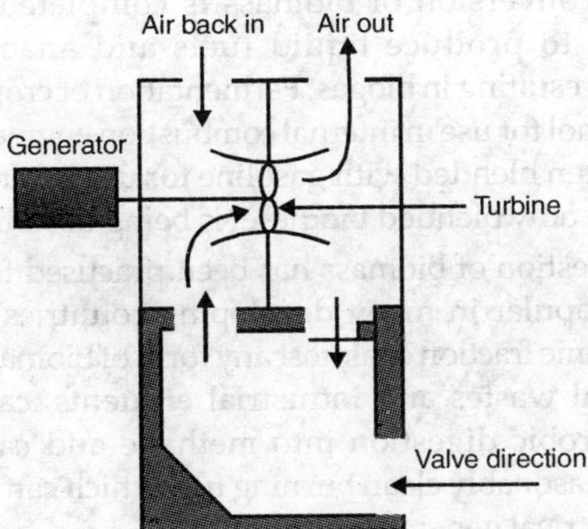

Fig. 5.8. Wave energy conversion.

5.11 BIOMASS

The Biomass is organic material which has stored sunlight in the form of chemical energy. All biomass is produced by green plants converting sunlight into plant material through photosynthesis. Biomass fuels include wood, wood waste, straw, manure, sugar cane, and many other byproducts from a variety of agricultural processes.

When burned, the chemical energy is released as heat. Wood had been the chief source of cooking and heating homes and other buildings for thousands of years. Biomass continues to be a major source of energy in much of the developing world even now.

Sugarcane is a good example of a biomass crop. The sugar is extracted from the cane by removing the juice; the remainder of the plant called bagasse produces heat when burnt.

The Biomass Resource. Main sources of biomass energy are: municipal and industrial wastes, agricultural crop residues and energy plantations. Industrial biomass wastes are primarily produced by the forest products, industry, and animal waste. Many different types of biomass can be grown for energy production. Crops that have been used for energy include: sugarcane, corn, husk, grains, grass, seaweed and many others.

Conversion Technology

(i) The simplest and most common method of obtaining energy from biomass is direct combustion. The heat of combustion is used to provide space or process heat, water heating or for producing electricity through the use of a steam turbine. In the developing world, many types of biomass such as dung and agricultural wastes are burned for cooking and heating.

(ii) Biochemical conversion of biomass is completed through alcoholic fermentation to produce liquid fuels and anaerobic digestion or fermentation, resulting in biogas. Fermentation of crops such as sugarcane produces ethanol for use in internal combustion engines has been common. Ethanol has been blended with gasoline for use in automobiles. Similarly in our country now blended biodiesel is being used by the railways.

(iii) Anaerobic digestion of biomass has been practised for almost a century, and is very popular in many developing countries such as China and India. The organic fraction of almost any form of biomass, including sewage sludge, animal wastes and industrial effluents, can be broken down through anaerobic digestion into methane and carbon dioxide. This "biogas" is a reasonably clean burning fuel which can be used for cooking, heating or electrical generation.

(iv) Wood and many other similar types of biomass which contain lignin and cellulose-agricultural wastes, cotton waste, wood wastes, peanut hulls etc. can be converted through thermochemical processes into solid, liquid or gaseous fuels. During the process biomass is heated in the absence of air and breaks down into a complex mixture of liquids, gases, and a residual char. If wood is used as the feedstock, the residual char is what is commonly known as charcoal.

5.12 BIO-GAS PLANT

In India a very large amount of cow dung is produced every year. Out of this only 50% is used for cooking purpose.

In the bio-gas plant, bio-gas is produced as a result of fermentation and decomposition of the dung.

Other than cow dung the material used in the production of bio-gas are corn husk, green leaves, dried leaves, vegetable waste and even human excreta.

Construction and Working. Figure 5.9 shows the schematic diagram of bio-gas plant.

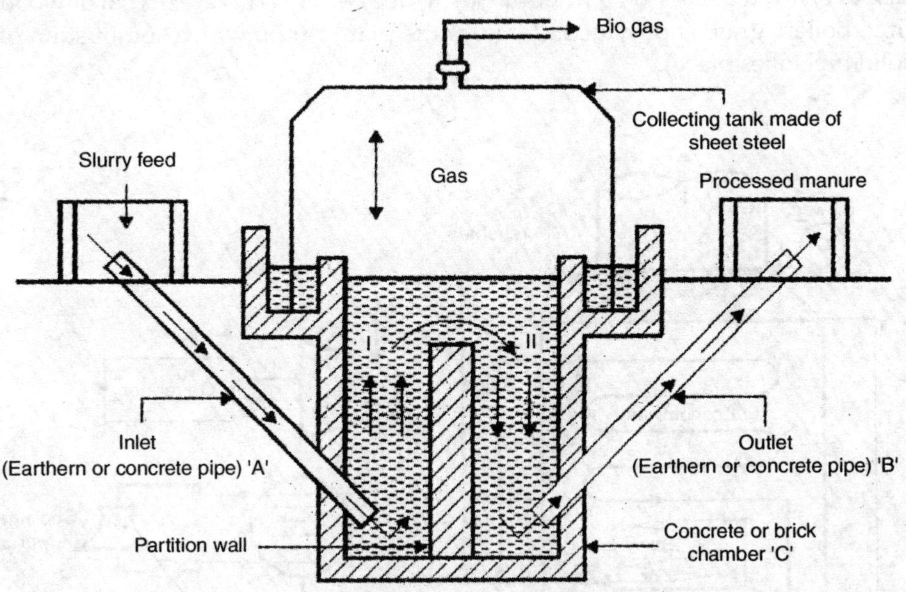

Fig. 5.9. Bio-gas plant.

It mainly consists of a very large concrete or brick work chamber 'C' which is divided into two separate compartments by a partition wall.

On the top of the chamber 'C', a gas collecting tank 'D' is provided. Also it is arranged with inlet pipe 'A' and outlet pipe 'B' as shown in Fig. 5.9.

Dung is mixed with sufficient amount of water so as to get dung slurry and this is then fed to the bottom of the chamber 'C' through the inlet pipe 'A'.

The fermentation process begins in the I–compartment. And as the atmospheric temperature increases, the decomposition of organic materials increases and the slurry rises up and over flows to the II–compartment, where further decomposition continues.

The gases produced during decomposition, bubble up through the slurry and get collected in the tank 'D'. The gas collected can be taken out through the valve.

From the II–compartment thick slurry will be thrown out through the outlet pipe 'B'. The calorific value of bio-gas is about 25 kJ / litre.

5.13 SIMPLE STEAM POWER PLANT (OR THERMAL POWER PLANT)

Figure 5.10 shows a simple steam power plant, it mainly consists of a boiler, steam turbine, condenser, feed pump etc.

1. Boiler. Boiler is a closed vessel in which water is heated and converted into steam, i.e. it is a steam generator. Boilers will be either oil fired or coal fired. In case of oil fired boilers, oil burners are provided, whereas in case of coal or wood fired boilers grate is provided. (Grate is the platform on which combustion of solid fuel takes place).

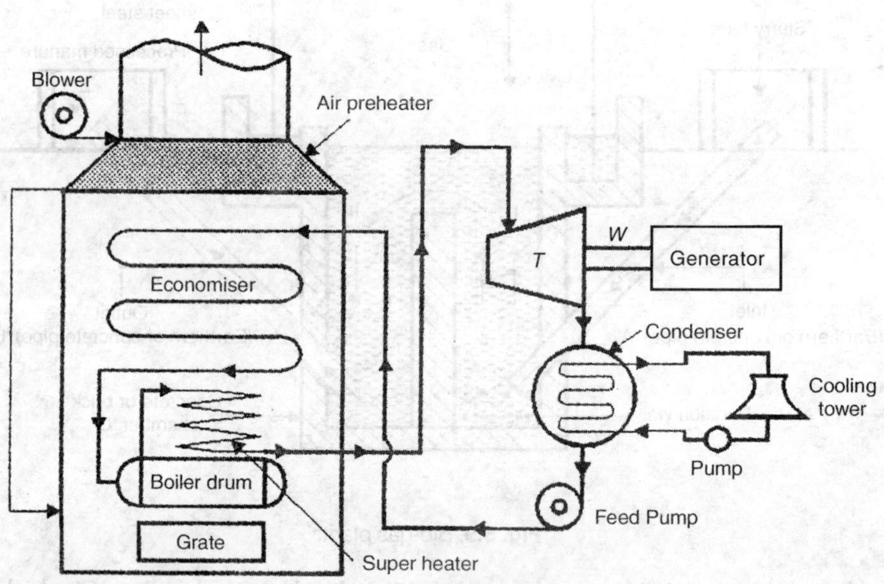

Fig. 5.10. Simple steam power plant.

Steam power plants using coal as the fuel are also called as *Thermal Power Plants*. As shown in Fig. 5.10 when the combustion of coal takes place over the grate, hot gases are generated. These hot gases heat the water in the boiler drum and convert it into steam. This steam generated in drum contains some water particles in suspension, so thus steam is called as *Wet Steam*. This wet steam is passed through the superheater. In superheater the temperature of steam is increased by further heating. Now the steam is known as *Superheated steam*.

2. Turbine. This superheated steam is then allowed to expand into the steam turbine. Turbine consists of an impeller or a cylindrical hub on which blades are fixed. This turbine is mounted on the shaft. When the steam expands on the blades it rotates the turbine impeller, which in turn drives the generator to generate electrical power.

3. Condenser. The steam after generating power is exhausted into the condenser and gets condensed into water (this water is also called as condensate). In the condenser, it comes in contact with the cold surface of water tubes.

4. Pump. The condensate is then pumped by means of the feed pump into the boiler drum through the economiser.

When the water passes through the economiser, it recovers some of the heat of the flue gases and becomes hot. Then this hot water will enter the boiler drum. In the boiler drum thus hot water is again heated and converted into steam and the cycle repeats.

Thus by using economiser some of the heat of the flue gases will be recovered, which otherwise would have been lost to the atmosphere. Thus by using economiser, the efficiency of the plant will increase.

Advantages of Steam Power Plant

1. Fuel used is cheaper.
2. Less space is required in comparison with that for hydro-electric plants.
3. As these plants can be set up near the industry, transmission costs are reduced.
4. Cheaper in production cost and initial cost compared with diesel power stations.
5. Can be located very conveniently near the load centre.
6. They can respond to rapidly changing loads with difficulty.

Disadvantages of Steam Power Plant

1. The cost of plant increases with increase in temperature and pressure.
2. Maintenance and operating costs are high.
3. Long time required for errection and putting into action.
4. A large quantity of water is required.
5. Greater difficulty experienced in coal handling.
6. Presence of troubles due to smoke and heat in the plant.

5.14 NUCLEAR POWER PLANT

Nuclear reactions produce energy either by fusion or by fission reaction. Fusion means union of lighter nuclei, such as two nuclei of Hydrogen, after fusion produce one nucleus of Helium and one neutron and energy. Atomic Fission is exactly opposite of fusion.

Here a heavy nucleus such as nucleus of U^{235} is bombarded by means of high energy neutrons, then it splits into number of fragments more or less of

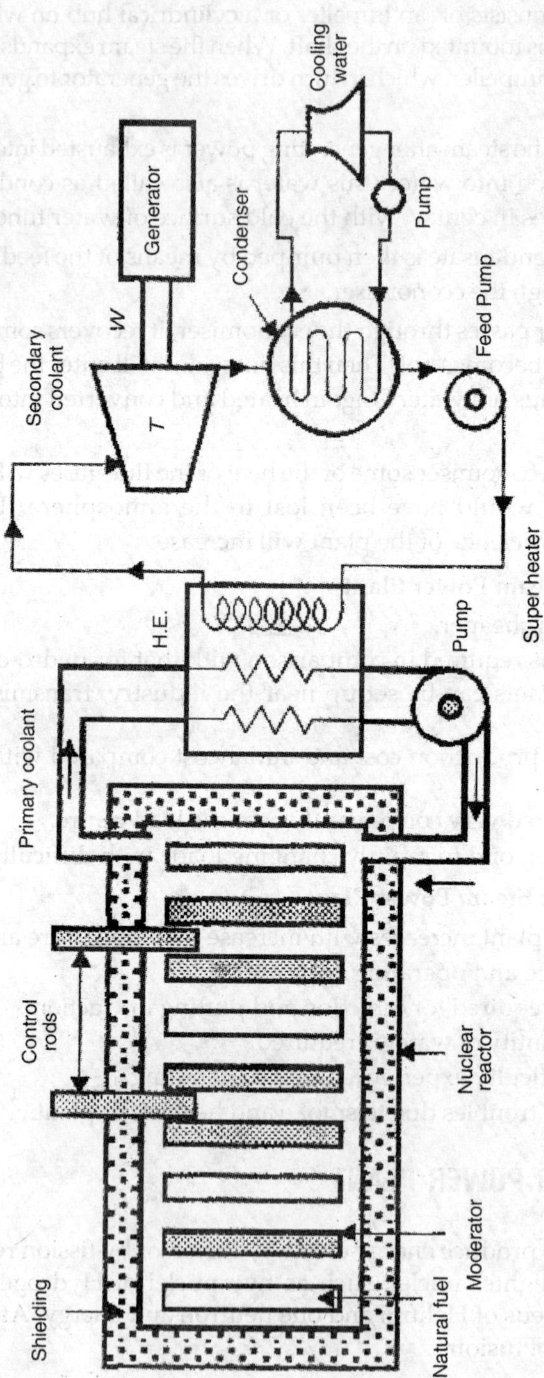

Fig. 5.11. Schematic of nuclear plant.

equal mass. The combined mass of fissioned products is slightly less than the parent nucleus. This disappearing mass is converted into energy according to $E = mC^2$ where C is velocity of light. This process is associated with the release of tremendous amount of heat energy.

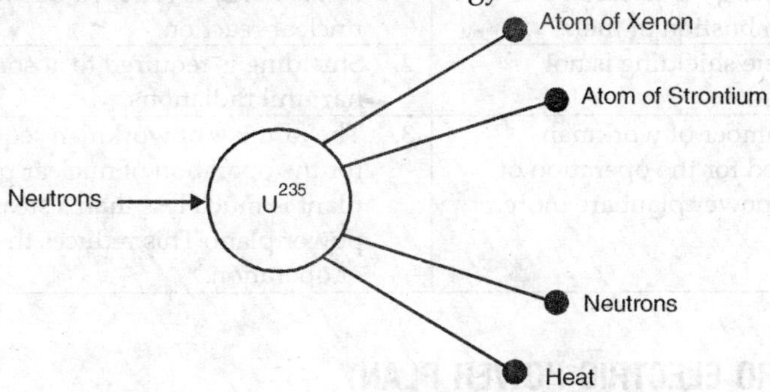

Fig. 5.12

The heat produced by the nuclear disintegration is taken up by the primary coolant (which may be Heavy water, Molten Sodium or an Organic liquid) and the primary coolant becomes hot. The heat of the primary coolant is transferred to the secondary coolant (i.e. water) in the heat exchanger. The water gets converted into steam. This steam then drives the turbine, which in turn drives the generator for generating power.

Note:

(i) **Moderators** are made of Graphite, Boron, Cadmium. These are used to slow down the fast neutrons to the right speed which is required for splitting up of U^{235}.

(ii) **Control rods** are made up of Boron or Cadmium. These are used to absorb excess neutrons released during fission process.

(iii) **Shielding** is provided to absorb harmful radiations like β- and γ-rays produced during fission reaction.

5.14.1 Advantages and Disadvantages of Nuclear Power Plants

Advantages	Disadvantages
1. Nuclear fuel has very high energy density i.e. very small quantity of nuclear fuel can produce very large amount of energy.	1. Problem of radiation hazard.
2. Cost of transportation and storage is much less as compared to the conventional fuels.	2. Problem of nuclear explosion.
3. Nuclear power virtually does not cause pollution.	3. Nuclear waste disposal poses problem.
4. Fast breeder reactors can yield Plutonium which can by used for military usages.	4. There are lots of international controls over the nuclear power plants.

5.14.2 Comparison between Steam Power Plant and Nuclear Power Plant

Steam Power Plant	Nuclear Power Plant
1. Heat energy is released due to the combustion of fuel.	1. Heat energy is released during nuclear reaction.
2. Concrete shielding is not required.	2. Shielding is required to absorb harmful radiations.
3. The number of workman required for the operation of steam power plant are more.	3. The number of workman required for the operation of nuclear power plant is much less than a steam power plant. This reduces the cost of operation.

5.15 HYDRO ELECTRIC POWER PLANT

Generation of power by hydroelectric stations is the utilisation of the part of hydrological cycle. Potential energy of rain falling on earth's surface, with respect to ocean is converted into mechanical energy by using water turbines. In order to generate energy by thus method economically, ample quantity of water at sufficient potential (head) must be available. Moreover, the past history of the place must be known in order to estimate the minimum and maximum quantity of water that can be made available for the purpose of power generation. As the availability of water depends on the natural phenomenon of rain, the maximum capacity of such plants is usually fixed on the basis of minimum quantity of water available. Usually storage reservoirs are constructed with such plant in order to store water during peak periods and to utilise the same during the off-peak periods. The civil engineering work involved in case of hydroelectric power station is considerable as it involves the construction of a dam and a large catchment area.

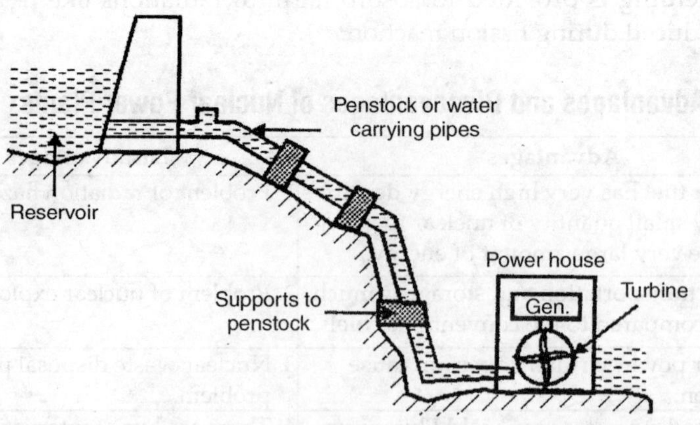

Fig. 5.13. Hydroelecteric power plant.

General arrangement of storage type Hydroelectric Power Plants is shown in Fig. 5.13 and are self-explanatory.

(a) The water stored in dams have potential energy which is used to run water turbines which in turn are connected to coupled to elect generators which produce electricity.

(b) Water of river stored in the reservoir which is between narrow valley is closed by a high rise dam.

(c) **Penstock.** The water is brought to the turbines by large size pipes known as penstock.

(d) The penstocks are held in position by anchor blocks.

(e) **Surge Tank.** To avoid damage to the penstock due to water hammer i.e. sudden variation in pressure of water surge tanks are provided as shown.

(f) **Water turbine.** It is a device that converts kinetic energy in water into mechanical energy (rotation).

(g) **Generation.** It is electrical device that converts mechanical energy (rotation) into the electrical power.

Advantages of Hydraulic Power Plant

(a) Operating cost low.

(b) No fuel required.

(c) Useful life 40–50 years as against 20–25 years for thermal plant.

(d) Fuel handling problem and ash problem absent.

(e) Cost of land is less (Remote place).

(f) Quick response.

(g) Effective change with time.

(h) No standby loss.

(i) Labour / operatorless.

(j) Can be used for irrigation and flood control.

Disadvantages

(a) Capital cost much more.

(b) Depends on rain (natural).

(c) Site is remote.

(d) Long period for selection of plant.

5.16 FUEL CELL

Fuel cells are electrochemical devices which directly convert hydrogen, or hydrogen-rich fuels into electricity without combustion. Thus process is much more efficient than traditional thermal power plants, converting up to 80% of the chemical energy in the fuel into electricity as compared to a maximum of 30–40% for conventional power plants. Although their structure is somewhat like that of a battery, fuel cells never need recharging or replacing and can consistently produce electricity as long as they are supplied with hydrogen and oxygen. Fossil fuels (coal, oil and natural gas), biomass (plant material) or pure hydrogen can be used as the source of fuel. If pure hydrogen is used, the emissions from a fuel cell are only electricity and water. Fuel cells are small and modular in nature and therefore fuel cell power plants can be used to provide electricity in many different applications, from electric vehicles to large, grid-connected utility power plants. First used in the U.S. space program in the 1950s, fuel cells are a developing technology with a few commercial uses today, but may emerge as a significant source of electricity in the near future.

Fuel Cell Electrochemistry. Fuel cells quite similar to a battery, with two porous electrodes separated by an electrolyte. Electricity is produced by chemical reaction between a hydrogen-based fuel and an oxidant—usually oxygen inside the fuel cell. Figure 5.14 is a diagram of the electrochemical reaction which takes place inside a fuel cell. Hydrogen (H_2) flows over the anode (the negative electrode) and splits into positively charged hydrogen ions, and electrons which carry a negative charge. The electrons flow through the anode to the external circuit, performing useful work (this is the electric current generated) while the hydrogen ions pass through the anode and into the electrolyte, moving towards the cathode (the positive electrode). The electrons eventually return to the cathode which is supplied with oxygen (O_2). At this point the electrons, hydrogen ions and oxygen react to form water (H_2O) and heat. In stationary fuel cell power plants thus heat can be captured and used for process heat in industries or space heating (co-generation). As long as the fuel cell is supplied with hydrogen and oxygen, this electrical production can continue indefinitely.

Electrolyte. The general design of most fuel cells is similar except for the electrolyte. Several different substances have been used as the electrolyte in fuel cells, each with their own advantages and disadvantages. Main types of fuel cells based on the electrolytes are:

(i) alkaline fuel cells

(ii) solid polymer fuel cells (also known as proton exchange membrane fuel cells)

(iii) phosphoric acid fuel cells

(iv) molten carbonate fuel cells and

(v) solid oxide fuel cells. Alkaline and solid polymer fuel cells operate at lower temperatures (50–260° C) and are mainly designed for use in transportation applications, while the other three operate at higher temperatures (up to 1000° C for solid oxide fuel cells and are being developed for use in cogeneration and large central power plants.

Sources of Fuel and Oxidant. Fuel cells operate at maximum efficiency, when operating on pure hydrogen and pure oxygen. Pure oxygen is very expensive, and thus air is used as the source of oxygen in most applications except where the extra cost can be justified, as in the space program. Pure hydrogen is expensive, and difficult to transport and store, therefore, like pure oxygen it is only used in special cases. Gaseous mixtures of hydrogen (H_2) and carbon dioxide (CO_2) which can be created by the 'processing' of fossil fuels or biomass are used instead of hydrogen in most commercial uses of fuel cells. The most economical sources of the necessary H_2/CO_2 fuel mixture have been found to be gaseous hydrocarbons such as natural gas and propane, light hydrocarbon liquids such as naphtha and methanol (from biomass), heavier hydrocarbon liquids such as fuel oil, and coal.

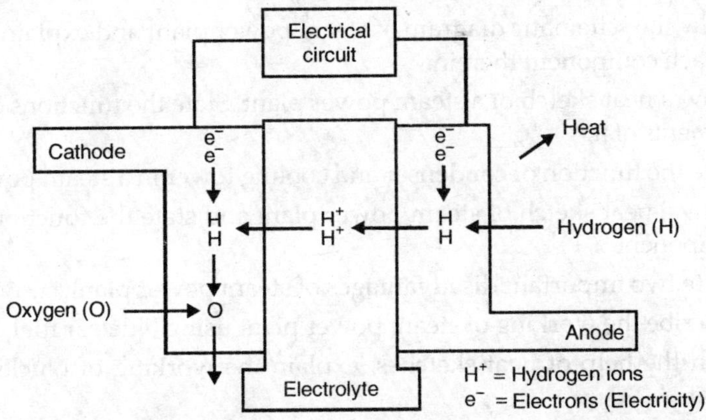

Fig. 5.14. Fuel cell.

Fuel Cell Power Plants. The fuel cell power section contains 'stacks' of one or more fuel cells. The individual fuel cells are small in size and produce between 0.5 and 0.9 volts of electricity. Therefore stacks of large number of fuel cells would be needed for power generation must have a large number of individual fuel cells. This eliminates the problems associated with large conventional power plants which can take up to ten years to build and have a fixed capacity. The fuel cells use air and hydrogen rich fuel and produce water, heat and direct current (DC) electricity.

The power conditioner section of a fuel cell power plant may also have an 'inventor' which converts the electricity to alternating current.

Uses of Fuel Cells. Fuel cells appear to be an important technology for the future as they.

THEORY QUESTIONS

1. Expiain various engineering applications of solar energy.
2. What are the merits and demerits of Renewable Energy sources./Compare renewable and non-renewable energy sources.

3. What are the types of energy sources? State various forms of renewable and non-renewable energy sources.
4. Explain with the help of block diagram working of a hydroelectric power plant. What are the advantages of hydroelectric power plant?
5. Draw a neat sketch of steam power plant and explain the function of each component.
6. With the help of a neat labelled diagram explain the working of Ocean Thermal Power Plant.
7. Draw the schematic diagram of steam power plant and explain functions of each component therein.
8. Draw a neat sketch of a steam power plant. State the functions of various elements of it.
9. State the function of condenser and cooling tower in a steam power plant.
10. Draw a neat sketch of steam power plant and state the functions of each component.
11. Write two important disadvantages of steam power plant.
12. Describe the working of steam power plant using Nuclear fuel.
13. With the help of neat sketches, explain the working of Nuclear power plant.
14. State the function of the following parts of a nuclear power plant.
15. (i) Nuclear reactor; (ii) Moderator; (iii) Control rod.
16. State the advantages and disadvantages of nuclear power plants.

CHAPTER 6

INTRODUCTION TO HEAT TRANSFER

6.1 INTRODUCTION

We have repeatedly used the term heat (Q) and heat exchanges between bodies without discussing the details of the phenomena by which heat is transferred. The purpose of this chapter is to give the student a general idea of the nature of the problems and to provide an introductory background for further study.

Heat transfer is a name of general process applied to any device that effects a transfer of heat from one substance to another.

The various devices that employ the principles of heat transfer are: (i) steam boilers (ii) surface condensers, evaporators, closed feed water heaters and the automobile radiator etc.

Whenever a temperature gradient exists within a system, or when two systems at different temperatures are brought into contact, energy is transferred. The process by which the heat energy transport takes place is known as heat transfer.

6.2 MODES OF HEAT TRANSFER

Heat transfer can be defined as the transmission of energy from one region to another as a result of a temperature difference between them. Since temperature differences exist all over the universe, the phenomena of heat flow over the universe, the phenomena of heat flow are as universal as those associated with gravitational attractions. Unlike gravity, however, heat flow is governed not by a unique relationship, but rather by a combination of various independent laws of physics.

The literature of heat transfer generally recognises three distinct modes of heat transmission:

(i) Conduction

(ii) Radiation and

(iii) Convection

Strictly speaking, only conduction and radiation should be classified as heat transfer processes, because only these two mechanisms depend for their operation on the mere existence a temperature difference. The last of the three, convection, does not strictly comply with the definition of heat transfer because it depends for its operation on mechanical mass transport also. But since convection also accomplishes transmission of energy from regions of higher temperatures to regions of lower temperatures, the term "heat transfer by convection", has been generally accepted.

6.2.1 Fourier's Law of Conduction

The basic relation for heat transfer by conduction was proposed by French scientist JBJ Fourier in 1882. It states that q_k the rate of heat flow by conduction in a material, is equal to the product of the following three quantities:

(i) K — the thermal conductivity of the material

(ii) A — the area of the section through which heat flows by conduction and

(iii) $\frac{dT}{dx}$ — the temperature gradient at the section i.e. the rate of change of temperature T with respect to distance in the direction of heat flow or

$$q_k = -K.A\frac{dT}{dx}$$

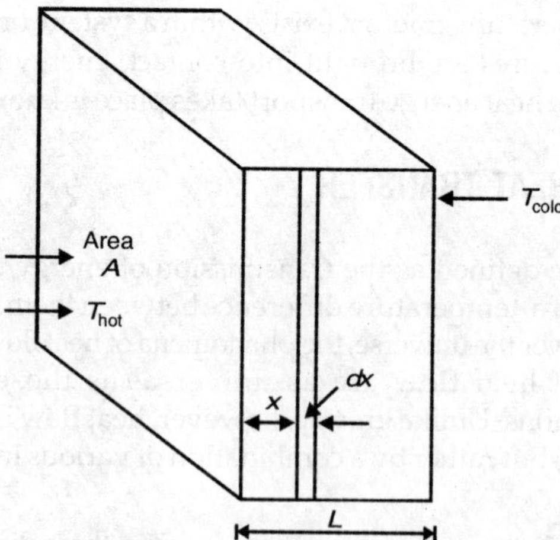

Fig. 6.1. One dimensional heat flow.

$q_k \rightarrow \text{kJ/hr}; \; A \rightarrow m^2; \; \frac{dT}{dx} \rightarrow °\text{C/m}$

$$K \rightarrow \frac{q_k}{A.dT/dx} \rightarrow \frac{\text{kJ}}{\text{hr}} \times \frac{1}{\text{m}^2} \times \frac{1}{°\text{C/m}} \rightarrow \frac{\text{kJ}}{\text{hr.m°C}} \text{ units of K}$$

Refer Fig. 6.1.

Consider a wall or slab of thickness *L*, let *A* be the area of heat flow across the wall. Let the temperatures on both sides of the wall be T_{hot} and T_{cold}. Consider a section at a distance *x* from hot surface and let *dx* and *dT* be the thickness of the strip and temperature drop across the strip as shown. Therefore, applying Fouriers Law, we get

$$q_k = \frac{-KA.dT}{dx}$$

We have
$$\frac{q_k}{A} = -\int K.\frac{dT}{dx}$$

$$\therefore \quad \frac{q_k}{A}\int_0^L dx = -K\int_{hot}^{cold} dT = K\int_{cold}^{hot} .dT$$

$$\therefore \quad \frac{q_k}{A}.L = K(T_{hot} - T_{cold})$$

or
$$\mathbf{q_k = \frac{AK}{L}(T_{hot} - T_{cold})}$$

We can write this also as,

$$q_k = \frac{AK}{L}.\Delta T = \frac{\Delta T}{L/AK}$$

Here we can compare with simple electric circuit as shown in Fig. 6.2.

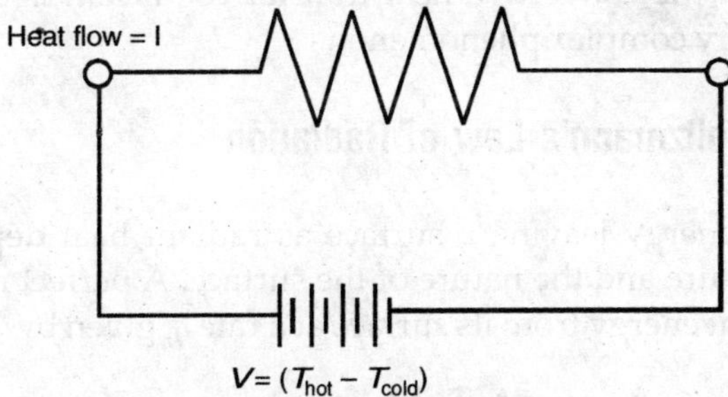

Fig. 6.2. Electric analog.

Here ΔT = Driving force for heat transfer similar to (Voltage)

$\frac{L}{AK}$ = Thermal resistance R_k, similar to (Electric Resistance)

q_k = Heat flow similar to (Electric current)

Thermal conductance is the reciprocal of thermal resistance

$$K_k = \frac{AK}{L} \rightarrow m^2 \times \frac{kJ/hrm^{\circ}C}{m} \rightarrow \frac{kJ}{hr^{\circ}C}$$

∴ This can be looked upon as an electric circuit as shown in Fig. 6.2.

6.2.2 Newton's Law of Cooling

The rate of heat transfer by convection between a surface and a fluid may be computed by the relation

$$q_c = \bar{h}_c A.\Delta T$$

where q_c = rate of heat transfer by convection (kJ/hr)

A = heat transfer area (m^2)

ΔT = difference between the surface temperature T_S and the temperature of the fluid $T\infty$ at some specified location (ºC).

$\bar{h}_c$ = average unit thermal convective conductance often called the surface coefficient of heat transfer or the convective heat transfer coefficient kJ/hr.m².ºC.

Thermal conductance K_c for convective heat transfer $K_c = \bar{h}_c . A$ and thermal resistance

$$R_c = \frac{1}{\bar{h}_c A}$$

The relation expressed by this equation was originally proposed by the British scientist, Issac Newton in 1701. This is also known as Newton's law of cooling. The evaluation of the convective heat transfer coefficient is difficult because convection is a very complex phenomenon.

6.2.3 Stefan Boltzmann's Law of Radiation

The quantity of energy leaving a surface as radiant heat depends upon the absolute temperature and the nature of the surface. A perfect radiator or black body emits radiant energy from its surface at a rate q_r given by

$$\mathbf{q_r = \sigma A_1 T_1^4 \ kJ / hr}$$

where A_1 is the surface area in sq.m. T_1 is the surface temperature in degrees Kelvin and σ is a dimensional constant and is known as Stefan–Boltzmann constant after two Austrian scientists, J. Stefan who in 1879 found the above equation experimentally and L. Boltzmann, who in 1884 derived it theoretically. Here 6 = 20.42 × 10^{-8} kJ/hr. m^2 T^4 = 5.671 W/m^2.T^4.

Stefan–Boltzmmnn Law of radiation states that any black body surface above a temperature of absolute zero radiates heat at a rate proportional to the fourth power of the absolute temperature.

While the rate of emission is independent of the conditions of the surroundings, a net transfer of radiant heat requires a difference in the surface temperature of any two bodies between which the exchange is taking place. If the black body radiates to an enclosure which completely surrounds it and whose surface is also black i.e. absorbs all the radiant energy incident upon it, the net rate of radiant heat transfer is given by,

$$\mathbf{q_r = \sigma A_1 \left(T_1^4 - T_2^4\right)}$$

where T_2 is the surface temperature of the enclosure in K.

Real bodies do not meet the specifications of an ideal radiator but emit radiation at a lower rate than the black-bodies. If they emit, at a temperature equal to that of a black body, a constant fraction of the black body emission at each wavelength, they are called *Gray Bodies*. The net rate of heat transfer from a gray body at T_1 to a black surrounding body at T_2 is

$$\mathbf{q_r = \sigma A_1 \varepsilon_1 \left(T_1^4 - T_2^4\right)}$$

where ε is the emissivity of the gray surface and is equal to the ratio of emission from the gray surface to the emission from a perfect radiator at the same temperature.

If neither of the two bodies is a perfect radiator and if two bodies possess a given geometrical relationships to each other, the net heat transfer by radiation between them is given by

$$\mathbf{q_r = \sigma A_1 F_{1-2} \left(T_1^4 - T_2^4\right)}$$

where F_{1-2} is a modulus which modifies the equation for perfect radiators to account for the emissivities and relative geometries of the actual bodies.

$$\mathbf{q_r = K_r \left(T_1^4 - T_2^4\right)}$$

By comparison the conductance K_r for radiation is

$$K_r = \frac{\sigma A_1 F_{1-2}\left(T_1^{\,4} - T_2^{\,4}\right)}{T_1 - T_2} \text{ kJ/hr°C}$$

Unit thermal conductance for radiation $\bar{h}_r$ is

$$\bar{h}_r = \frac{K_r}{A_1} = \frac{\sigma F_{1-2}\left(T_1^4 - T_2^4\right)}{T_1 - T_2} \frac{\text{kJ}}{\text{hr}} \text{m}^2/°\text{C or } \frac{\text{kJ}}{\text{hr.m}^2 °\text{C}}$$

Similarly, the thermal resistance for radiation is given by

$$\mathbf{R_r} = \frac{\mathbf{T_1 - T_2}}{\sigma \mathbf{A_1 F_{1-2}}\left(\mathbf{T_1^4 - T_2^4}\right)}$$

6.3 FILM COEFFICIENT OF HEAT TRANSFER

In heat exchangers, we are mostly concerned with walls separating liquids or gases from each other. In these cases we do not know the temperatures of both surfaces of the separating walls, but only the temperatures of the liquids or fluids on both sides of the walls. These temperatures are indicated as t_1 and t_2 as shown in Fig. 6.5. By measuring the temperature field in the liquids one obtains the curves shown.

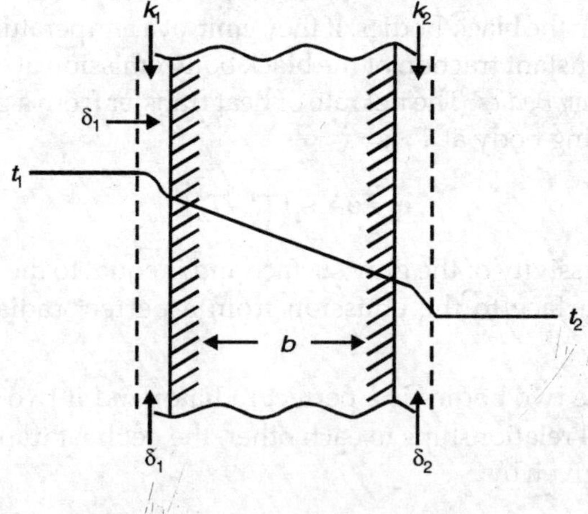

Fig. 6.3

The temperature gradient is confined to a relatively narrow layer of thickness δ quite close to the wall, whereas at a grater distance from the wall in most cases only small temperature differences exist. Simplified, the temperature curve can be replaced by the dashed broken line. This can be explained by assuming that a thin film of liquid (of the thickness δ') adheres to (1) the wall whereas outside thin film, temperature differences vanish as a result of mixing motions of the liquid. This picture oversimplifies the actual process considerably as we shall see later, but it brings out the salient features. Within the film, the heat transfer takes place by conduction, as in a solid wall. The temperature in the film, therefore, is again linear, and the flow of heat follows the equation

$$Q = \frac{K}{b}.A.(t_1 - t_2)$$

into which the thermal conductivity K of the liquid or of the gas and the thickness of the film δ' are to be inserted.

Thus, for the heat flow to the wall, the equation $Q = \frac{K}{\delta'} A(t - t_w)$ is obtained.

From this equation the rate of heat flow Q can be calculated if the film thickness δ' is known. The latter, however, depends to a great extent upon the external flow conditions, for instance upon the velocity with which the liquid flows along the wall, upon the shape of the wall, upon the structure of the wall surface and similar factors. In engineering practice it has become customary to calculate with the expression K/δ, rather (2) than directly with the film thickness δ'. This value is called film heat transfer coefficient and is designated by the letter h. Thus the expression

$$Q = hA\,(t - t_w)$$

as formulated by Issac Newton is obtained. The value $(1/hA)$ is called thermal resistance R_t, of the film heat transfer process

$$R_t = 1/hA$$

From Fig. 6.3 we have,

$$t_1 - t_{W_1} = \frac{1}{h_1 A}.Q$$

$$t_{W_1} - t_{W_2} = \frac{b}{kA}.Q$$

$$t_{W_2} - t_2 = \frac{1}{h_2 A}.Q$$

∴ Adding, we get

$$t_1 - t_2 = Q\left[\frac{1}{h_1 A} + \frac{b}{kA} + \frac{1}{h_2 A}\right]$$

$$= Q\left[R_{t_1} + R_c + R_{t_2}\right]$$

$$= R_0 Q$$

The sum of the individual resistances is the thermal resistance R_0 of the overall heat transfer process. We have

$$Q = UA(t_1 - t_2)$$

where $$U = \frac{1}{h_1} + \frac{b}{k} + \frac{1}{h_2}$$

The value U is called the **overall heat transfer coefficient.**

6.4 COMPOSITE WALLS

A composite wall, typical of the type used in a large furnace, is shown in Fig. 6.4. The inner layer, which is exposed to the high temperature gases, is made of firebrick. The intermediate layer consists of an insulating brick and is followed by an outer layer of ordinary red bricks.

The temperature of the hot gases is T_i and the unit surface conductance over the interior surface is $\bar{h}_i$. The atmosphere surrounding the furnace is at a temperature t_0 and the unit surface conductance over the exterior surface is $\bar{h}_0$.

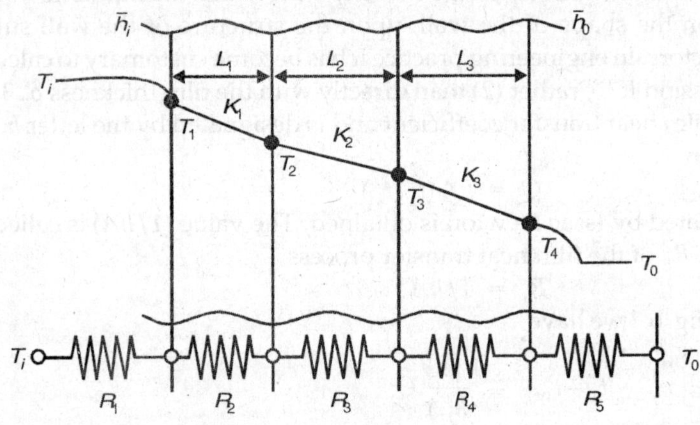

Fig. 6.4

Under these conditions there will be a continuous heat flow from the hot gases through the walls to the surroundings. Since the heat flow through a given area A is the same for any section, we obtain

$$q = \bar{h}_i A(T_i - T_1) = \frac{K_1 A}{L_1}(T_1 - T_2) = \frac{K_2 A}{L_2}(T_2 - T_3)$$

This equation can be written in terms of the thermal resistance of the various section as

$$q = \frac{T_i - T_1}{R_1} = \frac{T_1 - T_2}{R_2} = \frac{T_2 - T_3}{R_3} = \frac{T_3 - T_4}{R_4} = \frac{T_4 - T_0}{R_5}$$

where the resistances may be determined. Solving for the various temperature differences, we have

$$T_i - T_1 = qR_1$$

$$T_1 - T_2 = qR_2$$

$$T_2 - T_3 = qR_3$$

$$T_3 - T_4 = qR_4$$

$$T_4 - T_o = qR_5$$

Adding, we get

$$T_i - T_o = q(R_1 + R_2 + R_3 + R_4 + R_5)$$

$$\therefore \quad q = \frac{T_i - T_1}{R_1 + R_2 + R_3 + R_4 + R_5}$$

$$\therefore \quad \text{Heat flow} = \frac{\text{Overlap temp. potential}}{\text{Sum of resistance in the path}}$$

6.4 CONDUCTION OF HEAT THROUGH HOLLOW CYLINDER

The heat flow through a cylinder is considered along radial direction. Therefore the Fourier equation of conduction for cylinder can be written as

$$Q = -KA\frac{dT}{dr}$$

Figure 6.5 shows hollow pipe having inside radius R_1 and outside radius R_2. Let the inside temperature be T_1 and outside temperature T_2. The conductivity of the pipe material is say K.

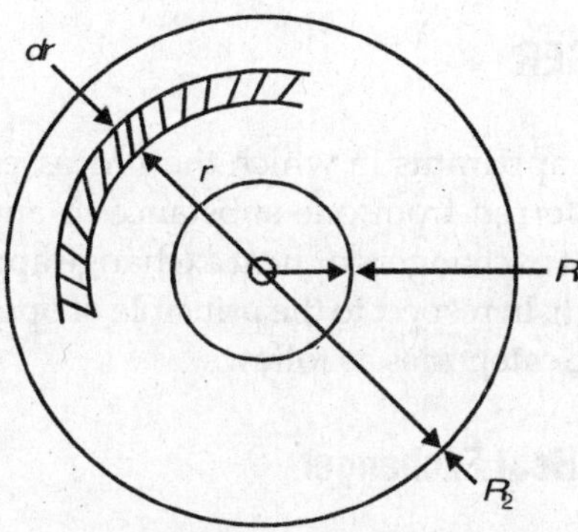

Fig. 6.5

Consider an elementary ring at a radius of r and having a thickness dr as shown in Fig. 6.5. The heat flow equation through an element dr under steady state condition can be written as

$$Q = -K.(2\pi r)\frac{dT}{dr}$$

where L is the length of pipe and $A = 2\pi rL$. Integrating this equation between the limits of R_1 and R_2 and T_1 to T_2 as the temperature changes from T_1 and T_2 through the thickness of the cylinder ($R_2 - R_1$),

$$\int_{R_1}^{R_2} Q.\frac{dr}{r} = \int_{T_1}^{T_2} -K2\pi L.dT$$

$$\therefore \quad Q.\log_e \frac{R_2}{R_1} = -2\pi LK(T_2 - T_1)$$

$$= 2\pi LK(T_1 - T_2)$$

$$\therefore \quad Q = \frac{2\pi LK(T_2 - T_1)}{\log_e \frac{R_2}{R_1}}$$

$$= \frac{T_1 - T_2}{\frac{1}{2\pi LK}\log_e \frac{R_2}{R_1}}$$

As before $\left[\frac{1}{2\pi LK}\right]\log_e \frac{R_2}{R_1}$ is known as thermal resistance of the pipe of length L.

6.5 HEAT EXCHANGER

A heat exchanger is an apparatus in which the processes of heating or cooling occur, i.e. heat is transferred from one substance to another. There is a large number of different heat exchangers or heat exchange apparatuses varying both in application and design. In respect to the principle of operation, heat exchangers may be divided in three categories as follows.

6.5.1 Direct Contact Heat Exchanger

The simplest type of heat exchanger is a container in which a hot and a cold fluids are mixed directly. In such a system both fluids will reach the same final temperature and the amount of heat transferred can be estimated by equating the energy lost by the hotter fluid to the energy gained by the hotter fluid to the energy gained by the cooler one. Open feed water heaters, desuperheaters, water cooling towers, scrubbers and jet condensers are the examples of heat transfer equipment employing direct mixing of fluids.

6.5.2 Recuperative Heat Exchangers

In heat exchangers of the recuperative variety, hot and cold fluids flow simultaneously through the heat exchanger and the heat is transferred through a wall separating the fluids. This group unites such heat transfer equipments such as steam boilers, water heaters, condensers etc. There are many forms of such equipment ranging from a simple pipe—within a pipe with a few sq-m of heat transfer surface upto complex surface condensers and evaporators with many hundreds of sq-m of heat transfrers surface. In between these extremes is a broad field of common shell and tube exchangers. These units are widely used because they can be constructed with large heat transfer surfaces in a relatively small volume, can be frabricated from alloys to resist corrosion, and are suitable for heating, cooling, evaporating or condensing all kinds of fluids.

6.5.3 Regenerative Heat Exchanger

A regenerative heat exchanger is an apparatus in which one and the same heating surface is alternately exposed to the hot and cold fluids. The heat carried by the hot fluid is taken away by and accumulated in the walls of the apparatus and is then transferred to the cold fluid flowing through the heat exchanger. Regenerators of open hearth and glass-melting furnaces, air preheaters of blast furnaces are some of the specimens of regenerative heat-exchange equipment.

The process of heat transfer in recuperative and regenerative heat exchangers is invariably bound with the surface of solid. That is why they are known as *surface exchanger*.

The special names given to heat exchanger are usually determined by their application and designation for example, steam boilers, furnaces, water heaters, evaporators, super-heaters, condensers etc. In spite of the great variety of shapes, layouts, principles of operation and working media, heat exchange apparatuses—ultimately serve one and the same purpose—transfer of heat from one, hot fluid to another cold fluid. The design fundamentals are, therefore common to all.

The complete design of a heat exchanger can be broken down into three major phases:

1. The thermal analysis
2. The preliminary mechanical design
3. Design for manufacture.

Here the emphasis will be on the thermal design. This phase of the design is primarily concerned with the determination of the heat-transfer surface area required to transfer heat at a specified rate for given flow rates and temperatures of the fluids.

The mechanical design involves considerations of the operating temperatures and pressures, the corrosive characteristics of one or both fluid, the relative thermal expansions and accompanying thermal stresses, and the relation of the heat exchanger to other equipment concerned.

The design for manufacture requires the translation of the physical characteristics and dimensions into a unit, which can be built at a low cost. Selection of materials, seals, enclosure, and the optimum mechanical arrangement have to be made and the manufacturing procedures must be specified.

6.6 BASIC TYPES OF HEAT-EXCHANGERS

The simple type of shell and tube heat exchanger is shown in Fig. 6.5. It consists of a tube or a pipe located concentricity inside another tube which forms the shell for this arrangement. One of the fluids flows through the inner tube and the other through the annulus formed between the inner and the outer tube. Since both fluid streams traverse the exchanger only once, this arrangement is called *single pass* heat exchanger. If both fluids flow in the same direction, the heat exchanger is a *parallel flow type*; if the fluids move in the opposite direction, the exchanger is of the *counter flow type*.

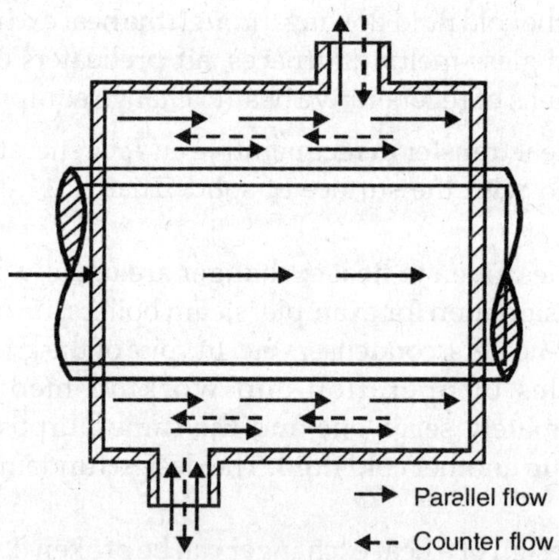

Fig. 6.5

The temperature difference between the hot and the cold fluid is, in general, not constant along tube, and the rate of heat flow vary from section to section. To determine the rate of heat flow one must therefore use an appropriate mean temperature difference.

When the two fluids flowing along the heat transfer surface move at right angles to each other, the heat exchanger is of the *cross flow type*. Three separate arrangements of this type of exchanger are possible. In the first case, each of the fluids is unmixed as it passes through the exchanger and therefore, the temperatures of the fluids leaving the heater section are not uniform, being hotter one side than on the other. A flat-type heater, a design used for turbine regenerators to claim the energy of the exhaust gases, or an automobile radiator approximates this type of exchanger.

In the second case, one of the fluids is unmixed and the other is perfectly mixed as it flows through the exchanger. The temperature of the mixed fluids will be uniform across any section and will vary in the direction of flow. An example of this type is the cross flow air heaters. The air flowing over the bank of tubes is mixed, while the hot gases inside the tubes are confined and therefore do not mix.

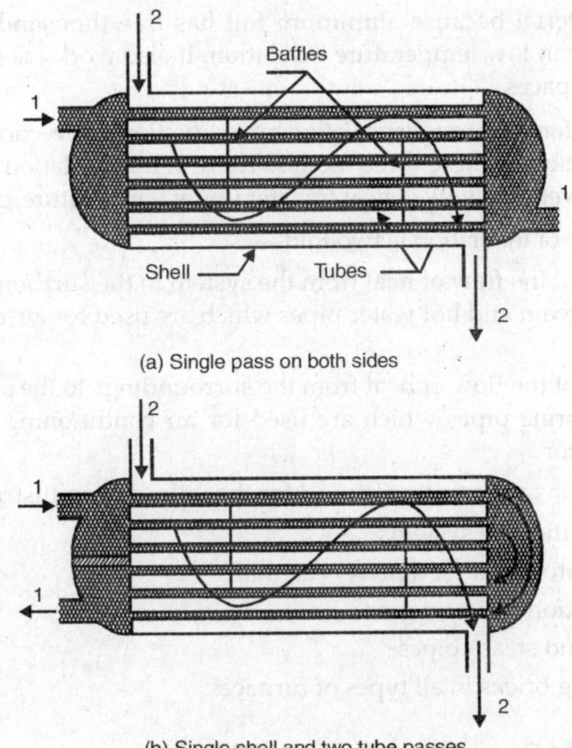

Fig. 6.7. Steam surface condenser.

In the third case, both of the fluids are mixed as they flow through exchanger, that is the temperature of both fluids will be uniform across the section, and will vary only in the direction of flow. Thus type of arrangement is less important than the other two.

In order to increase the effective heat transfer, optimum heat transfer surface area per unit volume is provided. Most commercial heat exchangers provide for more than a single pass through the tubes, and the fluid flowing outside the tubes in the shell is routed back and forth by means of baffles. Such an exchanger is known as *multi-pass* or *multiflow heat exchanger*. Figure 6.7 shows a three pass surface condenser.

6.7 INSULATION

6.7.1 Purpose of Insulation

The insulation is defined as a material which retards the heat flow with reasonable effectiveness. It is not always necessary to have low thermal conductivity for insulating material because aluminum foil has five thousand times greater conductivity than low temperature insulation. It also works as insulator when used with air spaces.

Heat is transferred through insulation by conduction, convection and radiation or by combination of these three modes. There is no insulation which is 100% effective to prevent the flow of heat transfer under temperature gradient.

The purpose of insulation is two fold—

(a) to prevent the flow of heat from the system to the surroundings as in the case of steam and hot water pipes which are used for air-conditioning in winter.

(b) to prevent the flow of heat from the surroundings to the system as in the case of brine pipes which are used for air conditioning in summer or refrigerator.

The insulations are commonly used for the following industrial purposes.

(i) Air-conditioning systems
(ii) Refrigerators and food preserving halls
(iii) Preservation of liquid gases
(iv) Boilers and steam pipes
(v) Insulating bricks in all types of furnaces.

6.7.2 Required Properties of Insulating Material

The desired properties of insulators are as mentioned below

1. Thermal conductivity. Some insulators have higher thermal conductivity than other. It is best to choose insulator having the lowest conductivity. The choice of the insulating material also depends upon outer practical requirements.

2. Structural strength. Some insulating materials tend themselves readily to various types of construction and easily handled under all circumstances.

Structural strength generally is obtained by the use of wood or steel frame work. For curved surfaces, material such as ground cork or mineral wool has been used extensively.

3. Weight. The weight of insulation for use in household cabinets is factor of small importance but when insulation under consideration is for use in truck, marine work, rail road, refrigerated car, aeroplane and similar application, weight often becomes a deciding factor.

4. Origin of insulation. Insulators derived from vegetable or animal sources are always subject to decomposition over a period of time. Such materials may develop odours unless special consideration is given to the method of applications.

5. Resistance to vermin. Insulators of the vegetable origin are regarded as food by certain forms of vermin. In tropical and semitropical countries, this difficulty may be trouble some on account of the presence of roaches, ants, mites and other forms of vermin, so that the vegetable type insulators need special protection.

6. Freedom from odour. Many foods absorb odours when they are exposed to odours. For this reason, the insulation used in refrigerator cabinets should be odourless when it is dry. In the presence of moisture, certain forms of vegetable insulation favour the growth of mold which produces odours and soon destroys the insulation. Insulation of this type when used in cabinets must be kept dry to avoid such difficulties.

7. Resistance to moisture absorption. An insulator which absorbs moisture deteriorates rapidly, especially if used on low temperature equipments. The rate of deterioration depends on the following factors:

(a) the thickness of insulation

(b) the temperature and humidity of atmosphere.

(c) method of application of the insulation

(d) the temperature difference between interior and exterior faces of the insulation.

8. Resistance to fire. A few insulators are extensively resistant to combustion and some burn very readily. Resistance to fire is not deciding factor, but in some application, such as, fire proof warehouse, is important.

6.8 TYPES OF INSULATIONS

Insulating materials are frequently classified in accordance with the temperatures at which they are used, such as low temperature (< 100ºC), moderate or steam temperature (100 – 600ºC) and high temperature (> 600ºC). This classification is not entirely satisfactory because many insulating materials can be properly employed in two or more of these temperature ranges. It is better to classify these insulators according to physical structure.

1. Loose Fill Insulators. This type of material is generally made in powered, granulated, cellular or fibrous form: (a) Mineral wool (b) Diatomaceous Earth (c) Venniculate (d) Silica Aerogal (e) Crushed insulating brick, asbestos fibres, granulated cork (f) Shredded redwood bark (g) Batts or Flexible Insulation.

2. Slab Insulation. Slab insulation consists usually of wood or sugarcane fibres, processed with an adhesive: (a) Cork board slabs (b) Rock cork (c) Balsa wood (d) Rubatex (e) Foam glass (f) Celotex (g) Hair felts (h) Kapok (i) Palco Bark (j) Insulating papers.

3. Insulating Fire-Brick. The insulating fire bricks are generally made by mixing an organic material with the refractory before moulding and the organic material is subsequently burned during firing leaving voids. The moulded brick is dried, fired. Standard size of brick is $6 \times 12 \times 25$ cm.

4. Reflective Insulation. The effectiveness of metallic surface for retarding heat transfer was well known to physicists early in the nineteenth century. Conduction and convection are practically eliminated by maintaining a high vacuum in the space.

(a) Aluminum Foil (Alfol)

(b) Low temperature reflective insulation

(c) Lead Tin alloy coated steel sheets.

6.9 EXTENDED SURFACES (FINS)

The equations of conduction heat transfer presented so far have not included the influence of heat losses or gains at the surfaces of the conducting body. When such concurrent heat flow must be considered, the surface coefficient of heat transfer h enters the basic equations. Also it is seen that the overall heat transfer resistance of a plane wall is determined principally by the greatest single resistance. If this resistance is one of the convection resistances, the heat flow through the wall can be increased by putting fins (extended surfaces) on the surface where this large resistance occurs. Such extended surfaces are widely used, for example, in economisers in steam power plants, convectors for steam and hot water heating systems, or electrical transformers, for the cylinders of air-craft engines, two wheeler engines etc.

There are many forms and shapes of these extended surfaces. Some of these are as follows.

1. A Thin Rod. This is transferring heat at its surface to a surrounding fluid and which is connected at its base to a heated wall. This system is shown in Fig. 6.8.

2. The Rectangular Fin. This is the simplest case for the plane finned surface. As long as the height of the fins on a tube is comparatively small with respect to the tube diameter, the formula derived from the plane wall can also be used for the tube. This system is shown in Fig. 6.9.

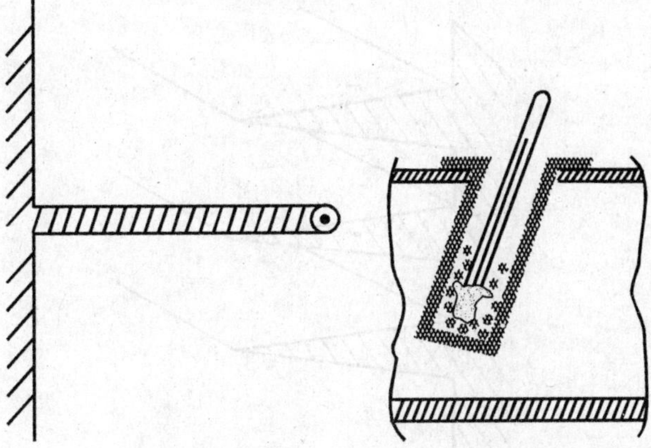

Fig. 6.8. Thermometer packet or well.

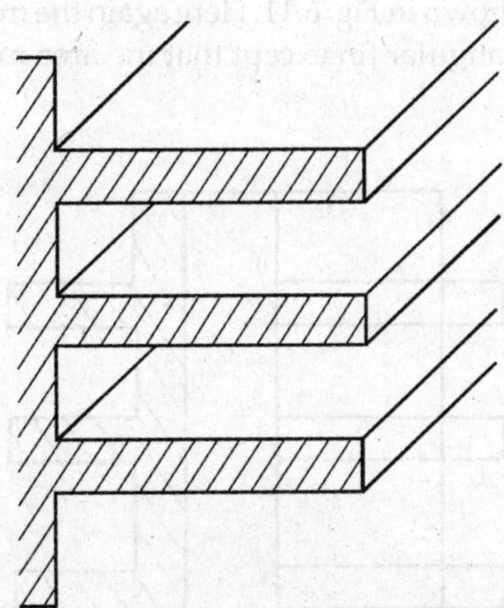

Fig. 6.9. Heating surface with rectangular fins.

3. Straight Fins of Triangular Profile. In determining the optimum fin, the question arises as to whether or not weight advantage can be gained by using a shape other that rectangular for the fin cross-section. Generally, a straight fin of triangular cross-section is considered. Such a fin is shown in Fig. 6.10.

The mathematical treatment in this case is similar to the case of the fin of rectangular cross-section except that the area normal to the heat flow is a function of the distance along the fin, decreasing as the fin length increases.

4. Cylindrical Fins. Fins which are arranged around tubes are called cylindrical fins and are quite important from an engineering point of view.

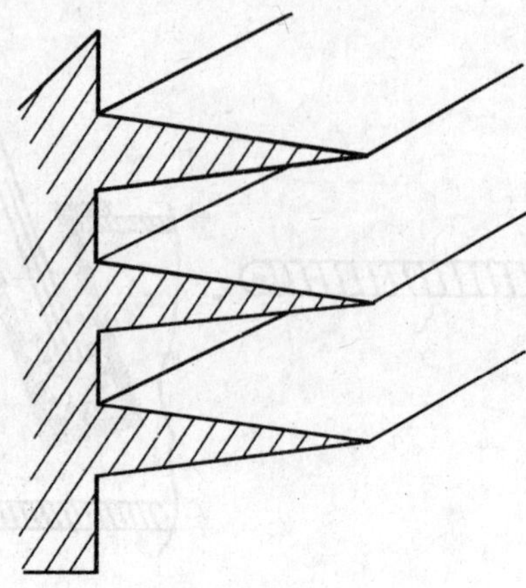

Fig. 6.10. Heating surface with triangular fins.

Such a fin system is shown in Fig. 6.11. Here again the treatment is substantially the same as for the rectangular fin except that the area must be allowed to vary with the radius.

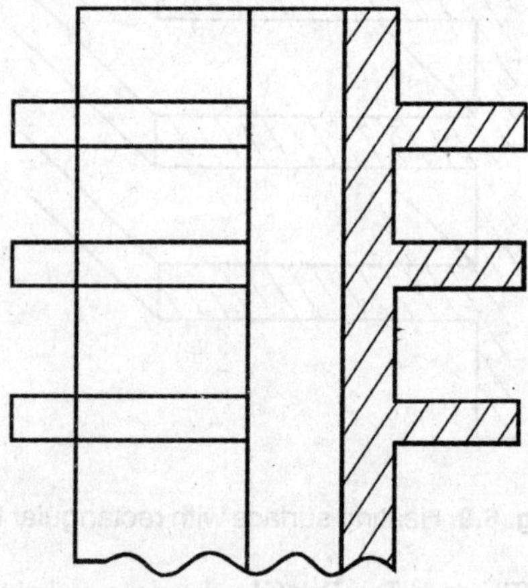

Fig. 6.11. Circular fins.

SOLVED EXAMPLES

EXAMPLE 6.1 *The inside surface of a furnace wall is at 270ºC and the outside surface is dissipating heat by convection into air at 20ºC. The wall is 40 mm thick and has thermal conductivity of 1.2 W/m-K. What is minimum value of heat transfer coefficient at the outside surface if outside surface temperature should not exceed 70ºC.*

Solution

Given:

$$t_1 = 270°C = t_2 = 70°C = t_3 = 20°C$$

$$t_2 = \text{Outside surface temperature}, K = 1.2\,W/m\,°K$$

Heat transfer through the furnace wall is given by

$$Q = \frac{KA(t_1 - t_2)}{L} = h_c A(t_2 - t_3)$$

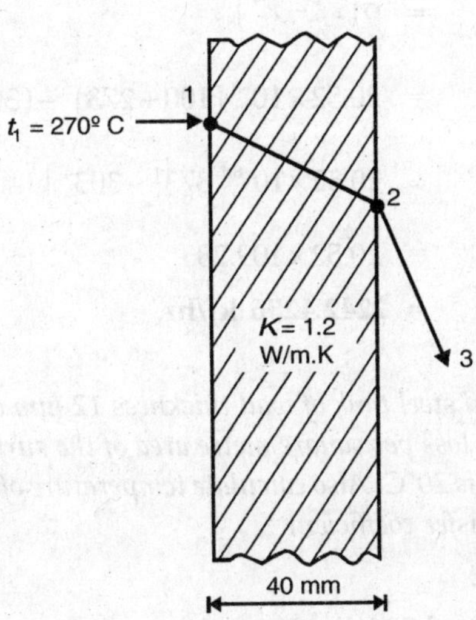

Fig. Ex. 6.1

Assume area A as 1

$$\therefore \quad \frac{1.2 \times 1 \times (270 - 70)}{0.04} = h_c \times (70 - 20)$$

$$\frac{1.2 \times 200}{0.04} = h_c \times 50$$

$$h_c = \text{heat transfer coefficient}$$

$$= \frac{1.2\times200}{50\times0.04}$$

$$= \mathbf{120\ W/m^2K}$$

EXAMPLE 6.2 *A black body at 30ºC is heated to 100ºC. Calculate the change in its emissive power.*

Solution

Emissive power of block body $= \sigma AT^4$

where $\sigma = 20.52\times10^{-8}\,\text{kJ/hrK}^4$

$\therefore$ For unit area, $A = 1\ \text{m}^2$

Increase of emmissive power

$$= \sigma\left(T_2^4 - T_1^4\right)$$

$$= 20.52\times10^{-8}\,(100+273)^4-(30+273)^4$$

$$= 20.52\times10^{-8}\left[373^4-303^4\right]$$

$$= 20.52\times109.28$$

$$= \mathbf{2242.4256\ kJ/hr.}$$

EXAMPLE 6.3 *(a) A mild steel tank of wall thickness 12 mm contains water at 100ºC. Calculate the rate of heat loss per square metre area of the surface of the tank when the atmospheric temperature is 20ºC. Also calculate temperature of the outside surface of the tank and overall heat transfer coefficient.*

Assume the following :

Conductivity of mild steel 50 W/m-K

Convection heat transfer coefficient

(i) on water side 2850 W/m² L

(ii) on air side 10 W/m² K

(b) A 60 W incandescent lamp has coil surface temperature 2500 K and room temperature 300 K. Estimate the surface area of the coil in mm².

(c) Derive an equation of conduction heat transfer rate in terms of temperatures dimensions and thermal conductivity for a thick, hollow cylinder.

Solution

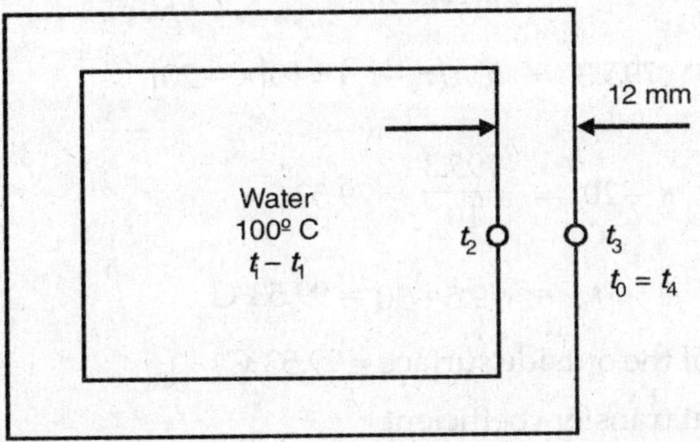

Fig. Ex. 6.3

Consider 1 sq-m of tank surface. The system is given in Fig. 6.13. Let Q be the heat rate

$$Q = h_i A(t_1 - t_2) \quad \text{(Water to inner surface)}$$

$$= \frac{KA(t_2 - t_3)}{x} \quad \text{(Inner to outer surface)}$$

$$= h_0 A(t_3 - t_4) \quad \text{(Outer surface to air)}$$

$$\therefore \quad t_1 - t_2 = \frac{Q}{h_i} \quad \text{for unit area}$$

$$t_2 - t_3 = \frac{Q.x}{K} \quad \text{for unit area}$$

$$t_3 - t_4 = \frac{Q}{h_0} \quad \text{for unit area}$$

$$t_1 - t_4 = Q.\left[\frac{1}{h_i} + \frac{x}{K} + \frac{1}{h_0}\right]$$

$$\therefore \quad 100 - 20 = Q\left[\frac{1}{2850} + \frac{12/1000}{50} + \frac{1}{10}\right]$$

$$80 = Q[0.0003509 + 0.00024 + 0.1] = Q \times 0.1006$$

$$Q = \text{Rate of heat transfer} / \text{m}^2$$

$$= \frac{\mathbf{80}}{\mathbf{0.1005989}} = \mathbf{795.3\ W}$$

Again $795.3 = h_0 A(t_3 - t_4) = 10[t_3 - 20]$

$$\therefore \quad t_3 - 20 = \frac{795.3}{10} = 79.53$$

$$\therefore \quad t_3 = 79.5 + 20 = 99.53^{\circ}\text{C}$$

Temperature of the outside surface = 99.53ºC

For overall heat transfer coefficient

We have $Q = UA(t_i - t_0)$

$$795.3 = U(100 - 20)$$

U = overall heat transfer coefficient

$$= \frac{\mathbf{795.3}}{\mathbf{80}} = \mathbf{9.94\ W/m^2/K}$$

(b) Heat transfer from the incandescent lamp is 60 W.

$$\therefore \quad 60 = \sigma(T_2^4 - T_1^4)$$

where $\sigma = 5.67 \times 10^{-8}\ \text{W/m}^2/\text{K}^4$

A = Area sq-m

T_1 = Lamp surface temperature

T_2 = Room temperature

$$\therefore \quad 60 = 5.671 \times 10^{-8} \times A \times (2500^4 - 300^4)$$

$$A = \frac{60 \times 10^8}{5.671 \times (2500^4 - 300^4)}$$

$$= \frac{60 \times 10^8}{5.671 \times 10^8 (25^4 - 3^4)} \text{m}^2 = \frac{10.58 \times 10^6}{(625 - 9)(625 + 9)} \text{mm}^2$$

$$= \mathbf{27.091\ mm^2}$$

= coil/surface area

(c) For derivation of the requrired equation, Refer Sec. 6.5.

EXAMPLE 6.4 *(a) Determine the rate of heat loss per hour through a 3.5 m² boiler plate if it is 20 mm thick with thermal conductivity 50 W/m-K the inner surface of the wall is covered with a layer of scale and deposits 2 mm thickness having a conductivity of 1 W/m-K. The hot gas temperature of the plate is 250ºC while that on the cold air (outside) is 200ºC. Also determine interface temperature.*

(b) Determine the loss by radiation from a steel tube 70 mm diameter and 3 m long at a temperature of 227ºC the tube is located in a huge room having wall temperature 27ºC.

Solution

The plate with scale and deposit is shown in Fig. Ex. 6.4

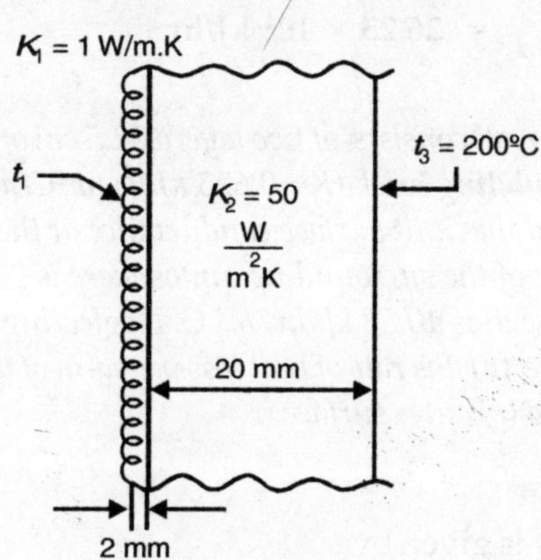

Fig. Ex. 6.4

With reference to this figure,

We write
$$Q = \frac{K_1 A_1 (t_1 - t_2)}{x_1} = \frac{K_2 A_2 (t_2 - t_3)}{x_2}$$

$$\therefore \quad t_1 - t_2 = \frac{Q.x_1}{K_1 A_1}$$

$$t_2 - t_3 = \frac{Q.x_2}{K_2 A_2}$$

$$t_1 - t_3 = \frac{Q}{A_1}\left[\frac{X_1}{K_1} + \frac{X_2}{K_2}\right]$$

$$250 - 200 = \frac{Q}{3.5}\left[\frac{2}{1000 \times 1} + \frac{20}{1000 \times 50}\right]$$

$$= \frac{Q}{3.5 \times 1000}[2+0.4]$$

$$\therefore \quad Q = \frac{50 \times 3.5 \times 1000}{2.4} = 72916.7 \text{ W}$$

= Rate of heat loss

= **72.92 kJ/s**

Rate of heat per hour

$$= 72.92 \times 3600 \text{ kJ/hr} = 26512 \text{ kJ/hr}$$

= $\mathbf{26.25 \times 10^4}$ **kJ/hr**

EXAMPLE 6.5 *A furnace wall consists of two layera 22.5 cm of brick (K = 4.984 kJ /hr.m ºC/m) and 12.5 cm of insulating brick (K= 0.623 kJ/hr.m ºC/m) the temperature inside the furnace is 1650 ºC and the unit surface conductance at the inside wall is 245.28 kJ/ hr.m² C. The temperature of the surrounding atmosphere is 27 ºC and the unit surface conductance at the outer wall is 40.88 kJ /hr. m² C. Neglecting the thermal resistance of the mortar joints. Estimate (a) the rate of heat loss per sq-m of wall and the temperatures at the (b) inter surface and (c) outer surface.*

Solution

(a) The rate of heat flow is given by

$$\frac{q}{A} = \frac{T_I - T_0}{\text{Total Resistance}}$$

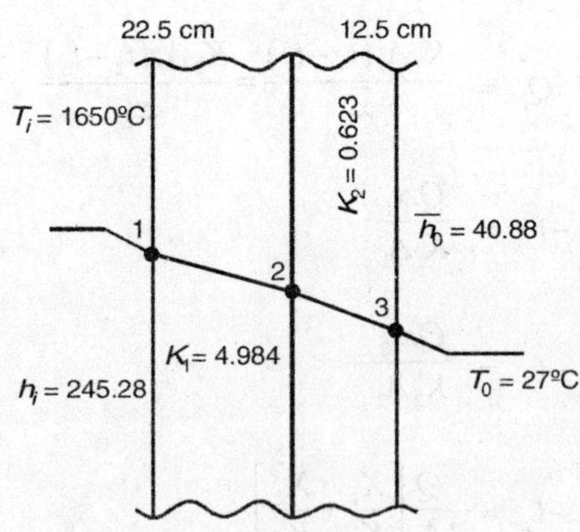

Fig. Ex. 6.5

$$\text{Total Resistance} = \frac{1}{h_I} + \frac{L_1}{K_1} + \frac{L_2}{K_2} + \frac{1}{h_0}$$

$$= \frac{1}{245.28} + \frac{22.5/100}{4.984} + \frac{12.5/100}{0.623} + \frac{1}{40.88}$$

$$= 0.00408 + 0.04514 + 0.20064 + 0.02446$$

$$= 0.27432$$

$$\therefore \quad q = \frac{1650-27}{0.27432} = \frac{1623}{0.27432}$$

$$= \mathbf{5916.9\ kJ/hr.m^2}$$

(b) We have $\frac{q}{A} = 5916.9\ \frac{T_i - T_1}{R_1} = \frac{1650 - T_1}{0.00408}$

$\therefore \quad T_1$ = Inner wall temperature

$$= 1650 - 5916.9 \times 0.00408 = 1650 - 24.14$$

$$= \mathbf{1625.86\ {}^\circ C}$$

(c) Again for outer surface temperatures

We have $\frac{q}{A} = \frac{T_i - T_3}{R_1 + R_2 + R_3} = \frac{1650 - T_3}{0.00408 + 0.04514 + 0.20064}$

$$5916.9 = \frac{1650 - T_3}{0.24986}$$

$$\therefore \quad T_3 = 1650 - 5916.9 \times 0.24986$$

$$= 1650 - 1478.38$$

$$= \mathbf{171.62\ {}^\circ C}$$

EXAMPLE 6.6 *A brick wall of a building is 30 cm thick and has an inside surface temperature of 24ºC and an outside surface temperature of –6ºC the wall is 2.75 m high and 6.1 m long. Estimate (a) the heat transfer by conduction through the wall per hour and (b) the conductance and resistance of the wall. The coefficient of conductivity of the brick material is 2.6 kJ/hr.m ºC.*

Solution

(a) Heat transfer by conduction

We have $q = \frac{KA\Delta T}{x}$

where $K = 2.6\ \text{kJ/hrm}\ {}^\circ\text{C}$

A = Area of the wall

$= 2.75 \times 6.1 = 16.8\ \text{m}^2$

x = Thickness of wall in m

$= \frac{30}{100} = 0.3\ \text{m}$

$$\Delta T = t_2 - t_1$$

$$= -(t_1 - t_2)$$

$$= -(24 + 6) = -30\ ºC$$

Substituting these values, we get

$$q = \frac{2.6 \times 16.8 \times 30}{0.3}$$

$$= \mathbf{4368\ kJ/hr}$$

(b) Conductance and resistance of the wall

$$\text{Resistance} = \frac{t_1 - t_2}{q} = \frac{30}{4368}$$

$$= \mathbf{0.00687\ C.hr/KJ}$$

$$\text{Conductance} = \frac{1}{\text{Resistance}}$$

$$= \frac{4368}{30}$$

$$= \mathbf{145.6\ kJ/m\ ºC}$$

EXAMPLE 6.7 *A temperature of the inner surface of a furnace wall is 450ºC and that of the outer surface 200ºC. The ambient temperature is 50ºC. It is required to reduce the heat loss from the outer surface by doubling the thickness of the brickwork. Assuming constant values convective heat transfer coefficient of air and thermal conductivity of wall material calculate percent decrease in heat loss.*

Solution

Let K be the conductivity of the wall material and $\bar{x}$ be the thickness.

Let T_0 be outside temperature and $\bar{h}_0$ be the unit coefficient of heat transfer for outer surface.

∴ Heat transfer

$$q = \frac{450 - T_0}{\frac{x}{K} + \frac{1}{h_0}} \text{ per unit area} \qquad \ldots\text{(i)}$$

When the wall is doubled in thickness, heat transfer.

$$q = \frac{450 - T_0}{\frac{2x}{K} + \frac{1}{h_0}} \qquad \ldots\text{(ii)}$$

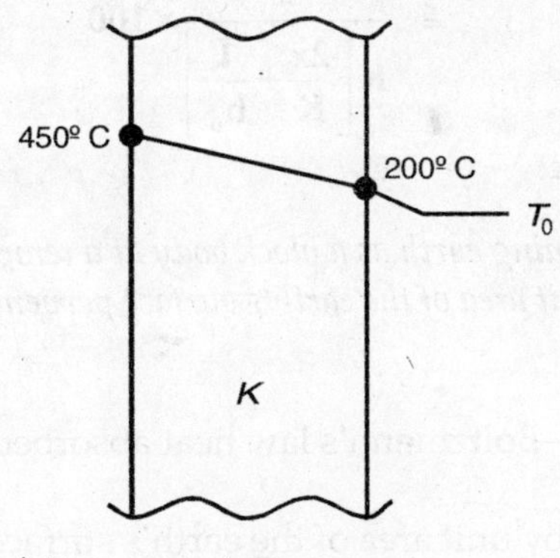

Fig. Ex. 6.7

Decrease in heat loss = (i – ii)

$$= (450 - T_0)\left\{\frac{450 - T_0}{\frac{x}{K} + \frac{1}{h_0}} - \frac{450 - T_0}{\frac{2x}{K} + \frac{1}{h_0}}\right\}$$

$$\text{Percentage decrease in heat lose} = \frac{(450 - T_0)\left\{\frac{1}{\frac{x}{K} + \frac{1}{h_0}} - \frac{1}{\frac{2x}{K} + \frac{1}{h_0}}\right\}}{(450 - T_0)\left\{\frac{1}{\frac{x}{K} + \frac{1}{h_0}}\right\}}$$

$$= 1 - \frac{\frac{1}{2x/K} + \frac{1}{h_0}}{\frac{1}{x/K} + \frac{1}{h_0}} = 1 - \frac{\frac{x}{K} + \frac{1}{h_0}}{\frac{2x}{K} + \frac{1}{h_0}}$$

$$= \frac{\frac{2x}{K} + \frac{1}{h_0} - \frac{x}{K} - \frac{1}{h_0}}{\frac{2x}{K} + \frac{1}{h_0}}$$

$$= \frac{\frac{x}{K}}{\frac{2x}{K} + \frac{1}{h_0}}$$

$$= \frac{x}{K\left[\frac{2x}{K}+\frac{1}{h_0}\right]} \times 100$$

EXAMPLE 6.8 *Assuming earth as a block body at a temperature of 300K. Estimate the energy received by unit area of the earth's surface perpendicular to the solar rays.*

Solution

According to Stefan–Boltzmann's law heat absorbed = Heat radiated

For a black body

∴ Heat received by unit area of the earth's surface perpendicular to the solar ray

$$Q = \sigma T^4 = 20.52 \times 10^{-8} \times (300)^4$$

$$= 20.52 \times 10^{-8} \times 10^8 \times 3^4$$

$$= 20.52 \times 81$$

$$= \mathbf{1662.12\ kJ/hr}$$

EXAMPLE 6.9 *Determine the loss of heat by radiation from a steel tube 70 mm diameter and 3 m long at a temp. of 227 ºC. The tube is located in a large room having wall temperature 27 ºC.*

Solution

$$Q = A.\sigma\left(T_H^4 - T_C^4\right)$$

$$A = 2\pi DL = 2\pi \times \frac{70}{1000} \times 3\text{m}^3$$

$$= \frac{42\pi}{100} = 0.42\,\pi\text{m}^2$$

$$= \frac{22}{7} \times \frac{42}{100} = \frac{132}{100} = 1.32$$

$$= 1.32 \times 20.52 \times 10^{-8}\left[500^4 - 300^4\right]$$

$$= 1.32 \times 20.52\left[5^4 - 3^4\right]$$

$$= 1.32 \times 20.52[625 - 81]$$

$$= 1.32 \times 20.52 \times 544$$

$= 14730 \text{ kJ/hr}$

$= \mathbf{4094 \text{ W or } 4.094 \text{ kW}}$

EXAMPLE 6.10 *The inside temperature of refrigerator is required to be 7 ºC the walls are constructed with two mild steel sheets 3 mm thick with 5 cm of glass wool insulation between them. The heat transfer coefficient on inner and outer surface of refrigerator are 10W/mºC and 12.5 W/m²C. Find the rate of heat removed from the refrigerator when it is kept in a kitchen room. The temperature in kitchen is 28ºC and K for mild steel is 40W/ºC and glass wool is 0.04 W/mºC.*

Solution

This is a problem of composit walls and the wall in problem is shown in Fig. Ex: 6.10,

$T_1 = 7°C$, $T_0 = 28°C$, $K_1 = 40$ W.mºC K_3; $x_1 = 3$ mm $= x_3$ and $x_2 = 50$ mm, $h_i = 10$ W/m²ºC and $h_0 = 12.5$ W/m²/ºC

Consider unit area and we write

$$Q = h_i A(T_i - T_1) \quad \therefore \quad T_i - T_1 = \frac{Q}{h_i A_i}$$

$$= \frac{K_1 A_1 (T_1 - T_2)}{x_1} \qquad T_1 - T_2 = \frac{Q \times x_1}{K_1 A_1}$$

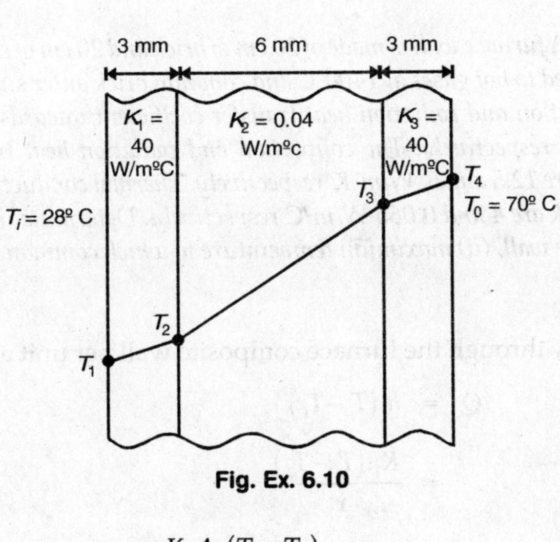

Fig. Ex. 6.10

$$= \frac{K_2 A_2 (T_2 - T_3)}{x_2} \quad \therefore \quad T_2 - T_3 = \frac{Q \times x_2}{K_2 A_2}$$

$$= \frac{K_3 A_3 (T_3 - T_4)}{x_3} \quad \therefore \quad T_3 - T_4 = \frac{Q \times x_3}{K_3 A_3}$$

$$= h_0 A_0 (T_4 - T_0) \quad \therefore \qquad T_4 - T_0 = \frac{Q}{h_0 A_0}$$

Adding these we get $T_i - T_0 = Q\left[\frac{1}{h_i A_i} + \frac{1}{\frac{K_1 A_1}{x_1}} + \frac{1}{\frac{K_2 A_2}{x_2}} + \frac{1}{\frac{K_3 A_3}{x_3}} + \frac{1}{h_0 A_0}\right]$

Assuming unit area of heat transfer

$$Q = \frac{T_i - T_0}{\frac{1}{h_i} + \frac{x_1}{K_1} + \frac{x_2}{K_2} + \frac{x_3}{K_3} + \frac{1}{h_0}}$$

$$= \frac{7 - 28}{\frac{1}{10} + \frac{0.003}{40} + \frac{0.05}{0.04} + \frac{0.003}{40} + \frac{1}{12.5}}$$

$$= \frac{-21}{1.43015}$$

$$= \mathbf{-14.684\ W/m^2}$$

The –ve sign indicates that the heat flows from the room to the refrigerator.

EXAMPLE 6.11 *A furnace wall is made of 20 cm of brick and 20 cm of common brick. The fire brick is exposed to hot gases at 1400ºC and common brick outer surface is exposed to 50ºC. The convection and radiation heat transfer coefficient towards gas side are 16.5 and 17.5 W/mºC respectively. The convection end radiation heat transfer coefficient towards outside are 12.5 and 6. W/mºK respectively. Thermal conductivity's of fire brick and common brick are 4 and 0.065 W/mºC respectively. Determine (i) heat loss par m² area of the furnace wall, (ii) maximum temperature to which common brick is subjected.*

Solution

(i) The heat flow through the furnace composite wall per unit area is given by

$$Q = h_i (T_i - T_1)$$

$$= \frac{K_1 . (T_1 - T_2)}{x_1}$$

$$= \frac{K_2 . (T_2 - T_3)}{x_3}$$

$$= h_0 (T_3 - T_0)$$

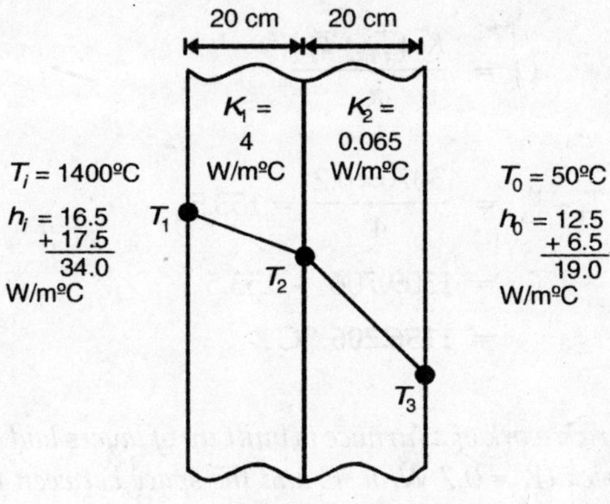

Fig. Ex. 6.11

$$\therefore \qquad Q = \frac{T_i - T_0}{\frac{1}{h_i} + \frac{x_1}{K_1} + \frac{x_2}{K_2} + \frac{1}{h_0}}$$

We take here $\quad h_i = 34 = \bar{h}_c + \bar{h}_r$

$$= 16.5 + 17.5 = 34$$

$$= \frac{1400 - 50}{\frac{1}{34} + \frac{0.2}{4} + \frac{0.2}{0.65} + \frac{1}{19}} = 1350$$

$$= \frac{1350}{0.0294 + 0.05 + 0.3077 + 0.0526}$$

$$= \frac{1350}{0.4397} = 3070 \text{ W/m}^2$$

$$= \mathbf{3.07 \ kW/m^2}$$

(ii) Maximum temperature to which common brick is subjected is T_2

$$\therefore \qquad Q = h_i (T_i - T_1)$$

$$\therefore \qquad 3070 = 34(1400 - T_1)$$

$$\therefore \qquad \frac{3070}{34} = 90.294 = 1400 - T_1$$

$$\therefore \qquad T_1 = 1400 - 90.294$$

$$= 1309.706\ ^\circ C$$

Similarly $$Q = \frac{K_1 (T_1 - T_2)}{x_1}$$

$$\therefore \quad T_1 - T_2 = \frac{3070 \times 0.2}{4} = 153.5$$

$$T_2 = 1309.706 - 153.5$$

$$= \mathbf{1156.206\ ^\circ C}$$

EXAMPLE 6.12 *A brick work of a furnace is built up of layers laid of fire clay ($K_1 = 0.93$ W/m ºC) and red brick ($K_3 = 0.7$ W/m ºC and the space between the two is filled with crushed diatomite brick ($K = 0.13$ W/m ºC). The thickness of fireday, diatomobile and red brick are 12 cm, 5 cm and 25 cm respectively.*

What should be the thickness of the red brick layer if the brickwork is to be laid without diatomite filling between the two layers so that the heat flow through the brick work remains constant.

Solution

With the diatomite filling, the heat flow will be given by (for unit area)

$$Q = \frac{T_1 - T_4}{\dfrac{x_1}{K_1} + \dfrac{x_2}{K_2} + \dfrac{x_3}{K_3}} = \frac{T_1 - T_4}{\dfrac{0.12}{0.93} + \dfrac{0.05}{0.13} + \dfrac{0.25}{0.7}}$$

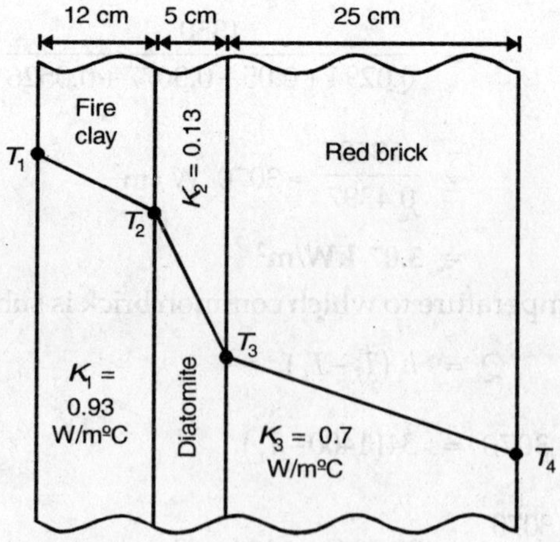

Fig. Ex. 6.12

$$= \frac{T_1 - T_4}{0.1290 + 0.3846 + 0.3571}$$

$$= \frac{T_1 - T_4}{0.8708} \text{ W/m}^2 \qquad \text{...(i)}$$

For the same Q-heat flow and without diatomite fillings, we have

$$Q = \frac{T_1 - T_4}{\frac{x_1}{K_1} + \frac{x_3}{K_3}} = \frac{T_1 - T_4}{\frac{0.12}{0.93} + \frac{x_3}{0.7}}$$

$$\therefore \quad \frac{T_1 - T_4}{0.8708} = \frac{T_1 - T_4}{0.1290 + \frac{x_3}{0.7}}$$

$$\therefore \quad \frac{1}{0.8708} = \frac{1}{0.1290 + \frac{x_3}{0.7}}$$

$$\therefore \quad 0.1290 + \frac{x_3}{0.7} = 0.8707$$

$$\therefore \quad \frac{x_3}{0.7} = 0.8707 - 0.1290$$

$$= 0.7417$$

$$\therefore \quad x_3 = \text{Thickness of the red brick layer}$$

$$= 0.7417 \times 0.7$$

$$= 0.51919 \text{ m}$$

$$= \mathbf{51.919 \text{ cm}}$$

EXAMPLE 6.13 *Hot air at a temperature of 656ºC is flowing through a steel pipe of 120 mm diameter. The pipe is covered with two layers of different insulating materials of thickness 60 mm and 40 mm and their corresponding thermal conductivities are 0.24 and 0.4 W/mºC. The inside and outside heat transfer coefficients are 60 and 12 W/m² ºC. The atmosphere is at 20 ºC. Find the rate of heat loss from a 60 m length of pipe. Neglect the resistance of the steel pipe.*

Solution

The pipe with insulations is shown in Fig. Ex. 6.13 we have

$$R_1 = 60 \text{ mm} = 0.06 \text{ m}$$

$$R_2 = 60 + 60 = 120 \text{ mm} = 0.12 \text{ m}$$

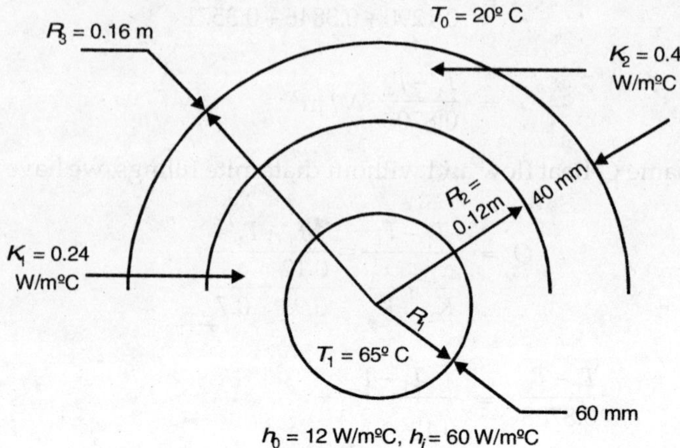

Fig. Ex. 6.13

$$R_3 = 60 + 60 + 40 = 160 \text{ mm} = 0.16 \text{ m}$$
$$K_1 = 0.24 \text{ W/m ºC} \; K_2 = 0.4 \text{ W/m ºC}$$
$$h_i = 60 \text{ W/m ºC} \; h_0 = 12 \text{ W/m ºC}$$

Length of the pipe to be considered is 60 m

Heat transfer through a circular, hollow pipe with raddi r_1 and r_2 is given by

$$Q = \frac{K.2\pi(t_1 - t_2)}{\log_e \frac{r_2}{r_1}} = \frac{t_1 - t_2}{\frac{1}{2\pi L}.\frac{1}{K}\log_e \frac{r_2}{r_1}}$$

Applying this equation to the problem, we have

$$Q = \frac{t_i - t_o}{\frac{1}{2\pi L}\left[\frac{1}{h_i R_1} + \frac{1}{K_1}\log_e \frac{R_2}{R_1} + \frac{1}{K_2}\log_e \frac{R_3}{R_2} + \frac{1}{h_0 R_3}\right]}$$

$$= \frac{65 - 20}{\frac{1}{2\pi \times 60}\left[\frac{1}{60 \times 0.06} + \frac{1}{0.24}\log_e \frac{0.12}{0.06} + \frac{1}{0.4}\log_e \frac{0.16}{0.12} + \frac{1}{12 \times 0.16}\right]}$$

$$= \frac{45}{0.002653[0.2778 + 2.888 + 0.7192 + 0.5208]}$$

$$= \frac{45}{0.002653 \times 4.4058}$$

$$= \frac{45}{0.01169}$$

$$= \mathbf{3.8494\ kW}$$

EXAMPLE 6.14 *A steam pipe of 16 cm inside diameter and 17 cm of outside diameter is covered with two layers of insulation. The thickness of first and second layers are 3 cm and 5 cm respectively and their corresponding conductivities are 0.15 and 0.08 W/m ºC respectively. The conductivity of the pipe material is 50 W/m ºC. The temperature of inner surface of the steam pipe is 300 ºC and that of the outer surface of insulation is 50 ºC. Find*

(i) the quantity of heat lost per meter length of the steam pipe.

(ii) if the steam is fully saturated at the entrance of the pipe find the quality of the steam coming out of one meter length of pipe. The quantity of steam flow is 40 gm/hr. Neglect the pressure loss.

Solution

The insulated pipe is shown in Fig. Ex. 6.14.

(i) Referring to this Fig. Ex. 6.14.

$R_1 = 8$ cm, $R_2 = 8.5$ cm, $R_3 = 11.5$ cm, $R_4 = 16.5$ cm, $T_1 = 300$ ºC, $T_4 = 50$ ºC

Length of pipe $= L = 1$m. $K_1 = 50$ W/m ºC, $K_2 = 0.15$ W/m ºC, $K_3 = 0.08$ W/m ºC.

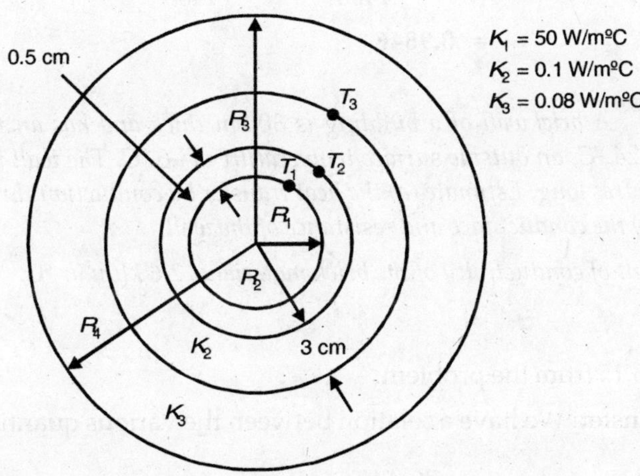

Fig. Ex. 6.14

The heat loss through a composite, cylinder is given by

$$Q = \frac{2\pi L(T_1 - T_4)}{\frac{1}{K_1}\log_e\frac{R_2}{R_1} + \frac{1}{K_2}\log_e\frac{R_3}{R_2} + \frac{1}{K_3}\log_e\frac{R_4}{R_3}}$$

$$= \frac{2\pi \times 1(300-50)}{\frac{1}{50}\log_e\frac{8.5}{8} + \frac{1}{0.15}\log_e\frac{11.5}{8.5} + \frac{1}{0.08}\log_e\frac{16.5}{11.5}}$$

$$= \frac{1570.8}{0.001212 + 2.0152 + 4.513}$$

$$= \frac{1570.8}{6.5291} = 240.58 \text{ W/m}$$

$$= 0.24058 \text{ kW} = \mathbf{866 \text{ kJ/hr}}$$

(ii) In enthalpy or total heat of dry and saturated steam at 300 ºC is obtained form steam tables and is 2751 kJ/kg total heat of steam at 1 m ahead of the entrance is given by

$$2751 - \frac{866}{40} = 2751 - 21.65$$

$$= 2729.35 \text{ kJ/kg}$$

$$= h_f + xhfg$$

$$= 1345 + x \times 1406$$

$\therefore$ x = Dryness fraction of steam

$$= \frac{2729.35 - 1345}{1406} = \frac{1384.35}{1406}$$

$$= \mathbf{0.9846}$$

EXAMPLE 6.15 *A brickwall of a building is 30 cm thick and has an inside surface temperature of 24 ºC, an outside surface temperature of –6 ºC. The wall is 2.75 metres high and 6.1 metres long. Estimate (a) the heat transfer by conduction through the wall per hour and (b) the conductance and resistance of the wall.*

The coefficient of conductivity of the brick material is 2.6 kJ/m.hr ºC.

Solution

Refer Fig. Ex. 6.15 from the problem.

(a) Heat transfer: We have a relation between the various quantities as

$$Q = \frac{-K.A\Delta T}{x}$$

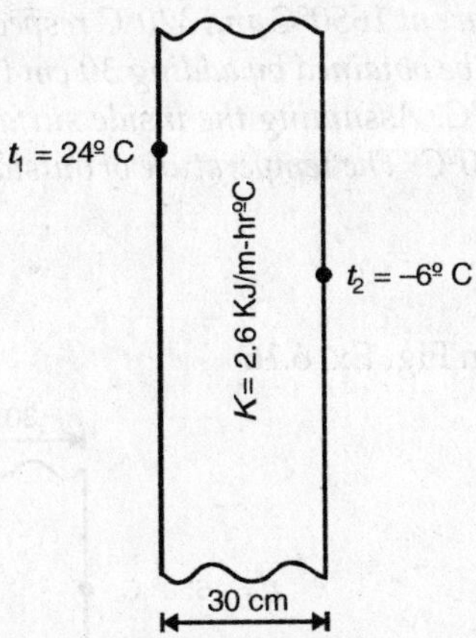

Fig. Ex. 6.15

where

$$K = 2.6 \text{ kJ/m.hr ºC}$$

$$A = \text{Area of wall in sq-m}$$

$$= 2.75 \times 6.1$$

$$= 16.8 \text{ m}^2$$

$$\Delta T = \text{Temperature difference across the wall}$$

$$= 24(-6) = 24 + 6$$

$$= 30 \text{ ºC}$$

$$x = \text{Thickness if wall in metres}$$

$$= \frac{30}{100} \text{ m}$$

Substituting these values in the above equation, we get

$$Q = \frac{2.6 \times 16.8 \times 30}{0.3} = 62 \times 16.8$$

$$= \mathbf{4368 \text{ kJ/hr}}$$

(b) Conductance and resistance of the wall:

$$\text{Resistance} = \frac{t_1 - t_2}{Q} = \frac{30}{1068}$$

$$= 0.00687 \text{ hr.C/kJ}$$

$$\text{Conductance} = \frac{1}{\text{Resistance}} = \frac{4368}{30}$$

$$= \mathbf{145.6 \text{ kJ/hr ºC}}$$

EXAMPLE 6.16 *The inside and outside surface of a furnace wall 30 cm thick of refractory bricks (K = 5.66 kJ/hr.m°C) are at 1650°C and 320°C respectively. Find the reduction in heat loss through the wall to be obtained by adding 30 cm thickness of insulating bricks for which K = 1.26 kh/hr m°C. Assuming the inside surface temperature of refractory bricks to remain fixed at 1650°C. The temperature of outside surface of the brick may be taken as 27°C.*

Solution

Arrangement are shown in Fig. Ex. 6.16

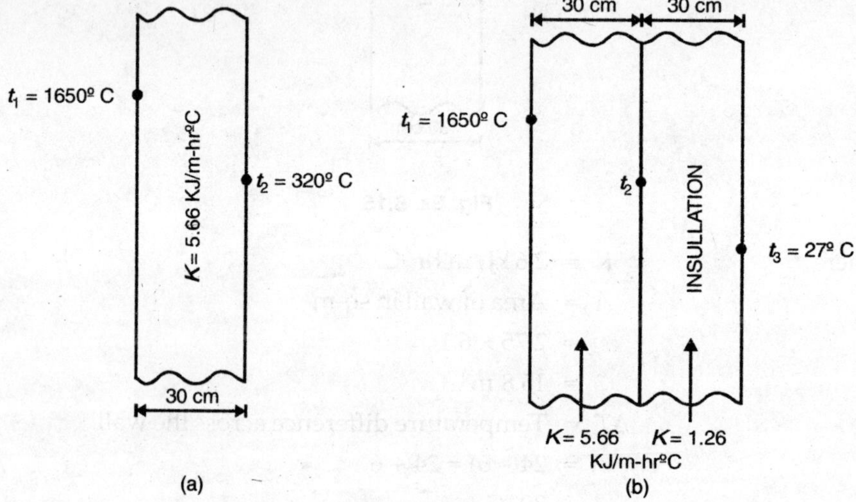

Fig. Ex. 6.16

(a) Without insulation: We have

$$Q = \frac{-K\,A\Delta T}{x}$$

Assuming area A as unity we get

$$Q = \frac{566 \times 1 \times (320 - 1650)}{0.3}$$

$$= 25093 \text{ kJ/hr/m}^2$$

(b) With insulation: For the refractory brick wall, we have

$$Q = -\frac{K_1 A \Delta t}{x_1} = \frac{5.66 \times (t_1 - t_2)}{0.3}$$

$$= \frac{5.66 \times (t_1 - t_2)}{0.3} = 18.87\,(t_1 - t_2)$$

or $$t_1 - t_2 = \frac{Q}{18.87} \quad \text{...(i)}$$

Similarly for insulating brick wall,

$$Q = \frac{-K_2 A(t_3 - t_2)}{x_2} = \frac{1.26 \times 1 \times t_2 - t_3}{0.3}$$

$$= 4.2(t_2 - t_3)$$

$$\therefore \quad t_2 - t_3 = \frac{Q}{4.2} \quad \text{...(ii)}$$

Adding Eqs (i) and (ii), we get,

$$t_1 - t_3 = Q\left[\frac{1}{18.87} + \frac{1}{4.2}\right] = Q[0.05299 + 0.2381]$$

$$= Q \times 0.29109$$

$$\therefore \quad Q = \frac{t_1 - t_3}{0.29109} = \frac{1650 - 27}{0.29109}$$

$$= 5575.6 \text{ kJ/hr/m}^2$$

$\therefore$ Reduction in heat loss through the wall

$$= 25093 - 5575.6 = \mathbf{1951.4 \text{ kJ/hr/m}^2}$$

EXAMPLE 6.17 *A cold room has one of the walls which measures 4.6 cm × 2.3 m constructed of bricks, 11.5 cm thick, insulated externally by cork slabbing 7.5 cm thick. The cork is protected externally by wood 2.5 cm-thick.*

Estimate the leakage through the wall per 24 hours if the interior temperature is –20°C and the exterior 18°C. The thermal conductivities of brick, cork and wood are 41, 1.9 and 7.5 kJ/hr.m°C respectively.

What will be the temperatures at the interfaces?

Solution

Figure Ex. 6.17 shows the arrangement of the various materials in the example.

For wood we have $$Q = \frac{KA(t_2 - t_1)}{x}$$

$$= 7.5 \times \frac{(4.6 \times 2.3)}{2.5/100}(t_1 - t_2)$$

$$= 3174(t_2 - t_1)$$

$$\therefore \quad t_2 - t_1 = \frac{Q}{3174}$$

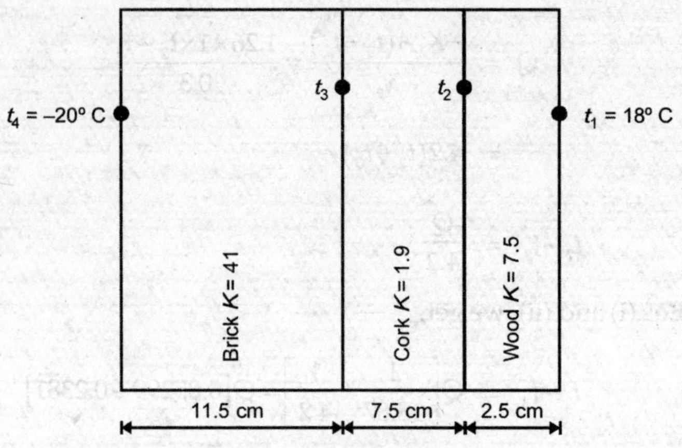

Fig. Ex. 6.17

$$= Q \times 0.000315$$

$$t_2 - t_3 = Q \times 0.003731$$

Similarly for cork we have

$$Q = \frac{1.9 \times 4.6 \times 2.3(t_2 - t_3)}{7.5/100}$$

$$= 268.03\ (t_3 - t_4)$$

$$\therefore \quad t_3 - t_4 = Q \times 0.00373 \qquad \text{...(ii)}$$

And for bricks, we have

$$Q = \frac{41 \times 4.6 \times 2.3(t_3 - t_4)}{11.5/100}$$

$$= 3772\ (t_3 - t_4)$$

$$\therefore \quad t_3 - t_4 = \frac{Q}{3772}$$

$$= Q \times 0.000265 \qquad \text{...(iii)}$$

Adding these three equations, we get

$$t_1 - t_4 = Q[0.000315 + 0.003731 + 0.000265]$$

$$= Q \times 0.00431$$

$$\therefore \quad Q = \frac{t_1 - t_4}{0.00431}$$

$$= \frac{18+20}{0.00431} = \frac{38}{0.00431}$$

$$= 8816.7 \text{ kJ/hr}$$

$$= \mathbf{211600.93 \text{ kJ/24 hr.}}$$

For interface temperatures

$$t_1 - t_2 = Q \times 0.000315$$

$$= 8816.7 \times 0.000315$$

$$= 2.7773$$

$$\therefore \quad t_2 = 18 - 2.7773$$

$$= \mathbf{15.2277\ °C}$$

Similarly $\quad t_2 - t_3 = \mathbf{8816.7 \times 0.00373}$

$$= 32.89$$

$$\therefore \quad t_3 = 15.2227 - 32.89$$

$$= \mathbf{-17.667\ °C}$$

EXAMPLE 6.18 *Prove that temperature distribution along the thickness of slab is uniform that is linear, if the thermal conductivity of the material of the slab is constant.*

Compute the heat loss per sq. metre of surface area for a furnace wall, 30 cm thick. The inner outer surface temperatures are 320° C and 38° C respectively. The variation of the thermal conductivity with the temperature is given by the following relation

$$K = \left(0.01256T - 4.2 \times 10^{-6} T^2\right) kJ / hr.m°C.$$

Solution

Consider a wall having a thickness as L and consider a plane surface at a distance x from the left face of the wall. Let the temperature at this surface be t and let dx be the elementary thickness of the slab at x. This is shown in Fig. Ex. 6.18.

$$Q = -\frac{KA.dt}{dx}$$

$$Q.dx = -KA.dt$$

Integrating this equation,

$$Q\int_0^L dx = -KA\int_{t_1}^{t} dt$$

Fig. Ex. 6.18

$$Q = -\frac{KA.dt}{dx}$$

$$Q.dx = -KA.dt$$

Integrating this equation,

$$Q\int_0^L dx = -KA\int_{t_1}^{t} dt$$

$$\therefore \quad QL = -KA(t-t_1)$$

$$\therefore \quad Q = -\frac{KA(t-t_1)}{L}$$

This shows that the temperature variation is linear one or uniform.

$$\frac{QL}{AK} = t_1 - t$$

or $$\mathbf{t = t_1 - \frac{QL}{AK}}$$

For the problem given, thermal conductivity of the material is varying.

$$\therefore \quad \text{We have } Q = \frac{AK.dt}{dx}$$

$$= -A\left[0.01256T - 4.2T^2 \times 10^{-6}\right]\frac{dT}{dx}$$

Integrating, we get

$$Q'L = A\left[\frac{0.01256T^2}{2} . \frac{4.2T^3}{3} \times 10^{-6}\right]_{38}^{320}$$

$$QL = A\left[0.00628\left(320^2 - 38^2\right) - \frac{4.2}{3} \times 10^{-6}\left(320^3 - 38^3\right)\right]$$

$$= 1\left[0.00628\left(358 \times 282\right) - 1.4 \times 10^{-6} \times 10^6\left(32.768 - 0.055\right)\right]$$

$$= 634 - 1.4 \times 32.713$$

$$= 634 - 45.8$$

$$= 588.2 \text{ kJ/hr/m}^2$$

$$\therefore \quad Q = \frac{588.2}{0.3}$$

$$= \mathbf{1960 \text{ kJ/hr/m}^2}$$

EXAMPLE 6.19 *A steam pipe line 11.5 cm outside diameter, is covered with two layers of different materials. The first layer is 5 cm thick and a thermal conductivity of 0.222 kJ/hr.m °C. The second layer is 3 cm thick and as a conductivity of 3.14 kJ/hr.m °C.*

Outside surface temperature of steam pipe is 235 °C and that of the outer surface of lagging is nm 38 °C. Calculate the heat loss per metre length of pipe per hour, and the temperature between the two layers of insulation.

Solution

Figure Ex. 6.19 shows the arrangement of the pipe with layers of different materials.

We have for cylindrical pipes,

$$Q = \frac{2\pi l \times K\left(t_1 - t_2\right)}{\log_e \dfrac{D}{d}}$$

For the inner layer,

$$\log_e \frac{D}{d} = \log_e \frac{11.5 + 10}{11.5}$$

$$= \log_e 1.87$$

$$= 0.6257$$

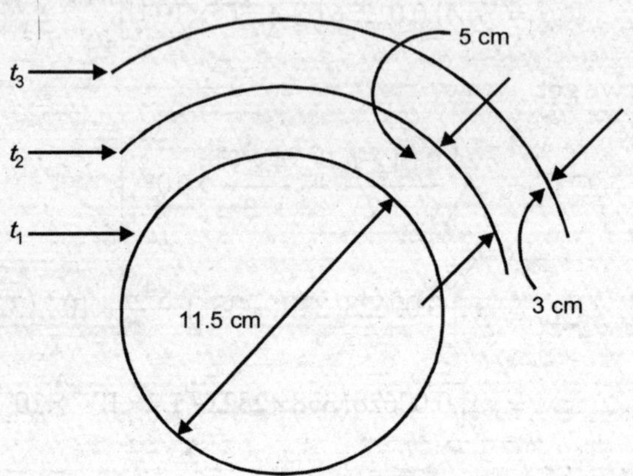

Fig. Ex. 6.19

Similarly for outer layer

$$\log_e \frac{D}{d} = \log_e \frac{21.5+6}{21.5}$$

$$= \log_e 1.28$$

$$= 0.2409$$

Taking the length of the pipes as 1 metre,

$$t_1 - t_2 = \frac{Q \times 0.6257}{2\pi \times 0.22}$$

$$= 0.4486\ Q \qquad \ldots\text{(i)}$$

Similarly, $$t_2 - t_3 = \frac{Q \times 0.2569}{2\pi \times 3.14}$$

$$= 0.0125\ Q \qquad \ldots\text{(ii)}$$

Adding Eqs (i) and (ii), we get,

$$t_1 - t_3 = Q \times [0.4486 + 0.0125]$$

$$= Q \times 0.4611$$

$$\therefore \quad 235 - 38 = 0.4611\ Q$$

$$\therefore \quad Q = \frac{\mathbf{197}}{\mathbf{0.4611}}$$

$$= \mathbf{427.24\ kJ/hr}$$

From Eq. (i) we have,

$$235 - t_2 = 0.4486 \times 427.24$$
$$= 191.66$$
$$t_2 = 235 - 191.66$$
$$= \mathbf{43.34\ °C}$$

= Temperature between the two layers of insulation.

EXAMPLE 6.20 A wall of refrigerated van of 1.5 mm of steel sheet at outer surface, 10 mm plywood at a inner surface and 2 cm of glass wool in between. Calculate the rate of heat flow if the temperature at the inside and outside surface are –15 and 24 °C. Take

$$K\,(steel) = 23.2\ W/m\ °C$$
$$K\,(glasswood) = 0.014\ W/m\ °C$$
$$K\,(plywood) = 0.052\ W/m\ °C$$

Solution

This is shown in Fig. Ex. 6.20. Consider unit area of flow of heat (1 m^2)

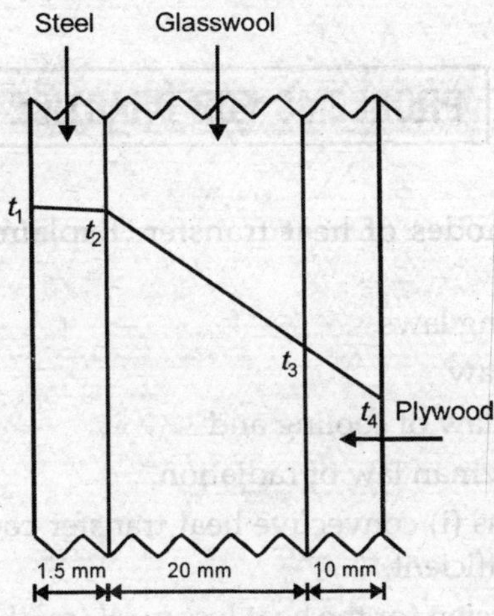

Fig. Ex. 6.20

With refrence to Fig. Ex. 6.20, we have

$$Q = \frac{KA(t_1 - t_2)}{x_1}$$

$$= \frac{23.2 \times 1(24 - t_2)}{1.5/1000}$$

$$\therefore \quad (24-t_2) = \frac{Q\times 1.5}{1000\times 23.2} = \frac{0.0646Q}{1000}$$

$$t_2 - t_3 = \frac{Q\times 20}{1000}\times\frac{1}{0.014}$$

$$t_3 - t_4 = \frac{Q\times 10}{1000}\times\frac{1}{0.052}$$

$$= \frac{Q\times 192.31}{1000}$$

Adding we get

$$24 + 15 = \frac{Q}{1000}[0.0646 + 1428.57 + 192.31]$$

$$39 = \frac{Q}{1000}1620.9446$$

$$Q = \frac{39000}{1620.9446} = \mathbf{24.06\ W/m^2}$$

PROBLEMS FOR PRACTICE

1. What are the modes of heat transfer. Explain very briefly each one mentioned.
2. State the following laws
   (i) Fouriers' law
   (ii) Newtons' law of cooling and
   (iii) Stefa–Boltzman law of radiation.
3. Explain the terms (i) convective heat transfer coefficient and (ii) overall heat transfer coefficient.
4. Derive an expression for the heat loss in W/m^2 through a composite wall of n layers.
   (a) without considering convective heat transfer coefficients and
   (b) with considering the convective heat transfer coefficient.
5. Prove that the heat flow through the hollow cylinder is given by

$$Q = \frac{2\pi LK(T_1 - T_2)}{\log_e\left(\frac{R_2}{R_1}\right)} \text{ where } T_1 > T_2$$

6. If the inner surface and outer surface temperatures of simple brickwall are 40°C and 20°C, calculate the rate of heat transfer per (metre)^2 of surface area of the wall having a thickness of 250 mm. Assume K for the brick is 0.52 W/m °C.

(*Ans.* 41.6 W/m^2)

7. Determine the rate of heat transfer/flow through the boiler wall of 20 nun thickness steel plate which is covered with an insulating material of 5 mm thick. The temperature at the inner and outer surfaces of the wall are 300°C and 50°C respectively. Assume K for steel = 58 W/m°C and for insulation as 0.116 W/m-K.

(*Ans.* 5.8 kW/m^2)

8. A steel tank of wall thickness 10 mm contains water at 90°C. Calculate the rate of heat loss per unit area of tank surface when the atmospheric temperature is 15°C. L for steel is 50 W/m°C and the heat transfer coefficients for inside and outside the tank are 2800 and 11 W/m^{2}K respectively. Calculate also the temperature of the outside surface of the tank.

(*Ans.* 820 W/m^2, 89.6°C)

9. Sheets of brass and steel, each 10 mm thick, are placed in contact with each other. The outer surface of brass is at 100°C and the outer surface of steel is at 0°C. What is the temperature of the common interface? Assume that the conductivities of brass and steel are in the ratio of 2 : 1.

(*Ans.* 66.7°C)

10. The wall of the furnace is made up of 250 mm of fire brick, K = 1.05 W/m°C; 120 mm of insulation brick, K = 0.85 W/m°C, and 200 mm of red brick, K = 0.85 W/m°C. The inner and outer surface temperatures walls are 850°C and 65°C respectively. Calculate the temperatures at the contact surfaces.

(*Ans.* 703°C, 210°C)

11. Explain the following
    (i) Fouriers' law of heat conduction
    (ii) Newton's law of cooling
    (iii) Stephan–Boltzman law of radiation heat transfer.
12. State different types of heat exchangers.
13. Explain the phenomenon of heat transfer by convection. State the Newton's Law of cooling.
14. State any three exchangers normally observed in thermal power plant, mentioning clearly the type of heat exchange.

(*Ans.* The heat exchangers that are normally used in thermal power plant are
1. Air preheater—Multipass, shell and tube type.
2. Economiser—Multipass, tube assembly kept in the flue gas passage.
3. Superheater—Steam carrying tubes kept in the flow gas passages)

15. Steam Generator:
    (i) Fire tube boiler furnace tubes through which gases flow and tubes are surrounded by water
    (ii) Water tube boiler-water tubes carrying water are placed in gas passage. Gases and water both are flowing.
16. Explain the following:
    (a) Conduction heat transfer and its associated property
    (b) Convection heat transfer and its associated property.

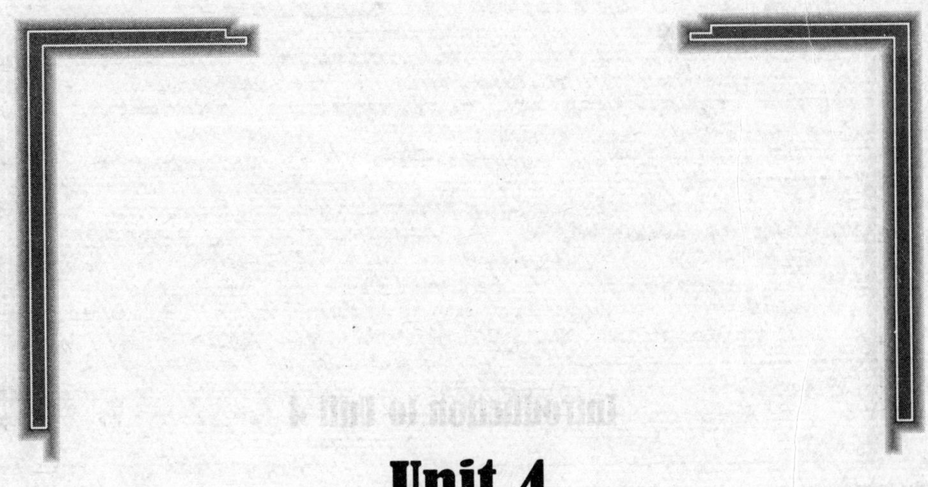

Unit 4

Metal Joining & Metal Joining Processes

Syllabus

Introduction to Metal Cutting Processes — Lathe, Drilling, Grinding and power saw machines, Lathe Machine Centre Lathe (Basic elements, working principle and types o operations)

Drilling Machine – Study of pillar drilling machine (operation only) Introduction to NC CNC M/c.

Introduction to Metal Joining Processes — Welding, soldering, Brazing, method and applications.

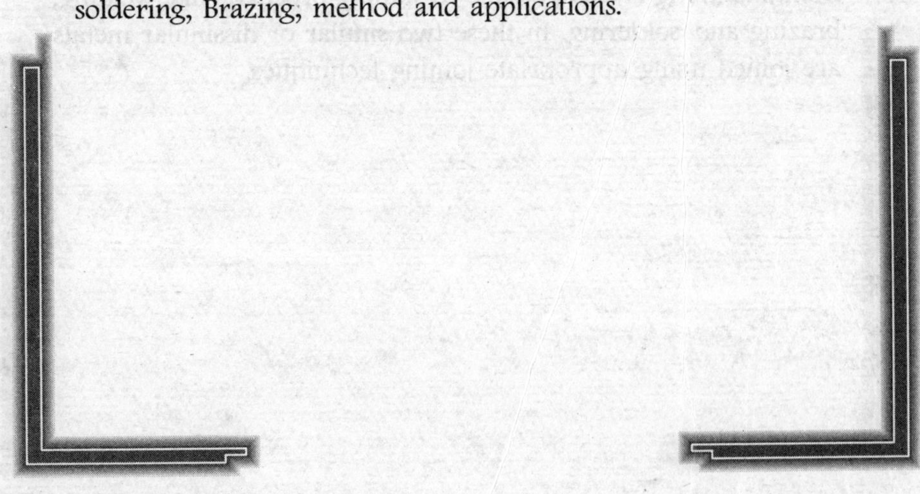

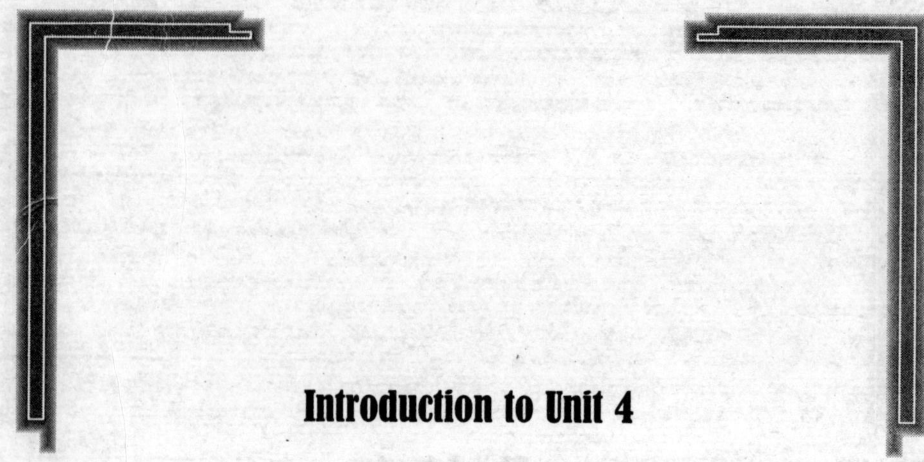

Introduction to Unit 4

All engineering and common use items, comes to exist only when they are manufactured. Hence the field of manufacturing processes has acquired a key place in mechanical engineering. Till today it is considered to be the most challenging field for developing engineering.

There are various methods of manufacturing which can be broadly classified into (a) Material cutting (b) Metal Joining

In the metal cutting operation the item is manufactured by removing excess material to get the required shape. For this purpose machines like lathe, grinding machines, power saws and the latest NC/CNC machines are used. In this section the other relevant aspects of these machines

and operations there on have been covered.

Manufacturing is also possible by joining processes like welding, brazing and soldering. In these two similar or dissimilar metals are joined using appropriate joining techniques.

CHAPTER 7

INTRODUCTION TO METAL CUTTING PROCESSES

7.1 MACHINE TOOL

A machine tool is a power driven device or apparatus to perform certain metal removing operations and produce a desired form on the surface being machined. A cylindrical or flat surfaces are generally required for the product.

Machine tools may be classified as standard or special purpose machine tools. A standard machine tool is that machine which can be used for a variety of work. As against this, a special purpose machine tool is required to perform specific work and cannot be used for any other purpose. Its application is a limited one.

A centre lathe is a very good example of a standard machine tool with the help of which variety of work can be performed, which include

(i) turning between centers

(ii) turning, facing, drilling and boring

In this chapter we will study the following machines:

1. Centre lathe
2. Drilling machine
3. Grinding machine
4. Power saw
5. NC and CNC machines (Introduction only)

7.2 CENTRE LATHE

The lathe is the father of all machine tools. It is a universal machine tool which can be used—with certain attachment—for many processes. For its development to the form in which we now know it, we owe much to Henry Maudlsey, who developed the sliding carriage, and in 1800 built a screw cutting lathe on which he turned screws having from 6 to 40 threads per centimetre and which were the best screws that had been made up to that time.

About 1830 Maudlsey constructed a lathe, with 3 metre face plate, which was used to bore large cylinders and turn flywheels.

In its operation the lathe holds a piece of material between two rigid supports called centers or by some other device such as a chuck or a face plate. The spindle carrying the work is rotated whilst a cutting tool supported in a tool post is caused to travel in a certain direction, depending upon the form of the surface required. If the tool moves parallel to the axis of rotation of the work a cylindrical surface is produced Fig. 7.1(a) whilst if it moves perpendicular to this axis it produces a flat surface as at Fig. 7.1(b)

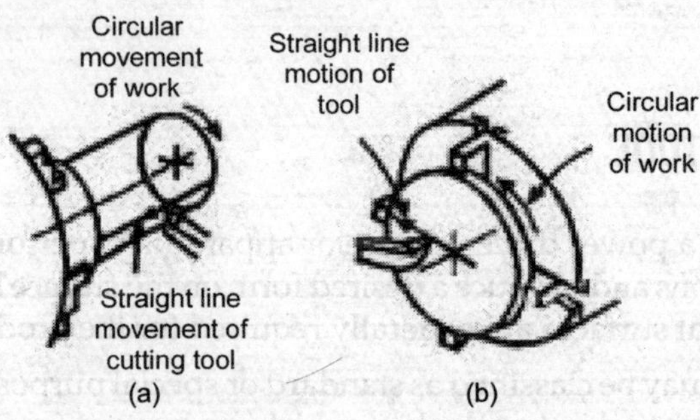

Fig. 7.1. Surface formation of the lathe.

Figure 7.2 shows a simple centre-lathe or general purpose lathe showing its main components. The components are as follows:

(i) Bed. The bed of the lathe is made of a good quality gray iron casting or alloy and is supported on legs or a bench. It has accurately machined hardened and ground flat or inverted Vee-shaped inner and outer surfaces to guide the carriage head stock and tail-stock and to ensure accurate alignment.

(ii) Head Stock. It is the powered end of a lathe which is securely fixed on the inner ways of the bed. It is always on the left side of the operator. It houses the speed changing gears and the driving spindle. Normally a built in electric motor supplies power to the spindle. The centre of the spindle is hollow so that long bars can be put through and be held by chucks for machining. It is made of cast iron.

(iii) Tail Stock. It is supported on the inner ways of the bed and is located on the right side of the operator. The main purpose of the tail-stock is to support the free end of the work piece when it is machined between centers. It is also used to hold tools when operations such as drilling, reaming, tapping etc. are performed on a lathe. (Ref. Fig. 7.3).

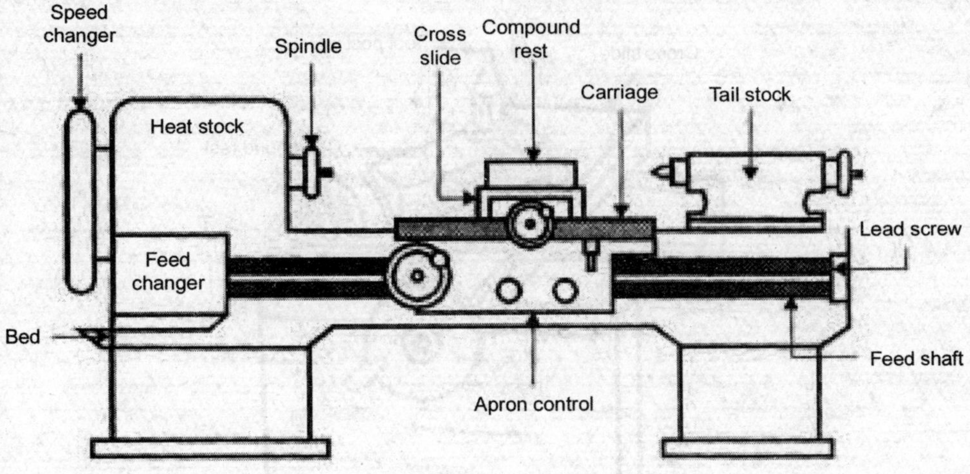

Fig. 7.2. Centre lathe.

Fig. 7.3. Tail stock.

(iv) Carriage. It consists of three main parts—a saddle, a cross slide and apron. It is used for moving the cutting tool along the lathe bed. (Fig. 7.4)

The saddle, an H-shaped casting, is bridged across the lathe bed and carries a cross-slide, a compound rest, a top slide and a tool post. The cross slide can be moved at right angles to the lathe bed either by hand or by power. The compound rest permits the swivelling of the tool to the required angle.

The apron is fastened to the saddle and contains the gears and clutches for transmitting motion from the feed rod to the carriage, and also contains the split nut (or half nut) which engages with the lead screw while cutting threads.

The compound slide: On most lathes a compound slide is interposed between the Tool post and the cross-slide. This slide may be swivelled to any angle and its use enables the tool to be moved in directions other than those permitted by the carriage and cross-slide. The compound slide is useful for turning and boring short tapers, chamfers and other jobs requiring an angular movement of the tool.

(v) The tool post is located on the top of the compound-rest. The purpose of tool post is to hold the tool and to enable it to be adjusted to a convenient working position.

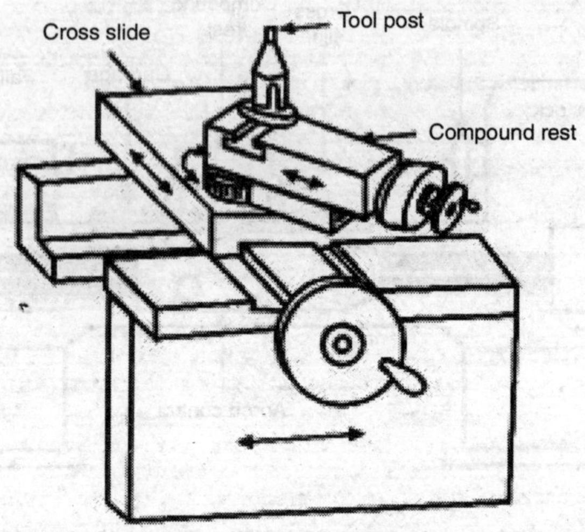

Fig. 7.4. Lathe carriage.

(vi) Chucks may be of the 4-jaw independent, or a 3-jaw self-centring patterns. (Fig. 7.5)

Fig. 7.5. Lathe chucks.

Each jaw of the 4-jaw chuck is operated by a separate square-threaded screw, whilst in the 3-jaw type all the jaws close in together, actuated by the scroll, which is a spiral groove cut on the face of a flat disc. The principle is similar to the movement of a nut by a screw except that the screw is cut on the face of a disc. A 3-jaw Chuck is easier to operate, but the gripping efficiency is much less than that of the 4-jaw. The chuck is adapted to the nose of the lathe by being screwed to a back-plate which screws on to the machine nose. A Magnetic Chuck is available in several designs, most common being a rotary type.

An Air or Hydraulic Chuck is often used on lathes and other turning machines engaged in a mass production work. It is quick acting and grips the work strongly. The chuck can be worked by compressed air or by a fluid.

7.3 TYPES OF LATHES

Lathes are manufactured in such a large number of types and sizes, and for such variety of purposes that it is quite impossible to attempt within the scope of a volume like this, to illustrate completely the different classes of lathes. Some of the different types of lathes are as follows:

1. Speed lathe. Normal lathe 1200–3600 rpm. These lathes are used for polishing, metal spinning and wood turning Hand-operated tool.

2. A bench lathe. A small lathe which can be mounted on a work bench. All operation carried on centre lathe can be there with this lathe.

3. A precision lathe. It is a bench-type lathe capable of giving very accurate work.

4. A centre lathe. It is a general purpose lathe.

5. A tool room or tool maker's lathe. Similar in appearance to centre lathe, but is very accurate. Used for precision work on tools, jigs, dies etc.

6. A production lathe. Simple lathe for medium run production work. No lead screw, no compound rest.

7. A single spindle automatic lathe. It is used to produce components in large quantities.

8. A gap bed lathe. It has a bed with a removable section adjacent to the head-stock so that a large work can be swung.

9. Special purpose lathe. It has such characteristics that they are meant for only certain types of work. •

7.4 LATHE OPERATIONS

The operations that can be performed on a lathe are known as turning operations. The word turning usually refers to machining external surfaces while machining internal surfaces is termed boring.

The operations carried out on a lathe can be grouped into the following categories:

(1) Plain turning between the centres
(2) Facing in chucks
(3) Taper turning
(4) Screw cutting
(5) Eccentric turning
(6) Drilling and boring
(7) Knurling
(8) Form turning

7.5 DRILLING MACHINES

Drilling is a manufacturing process by means of which cylindrical holes are produced in the work piece. The cutting tool used for this purpose is called a Drill.

Drills are of two types:

(1) Flat drill and (2) Twist drill.

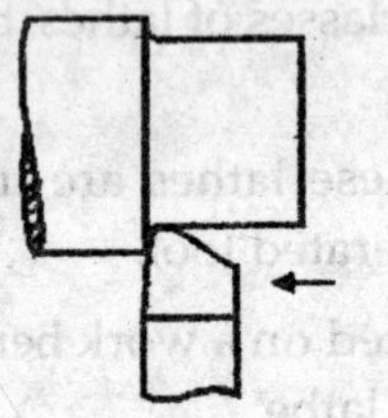

Fig. 7.6. Plain or rough turning.

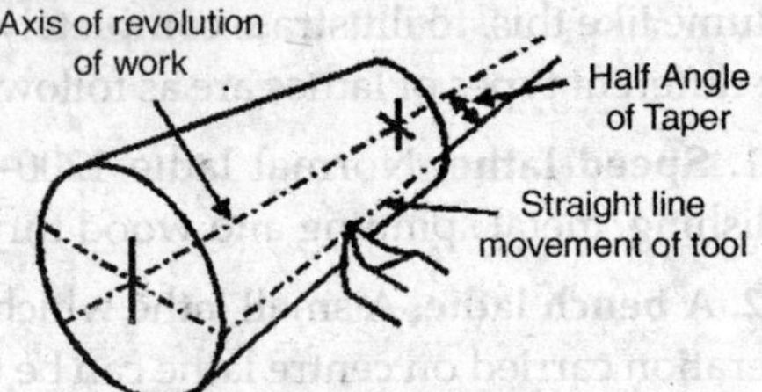

Fig. 7.7. Taper turning.

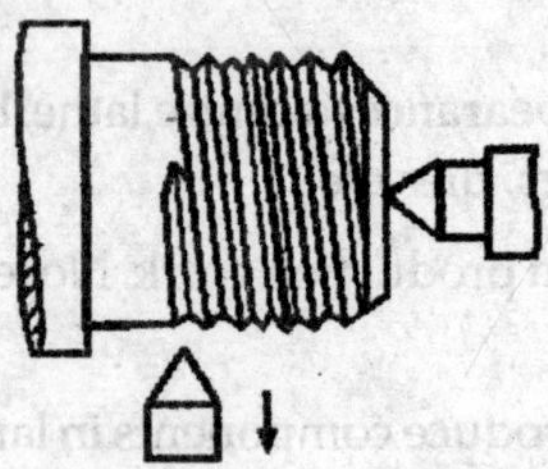

Fig. 7.8. Thread cutting.

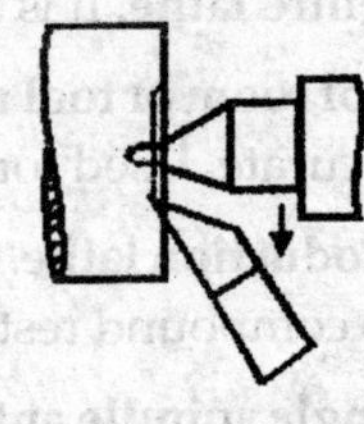

Fig. 7.9. Facing.

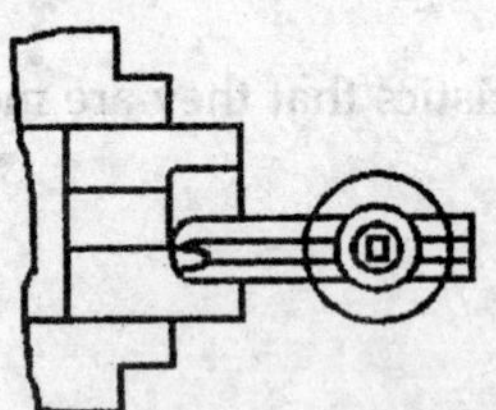

Fig. 7.10. Drililng and boring.

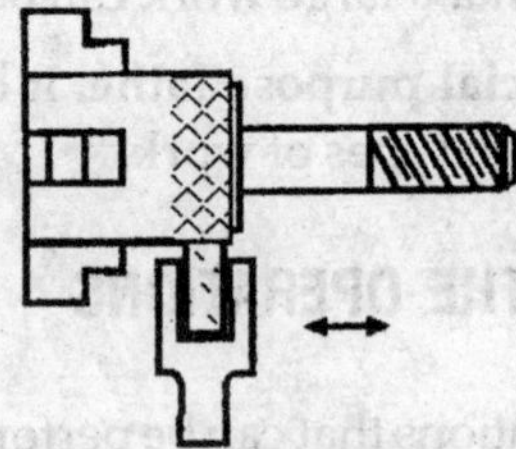

Fig. 7.11. Knurling.

1. Flat Drill. The flat drill, which is the forerunner of the twist drill is shown in Fig. 7.12. This drill is easily forged and ground to any required size, but the results it gives are not comparable with those obtained from a twist drill; this is mainly because the twist-drill point is backed up and kept true by the body of the drill following behind, while the point of the flat drill has no such influence to guide it.

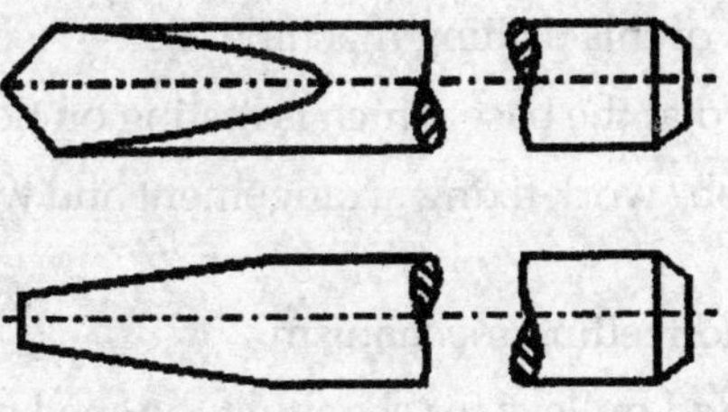

Fig. 7.12

The principal present day use of the flat drill is for boring very long holes, in which process a cutter in the form of a flat drill is carried at the end of a long bar. The arrangement gives the best results when the work rotates, the drill and bar being held stationary and fed into the hole (as in lathe). Generally the bar which carries the flat cutter has tube sunk into it for conveying cutting lubricant to flush away the swarf and avoid choking up of the drill.

2. Twist Drill. The twist drill is the most commonly used variety of drill, and twist drills are specified according to the end by which they are held, called the Shank. Drills may be taper Shank, parallel shank or jobber's drills and are shown in Fig. 7.13.

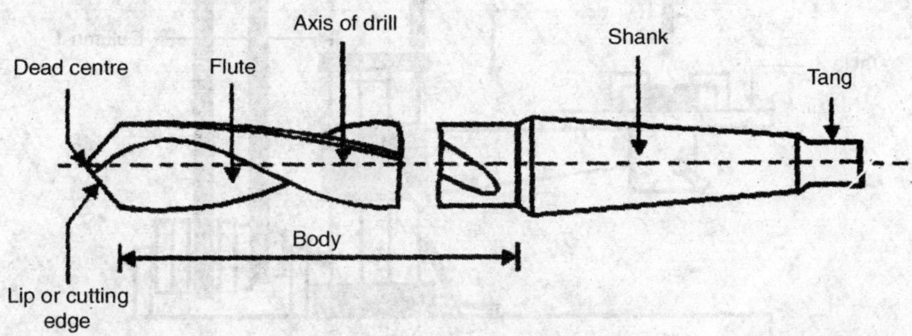

Fig. 7.13. Twist drills.

The drills are held in the machine by special chucks. Drills may be made of either carbon steel or high speed steel.

7.6 DRILLING MACHINES

The three principal types of power machine used for drilling are briefed as below.

(a) The sensitive machine. This is a high form of machine used for holes upto about 1/2" (12 mm) diameter.

(b) The upright or pillar machine. The machines are in sizes upto that which can drill holes up to 2–3". diameter (50–75 mm).

A diagram of a medium upright or pillar drilling machine shown in Fig. 7.14. The main components of this drilling machine are:

(i) A column fixed at the base which is resting on floor.

(ii) A table with job / work fixing arrangement and which can be moved up and down.

(iii) Drill chuck and feeding mechanism.

(iv) A set up belt and pulleys for changing the sped of drill.

(v) Base frame on which machine is mounted.

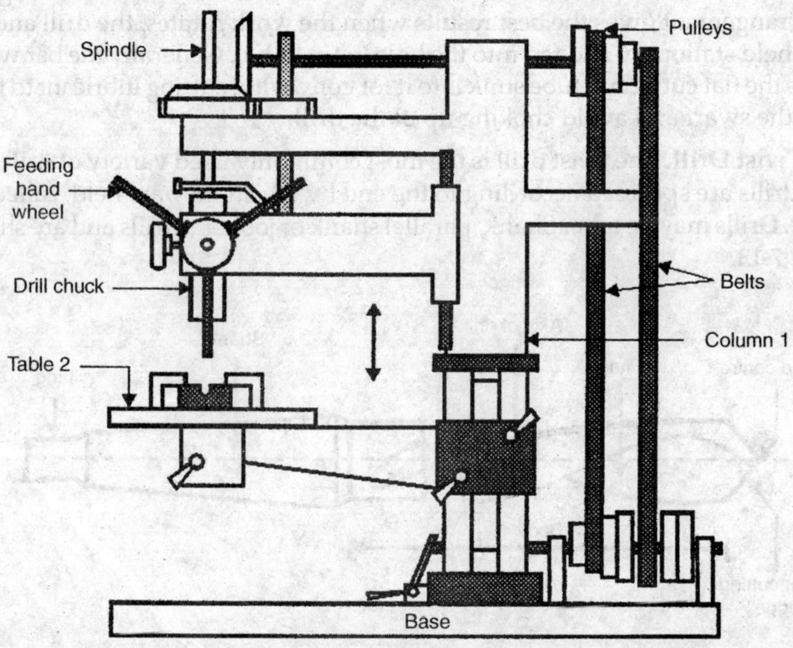

Fig. 7.14. Pillar/upright drilling machine.

Working. Work piece is held on the table by means of fixtures. Table can be moved so that drilling can be possible by rotating the hand wheel connected for feeding the drill.

The speed of the drill can be changed from 470–1440 rpm. This variation in speed can be achieved by providing an electrical pole changing motor.

(c) The radial drilling machine. In this machine the drill head is on an arm and may be swung or moved over the area of the machine table or base.

Radial drilling machine may be of the sensitive or heavy types.

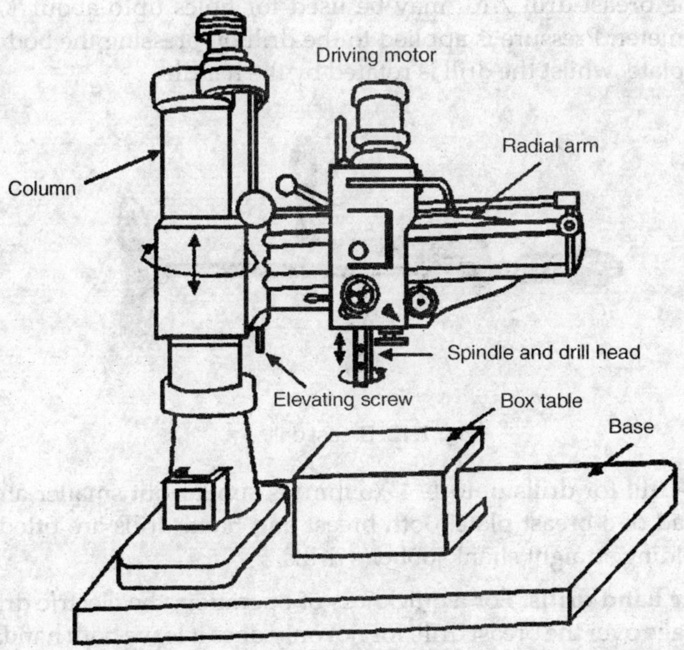

Fig. 7.15. (a) Schematic diagram of a radial drilling machine.

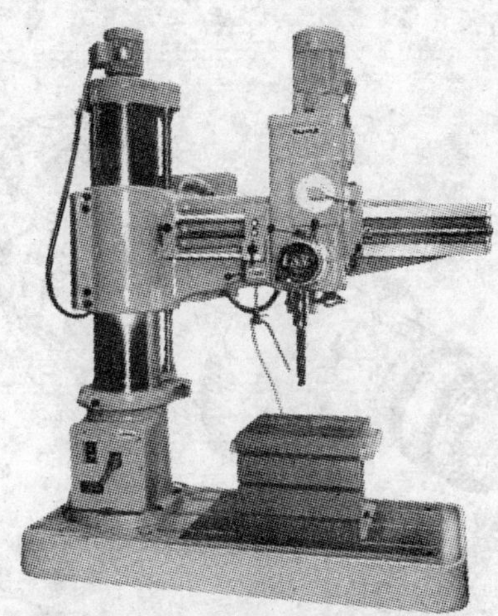

Fig. 7.15. (b) Radial drilling machine.

(d) Hand-drilling machine. The fitter will have occasion to use various hand-drilling appliances for holes where it is not possible to bring the work up to a machine. The breast-drill 7.16 may be used for holes upto about 1/2 inch. (12 mm) diameter. Pressure is applied to the drill by pressing the body on the shaped end plate, whilst the drill is rotated by the handle.

Fig. 7.16. Breast drill.

The hand-drill for drills upto 1/4" (6 mm) is similar but smaller and has a handle instead of a breast plate. Both breast and hand drills are fitted with a chuck for holding straight shank jobber's drills.

(e) Electric hand drills. For a quickness of operation, the electric drill has a great advantage over the breast drill, for not only does it leave both hands free to guide and feed the drill, but also rotates the drill with more uniformity and possesses that extra amount of weight necessary to give improved balance and manipulation. A switch is incorporated convenient for the hand so that the drill may be positioned before it is started-up. Electric drills of the breast type may be obtained to take drills upto about 5/8" (16 mm) in diameter, and an example of one is shown in Fig. 7.17.

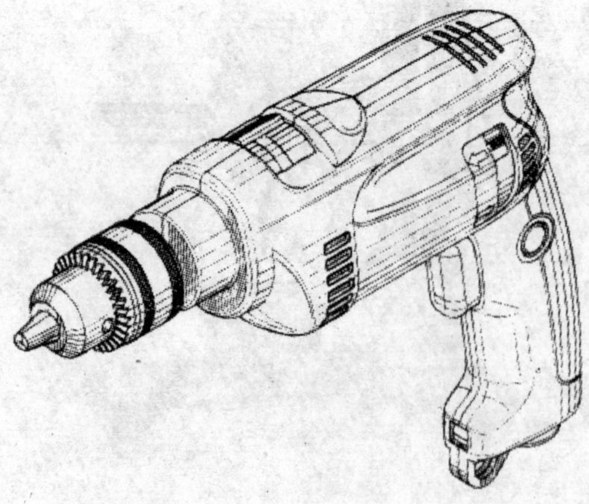

Fig. 7.17. Electric hand drill.

Where extremely high speeds are necessary, small hand drills driven by a compressed air motor may be used.

7.7 DIFFERENT DRILLING OPERATIONS

Other than drilling, following operations can be performed on the drilling machine.

1. Reaming. A drill cannot be relied upon to produce a hole having sufficiently good qualities of finish and accuracy for many purposes, and when accurate holes are required a reamer is a must for finishing to size. The reamer does not originate the hole in the same way as the drill but merely imparts to the previously drilled hole the necessary smoothness, parallelism, roundness and accuracy in size. Reamers may be made of cast steel, case-hardened mild steel. Reamers are shown in Fig. 7.18.

Fig. 7.18. Reamers.

2. Countersinking. Countersunk head screws and wood screw require a 90º chamfer cut round their hole as a seating for the under side of the head. This is cut by means of a countersinking cutter as shown.

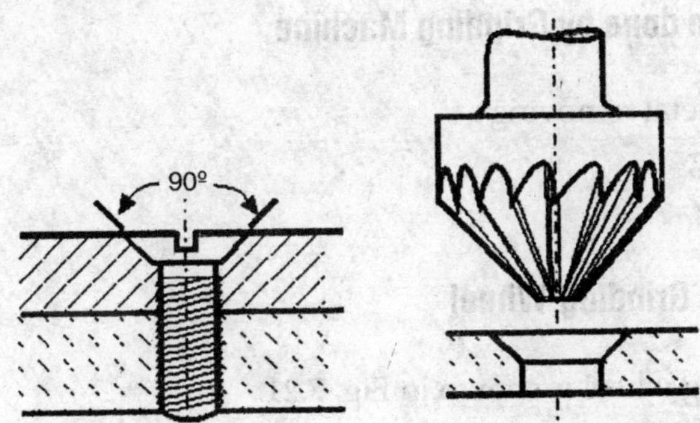

Fig. 7.19. Countersinking.

3. Counter boring. The preparation of holes for certain purposes involves increasing the diameter of the hole for a certain distance down. This is called Counterboring, and is done with a cutter of the type shown in Fig. 7.20.

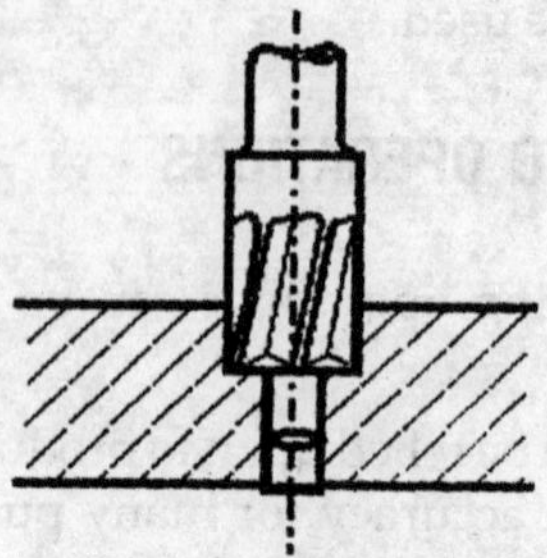

Fig. 7.20. Counter boring.

7.8 GRINDING MACHINE

Introduction. Grinding is a process of metal cutting or metal removal, to obtain better finish by a rotating abrasive wheel called grinding wheel.

It removes very little quantity of metal and we get a very good surface finish.

7.8.1 Applications

1. To sharpen the cutting tools
2. To grind the gears
3. For thread grinding
4. To obtain better surface finish
5. For removing any excess material to get the required dimensions in very close tolerances.

7.8.2 Operation done by Grinding Machine

1. Excess metal removing
2. Finishing
3. Buffering

7.8.3 A Typical Grinding Wheel

A typical grinding wheel is shown in Fig. 7.21.

7.8.4 Types of Grinding

(A) Rough Grinding. When the surface finish and accuracy are not of prime importance, then this method is used. In this the job is pressed against the rotating grinding wheel (or vice-versa), so as to remove excess metal.

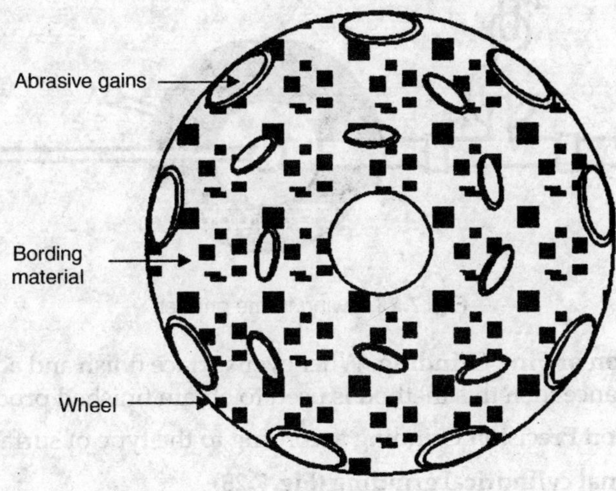

Fig. 7.21

Following are the Rough Grinding machines.

1. Stand Grinder. Figure 7.22 shows a stand grinding machine. In this case on the cast stand motor and grinding wheel are mounted. Grinding wheel will be rotating and the job in pressed as shown to remove excess metal.

2. Portable Grinding Machines. Figure 7.23 shows a portable grinding machine (The figure is self explanatory).

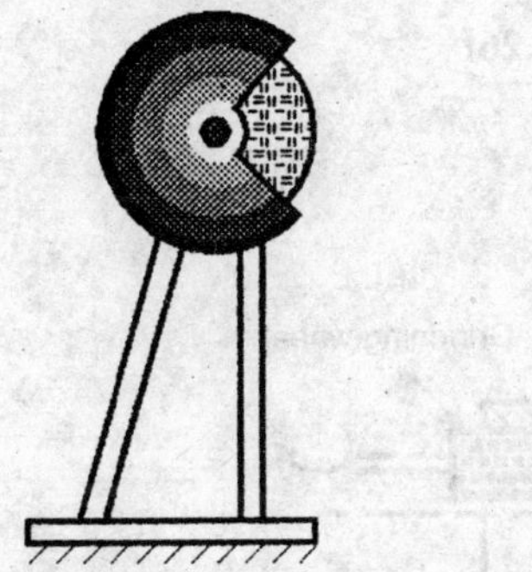

Fig. 7.22. Stand grinder machine.

Fig. 7.23. Potable grinder.

3. Swing Frame Grinders. It has a horizontal arm 'AOB'. The length of the arm will be around 2–3 m. 'O' in the centre of gravity where it is used. It can move within the specified area for grinding purpose. For grinding we have to apply the force at end 'B', so that the rotating grinding wheel is pressed against the job and excess metal is removed.

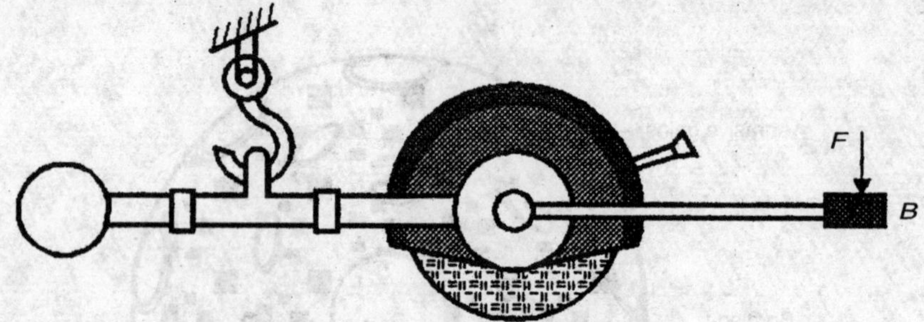

Fig. 7.24. Swing frame grinder.

(B) Precision or Fine Grinding. Whenever surface finish and accuracy are of prime importance then this method is used to obtain finished products.

Classification Precision Grinding according to the type of surface:

(i) External cylindrical grinding (Fig. 7.25)

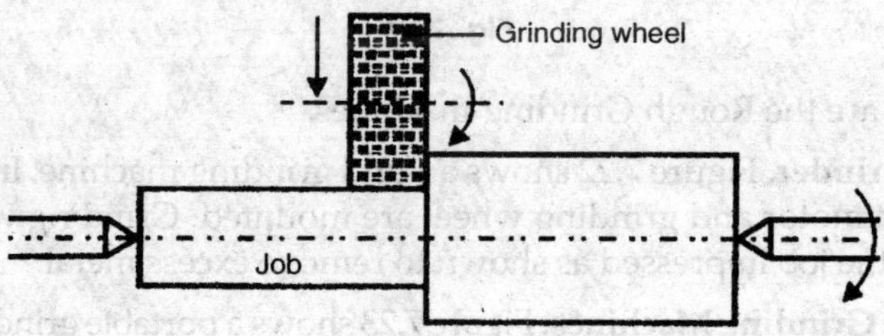

Fig. 7.25

(ii) Internal cylindrical grinding (Fig. 7.26)

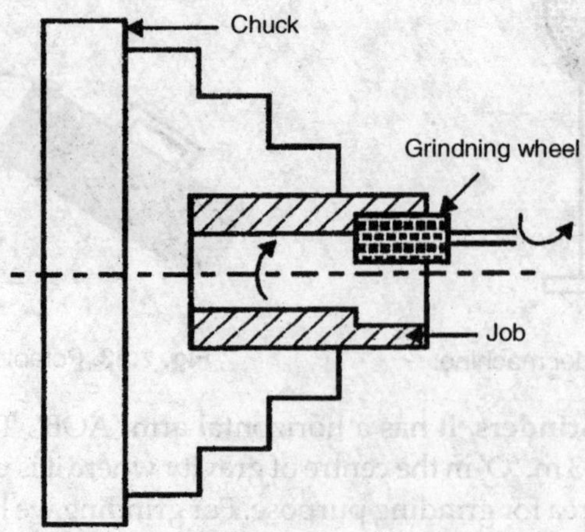

Fig. 7.26

(iii) External taper griding (Fig. 7.27)

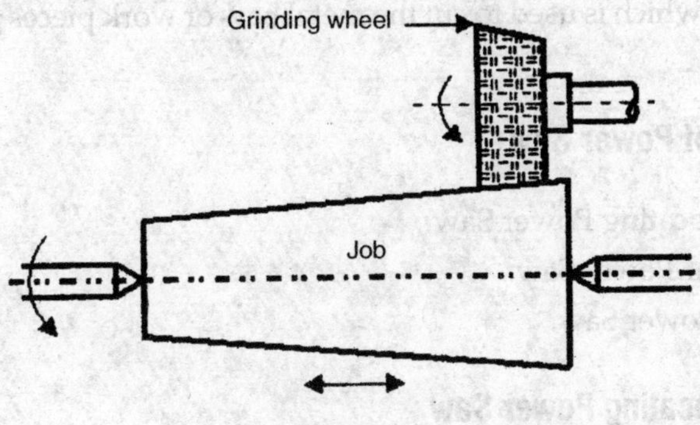

Fig. 7.27

(iv) Surface griding (Fig. 7.28)

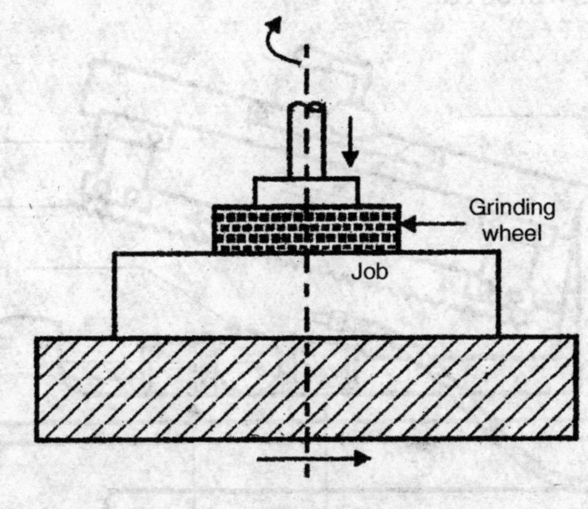

Fig. 7.28

(v) Face grinding (7.29)

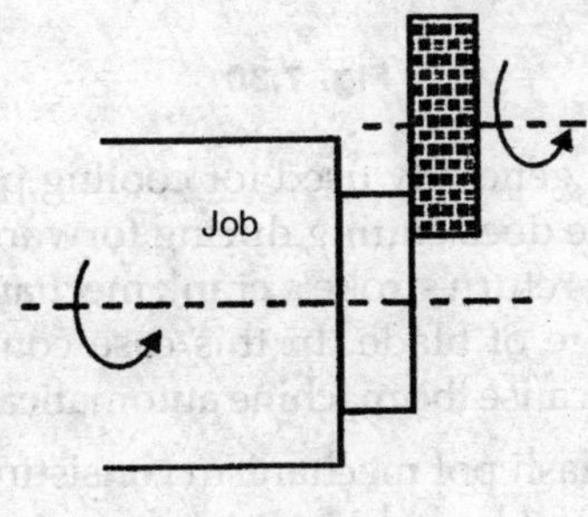

Fig. 7.29

7.9 POWER SAW

It is a machine, which is used to cut the metal bars or work pieces to the required lengths.

7.9.1 Types of Power Saw

(i) Reciprocating Power Saw.
(ii) Circular Power Saw.
(iii) Band Power Saw.

7.9.2 Reciprocating Power Saw

Figure 7.30 shows the horizontal power saw. In this case the work piece will be held by means of the clamps in the correct position. As the ram with blade reciprocates the job will be cut.

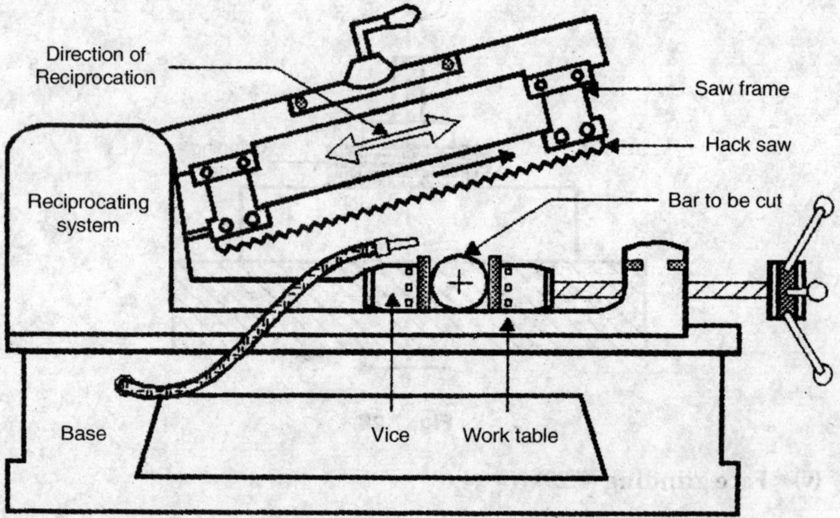

Fig. 7.30

In this case a coolant is generally used for cooling purpose since it increases the blade life. The machine does cutting during forward stroke only and return stroke will be idle. During return stroke a crank mechanism in provided to raise the blade to avoid damage of blade. In this case continuous attention is not required during cutting because the machine automatically shuts off after cutting.

It is to be noted that a dash pot mechanism consisting of piston and cylinder filled with oil is provided on the machine. It produces reciprocating motion of the ram and blade when the power is supplied.

The average thickness of cut of power saw is about 3 mm during each forward stroke.

7.10 INTRODUCTION TO NC AND CNC MACHINE

Introduction. In machine tools say a lathe machine-metal is removed from a job to create required shape-using cutting tools. For this purpose travel of the tool both along the bed of the lathe and the depth of the cut, speed etc. are to be controlled. This is done by skilled worker who controls these variables manually.

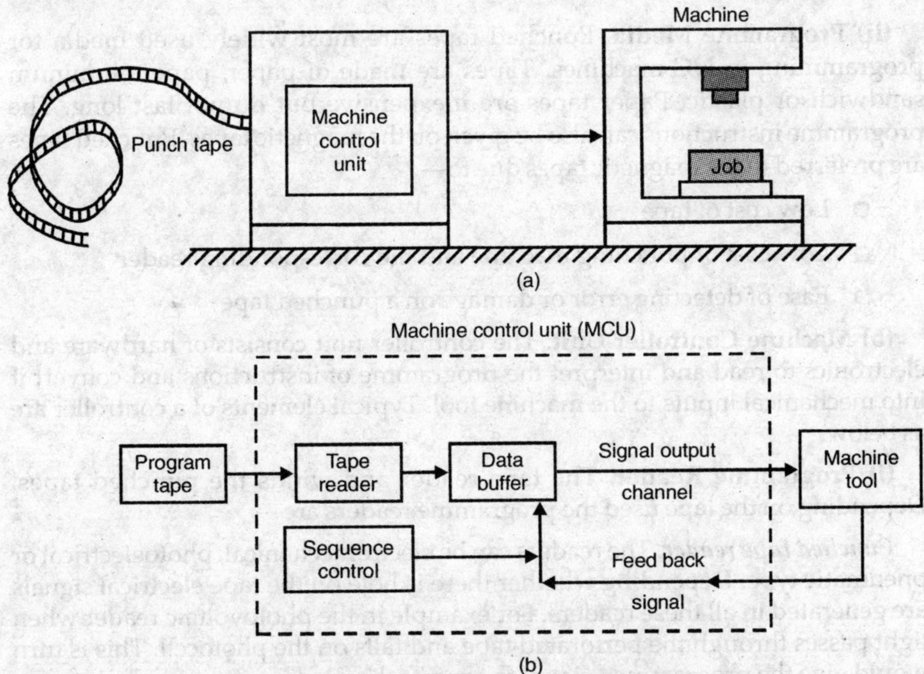

Fig. 7.31. Block diagram of NC machine.

Manual operations have its limitations like accuracy, low production and high rejections. In numerically controlled machines most of the operations can be done automatically. A machine tool which operates automatically or semi-automatically as per coded instructions given to it is called *numerically controlled* machine (NC machine). In NC machines servos motors replace human operators in positioning of the work piece and in positioning/operation of the cutting tools.

For the NC machines first programme to do the required operation has to be written and punched on the tapes. The reader reads the tapes and sends required electrical signals to the controller to execute the job. Block diagram of an NC machine is given below.

7.11 COMPONENTS OF NC MACHINE

Main components of an NC machine are as follows:

(a) Programme of Instruction. (i) It is part programme which contains coded instructions to execute the operation in a cycle of operation. Numbers, letters and symbols logically organized to direct a machine tool to the required job are called NC programme. The instructions are in the form of numerals, letters and symbols.

(ii) Programme Media. Punched tapes are most widely used media for programming in NC machines. Tapes are made of paper, paper-aluminum sandwich or plastic. Paper tapes are inexpensive but do not last long. The programme instructions can also be given on the magnetic tapes. Punched tapes are preferred to the magnetic tapes due to—

- Low cost of tape
- Low cost of punching machine and the corresponding reader
- Ease of detecting error or damage on a punched tape

(b) Machine Controller Unit. The controller unit consists of hardware and electronics to read and interpret the programme of instructions and convert it into mechanical inputs to the machine tool. Typical elements of a controller are as below:

(*i*) Programme Reader. The tape reader and winds the punched tapes. Depending on the tape used the programme readers are—

Punched tape reader. The readers can be electromechanical, photoelectrical or pneumatic type. Depending whether there is hole on the tape electrical signals are generated in all these readers. For example in the photovoltaic reader when light passes through the perforated tape and falls on the photocell. This is turn would give the programmed signal to operate the machine.

Magnetic tape reader. It consists of a magnetic head which reads as well encodes the instructions. For reading propose the tape is moved across the magnetic head. An emf would be induced in the winding which is used for controlling.

(c) Machine Controller. The controller gets signals from the reader and controls various activities of the machine to execute the instructions. The controller consists of control panel. Decoder, feed rate generator, interpolator and auxiliary rate function.

(d) Decoder. It converts the signals from the reader to operate or switch on a particular machine tool function. The decoders have relays to operate the switches.

(e) Buffer Storage. If the tape speed is not sufficient it is necessary to have memory storage device for execution of the programme. Buffer storage is used to temporarily store data received from the decoder.

(f) Interpolator. As the name suggests it is to interpolate the data when needed. Interpolation is required for computing the intermediate points of a curve while machining on NC machines.

7.12 ADVANTAGES OF NC MACHINES

Advantages of NC machine are as below.

1. Increased productivity
2. Fewer rejections
3. Job accuracy
4. Lower tooling cost
5. Easy design changes
6. Less number of jigs and fixtures needed

7.13 DISADVANTAGES OF NC MACHINES

1. Higher initial investment
2. Higher maintenance cost
3. Need for having programmer

7.14 USES OF NC MACHINES

It is possible to have NC and CNC machines from simple single spindle drilling to complex machining centers like:

(a) Metal cutting operations. Turning, drilling, boring, grinding, milling etc.

(b) Production job. Where geometry is complex and need many operations, works where high accuracy is needed.

7.15 ILLUSTRATIVE EXAMPLE OF NC PROGRAMMING

To operate the NC machine tool as programmed apart from the coded instructions—electronic / electrical and mechanical interfacing is required to move and control the cross slide, tool post, control speeds etc. As way of illustration a simple turning job where 2 mm material is required to be removed from a shaft of 50 mm length, then

Step (a) Movement of the tool post through 2 mm along Z-axis

Step (b) Move of the tool by 50 mm in X-direction

Step (c) Move the tool post by 2 mm in Z-axis

Step (d) Move the tool post by 50 mm along X-axis

7.16 CNC MACHINES

Computer Numerically Controlled machines (CNC machines) are extensively used these days for precision machining. There is hardly a facet of manufacturing that is not in some way touched by what these innovative machine tools can do. It incorporates a dedicated computer or microprocessor. In the CNC machines a computer receives information from a tape directly or from a computer storage and controls the machine.

In the NC machines the physical components are connected through hardware whereas in the CNC the physical components are soft wired. There is lot of flexibility in the CNC machines as compared to the NC machines. The block diagram of a CNC and DNC machine along with that of NC machine is given below.

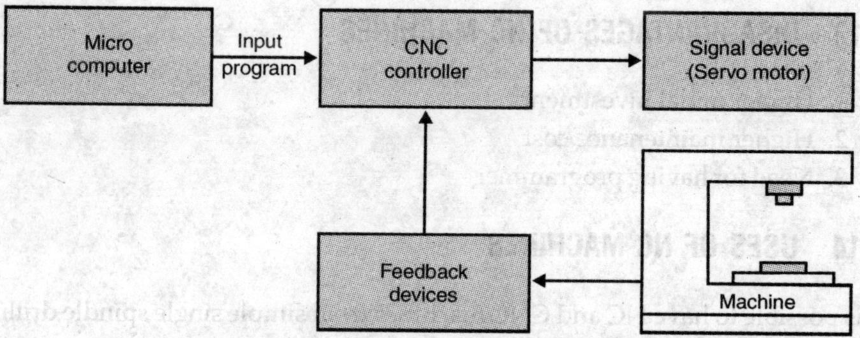

Fig. 7.32. Block diagram of CNC machines.

THEORY QUESTIONS

1. Explain centre lathe and its components.
2. Explain pillar drilling machine with a sketch.
3. Explain different types of grinders.
4. Explain power saw.
5. Explain in detail NC/CNC machine in detail.

CHAPTER

8

Introduction to Metal Joining Process

8.1 WELDING

In olden days, in order to join two pieces of metals, rivetting was done. But nowadays for joining two pieces of metals welding is done.

So, welding is a process of joining two pieces of metals by heating them.

Nowadays welding is used in almost all the industries since it is the easiest and rapid method of joining the metals.

Nowadays methods have also been developed to weld dissimilar metals. One important feature of welding is that the strength of the welded joint is more than 100% when compared to the metals.

8.2 WELDING APPLICATIONS

It is used in joining metals in the following fields:

(a) In fabrication of tanks, vessels, boilers.

(b) In automobile industry.

(c) In structural work, i.e. to fabricate trusses, frames.

(d) In pipe line fabrication

(e) In ship building, motor building i.e. in general, in all the types of joining metals, welding is used.

8.3 REQUIREMENTS OF GOOD WELDED JOINTS

Requirements of good welded joints are described briefly as follows.

(i) For good welded joints, the surfaces should be clean. The cleaning can be done by using wire brush.

(ii) Proper edge preparation should be done as discussed below otherwise the joint may not have required strength. The edges must be prepared depending upon the thickness.

Various types of edge preparations for butt welds (Fig. 8.1)

- Square butt weld
- Single V or single U butt weld
- Double V or Double U butt weld

The square butt weld is used for the sheets having thickness between 1–5 mm.

Single V or Single U butt welds are used for the plates having thickness between 5–15 mm.

Double V or Double U butt welds are used for welding plates above 15 mm thickness.

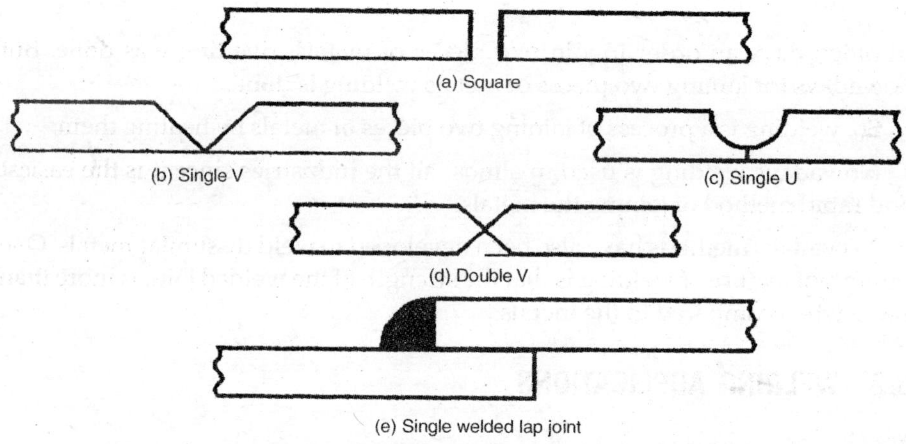

Fig. 8.1

8.4 WELDING CLASSIFICATION

Welding is broadly classified into two types:

(i) Plastic Welding. In this case the two pieces of metals to be joined are heated to plastic state and welding is completed by forcing them. In this case no filler material is added.

(ii) Fusion Welding. In this case the two pieces of metal are heated and brought to molten state and filler material is added and allowed to solidify.

Plastic and fusion welding processes are further classified as under.

(I) Plastic Welding (Under pressure without filler material)

(a) Forge Welding
(Heat is created by Blacksmith's fire)

(b) Resistance welding
(Heat is created by electric current)

- Lap welding
- Butt welding
- Spot welding
- Projection welding
- Seam welding

(c) Thermit welding with pressure
(Heat is created by chemical reaction)

(II) Fusion Welding (No pressure but filler material is added)

(a) Gas welding
(Heat created by gas)

- Oxy – Acetylene welding
- Air – Acetylene welding

(b) Arc welding
(Heat is created by electric arc)

- Metal arc
- Submerged arc
- Tungusten Inert gas welding
- Gas metal arc welding.

(c) Thermit welding without pressure.

Now we will study the welding processes in detail.

(I) (A) Forge Welding

Fig. 8.2

It is the oldest method of joining two pieces of metals. In this case heat is created by blacksmith's fire and the two pieces of metals to be joined are heated and brought to plastic state. Then they are superimposed and hammered together to form the joint.

(I) (B) Resistance Welding. In this case a heavy electric current is passed through the pieces of metals to be joined. Then because of electric resistance, the metals are heated to plastic state and by applying force, welding can be completed.

In this case, no additional filler material is required and the metal pieces are pressed in the two copper electrodes.

(i) Spot Welding. It is used for welding ferrous and non-ferrous sheets upto a thickness of 8 mm. The sheets to be welded are held between the fixed and movable electrodes as shown. Then the electrodes are pressed by pressing the foot lever. When the electrodes are pressed, a current flows and the two parts at

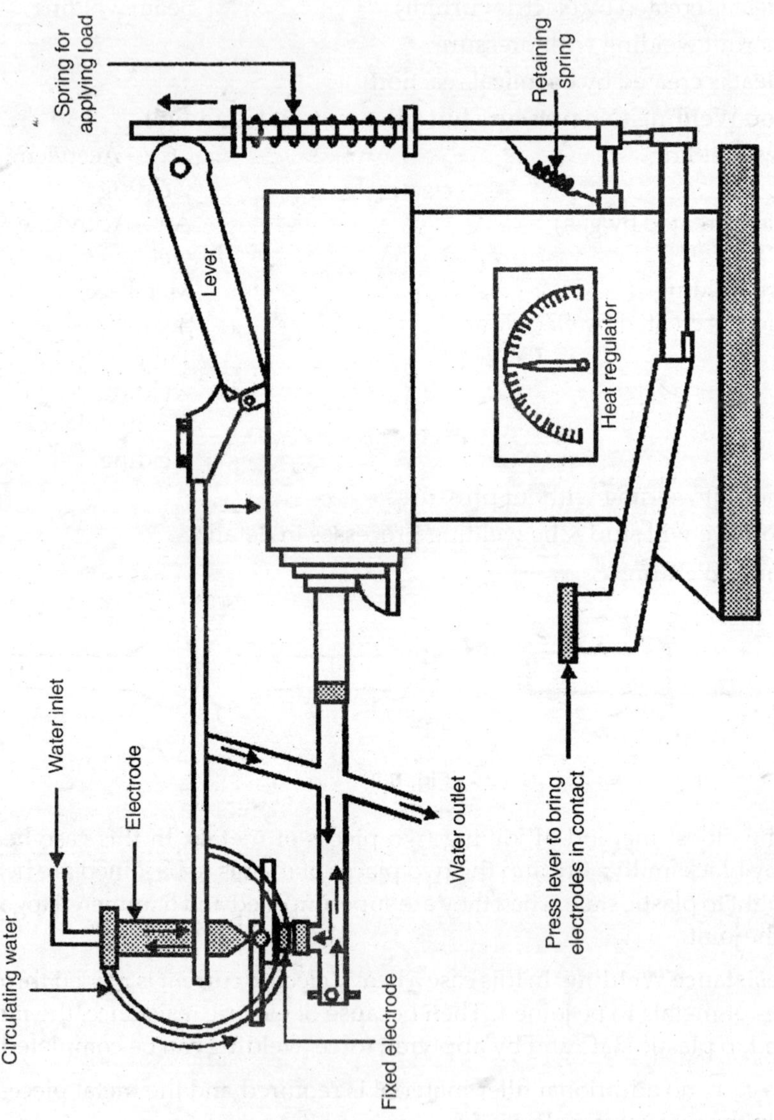

Fig. 8.3. Spot welding machine.

the pressed points are brought to plastic state. Then, this plastic metal mixes, solidifies to form the joint.

(ii) Projection Welding. It is the slight modified form of spot welding process. In this case current and pressure are localised at the welding points by making projections for the upper sheet of metal as shown in Fig. 8.9. The two metal sheets are held in position between fixed arm and upper movable arm. Then the current is passed and welds at all the points of projections are obtained due to flattering of projections.

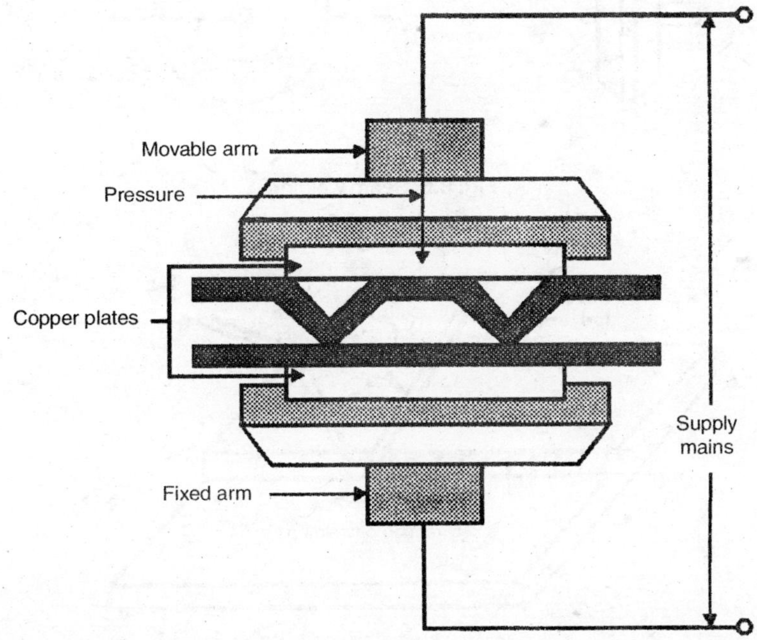

Fig. 8.4

(iii) Seam Welding. For producing continuous welds between two over lapping sheets this seam welding is used. In this case the two sheets to be seam welded are held as shown in Fig. 8.5 between the electrodes. When the electrode wheels rotate continuous weld will be produced.

(II) (a) Gas Welding (Fig. 8.6). In gas welding heat energy required to heat the surfaces is obtained by the combustion of mixture of two gases. Generally oxygen and acetylene are used for gas welding purpose. These gases are mixed in proper proportions in the welding torch. This welding torch is provided with two regulators to regulate the gases.

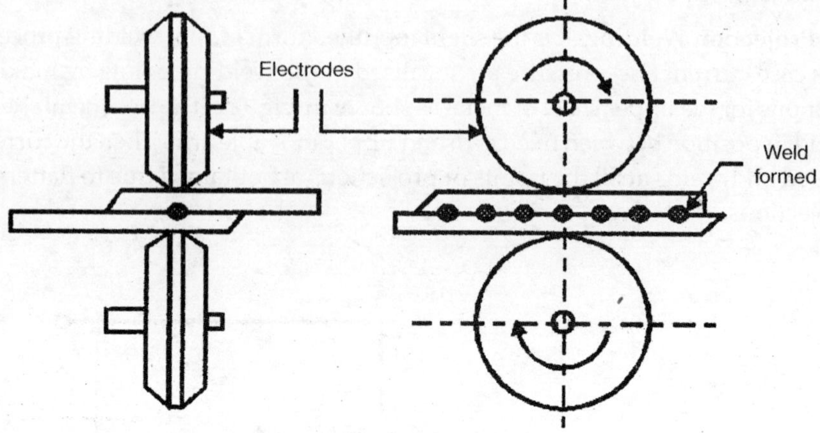

Fig. 8.5. Seam welding.

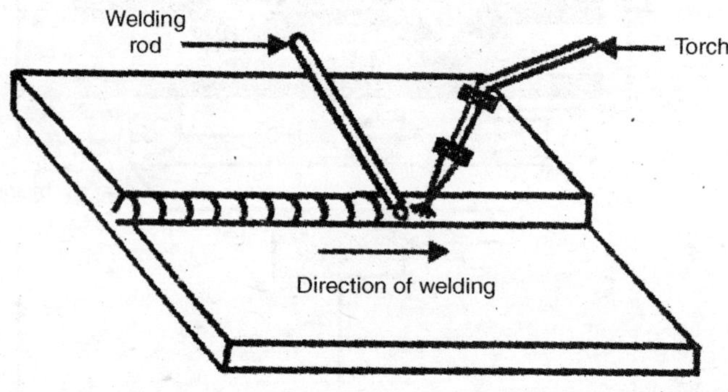

Fig. 8.6

The flame produced due to the combustion is used to heat the metal surfaces to plastic state and welding is completed by adding the filler material as shown.

(b) Oxyacetylene Welding. As the name implies in this case the two gases used are oxygen and acetylene. It is to be noted that oxygen and acetylene are commercially available in cylinders. If required acetylene can be produced by the chemical reaction between calcium carbide and water as under:

$$CaC_2 + 2\,H_2O \longrightarrow \underset{\text{Acetylene}}{C_2H_2} + Ca\,(OH)_2$$

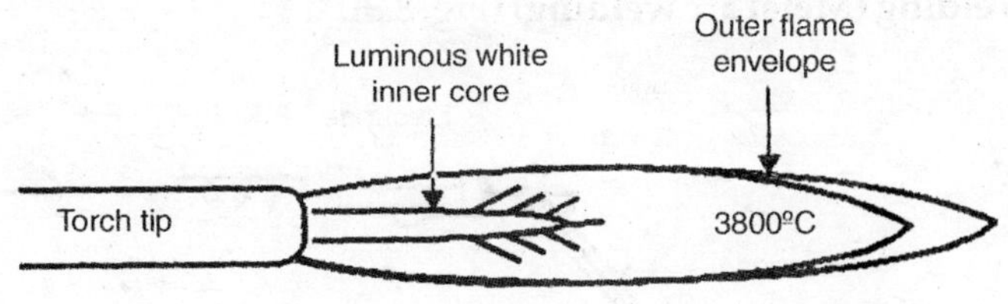

Fig. 8.7

And the chemical reaction for the combustion of acetylene is,

$$\underset{\text{(Acetylene)}}{2\,C_2H_2} + 5O_2 \longrightarrow 4\,CO_2 + 2\,H_2O$$

When the combustion takes place, we get the flame at the tip of the torch as shown in Fig. 8.7.

There are mainly three types of flames:

(i) Natural flame; (ii) Carburising flame; (iii) Oxidising flame.

(i) Neutral Flame. It is obtained by mixing equal quantities of acetylene and oxygen. It is used for welding all the metals like ferrous metals, Cu, and Aluminium alloys.

When we change the proportions of oxygen and acetylene. Carburising and oxidising flames can be obtained.

(ii) Carburising Flame. It is obtained by more quantity of acetylene. It is very much suitable for welding of steel as the rate of welding is faster by this flame.

(iii) Oxidising Flame. It can be obtained by more quantity of oxygen. It is mainly used for welding of brass. It is also very much suitable for cutting operations.

Advantages

(1) Gas welding is more suitable for thin sheets.

(2) The equipment is portable, so suitable for out door repair works.

(3) By changing the nozzle in the torch, the torch can be used for gas cutting.

Limitations

(1) It is a slow process compared to arc welding.

(2) Gases used in the gas welding are more costly.

Note that we will be studying welding torch, and other welding equipments while studying gas cutting.

Air Acetylene Welding. As the name implies in this case air and acetylene are used as the gases to produce the flame. In this case temperatures produced are low compared to other gas welding processes. This method is generally used for Lead welding.

(b) (i) Arc Welding (Metal arc welding) (Fig. 8.8)

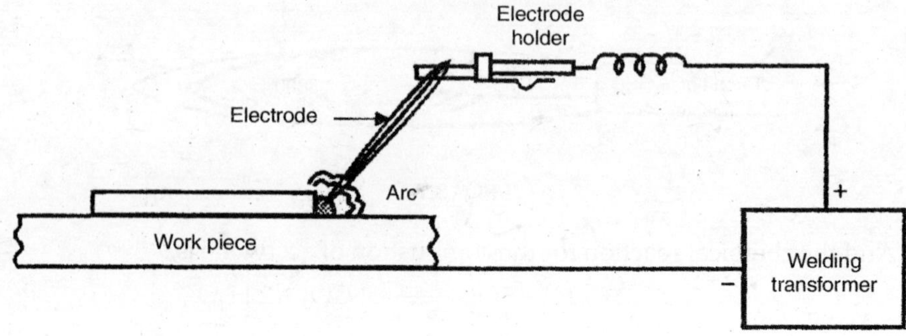

Fig. 8.8

Figure 8.8 shows an arc welding circuit. It mainly consists welding rectifier, electrode (or welding rod) holder, welding rod, and work piece.

When the welding rod touches the work piece, an arc is produced and tremendous amount of heat is liberated. The temperature of the arc is about 3600° C. This heat energy is utilised for melting work piece and welding rod. So, a small pool of molten metal is formed. This molten metal is agitated by the action of arc and the metal is perfectly mixed and after cooling it produces a sound joint.

Note:

(i) Arc initiation voltage is 60–100 V.

(ii) Arc maintenance voltage is 25–45 V.

(iii) Power source may be a.c. or d.c.

(iv) The gap between welding rod and work piece is to be 3 mm.

(v) Molten metal in the pool when exposed to air forms oxides and nitrides in the steel, when it reacts with oxygen and nitrogen in air. This weakens the welded joint, also it reduces resistance to corrosion. So in order to avoid this, the welding rods are provided with flux coating. This flux after welding forms a slag covering on the welded joint and prevents from oxidation. This slag has to be chipped off afterwards.

Application. This arc welding is most commonly used in the fabrication of tanks, vessels, trusses, frames, boilers, automobile chassis and body buildings etc.

II (b) (ii) Submerged Arc Welding (SAW). It is the improved arc welding process, which is used for the production of butt welds of thick steel plates.

In this case the arc produced is submerged (covered) in the flux, hence the name submerged arc welding.

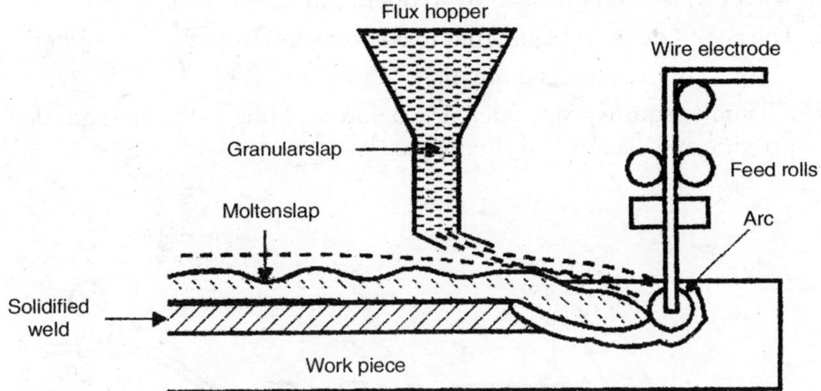

Fig. 8.9

As shown in Fig. 8.9 the two plates to be welded are placed in position on a back up plate. Through the flux hopper, flux is supplied and wire feeder feeds the bare wire continuously.

In this case arc produced is submerged in the flux. Because of heat of arc the wire and parent metals melt and form a pool of molten metal. Thus molten pool produces a welded joint after cooling.

II (b) (iii) Gas Tungsten Arc Welding (GTAW). It is also known as Tungsten Inert Gas welding (TIG).

It is a fast process, produces clean welds and it can weld metals considered to be impossible to weld.

TIG welding utilises a non-consumable electrode in a special holder, a separate filler material and an inert gas i.e. argon gas cylinder, power supply source as shown in Fig. 8.10.

When the tungsten electrode strikes the work piece, an arc will be produced. Around the arc inert gas shielding is formed as the gas is coming out from the torch. Because of heat of arc, work piece and filler material melt and form a molten pool. This molten pool after cooling forms a sound welded joint in the shielding of inert gas.

Advantages of TIG Welding

1. Welded joints are stronger more ductile and corrosion resistant than the other weld made by other methods.
2. The welding of non-ferrous metals is simplified as no flux is required.
3. Dissimilar metals can also be welded easily.
4. Due to presence of inert gas, there is less smoke.
5. The arc is transparent due to shielding inert gas, so the welder gas can clearly observe the weld as it is being made.

Limitations of TIG Welding

1. The process is relatively slow in operation.
2. The back side of weld joint has to be protected from the atmosphere.
3. The cost of inert gas is quite high.
4. All joints require proper cleaning before welding as the inert gas does not provide any cleaning or fluxing action.

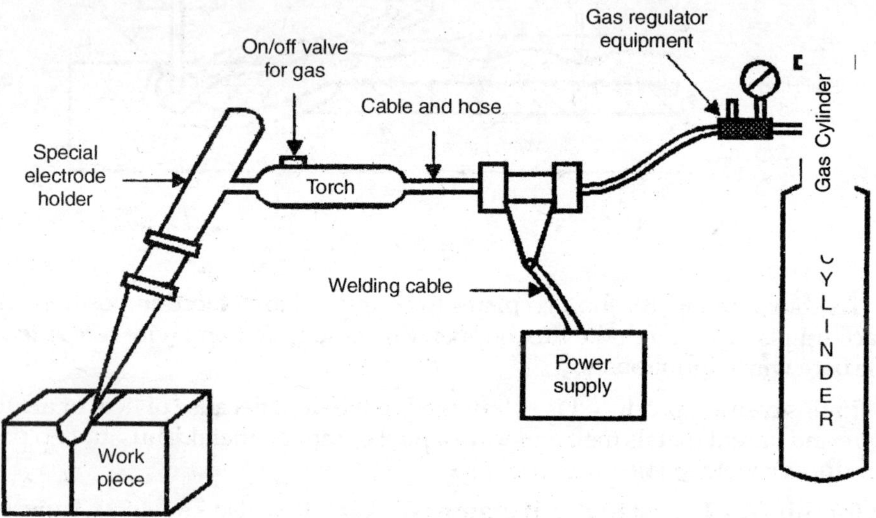

Fig. 8.10. TIG welding.

II (b) (iv) Gas Metal Arc Welding (GMAW). It is also known as Metal Inert Gas (MIG) welding. In MIG welding we get clean and good welds also fast filler metal deposition rates. It uses high welding current which is used to break the globules of molten metal into fine spray.

Figure 8.11 shows schematic diagram of MIG welding. MIG uses a consumable electrode. It is supplied through an electrode holder into the arc. Then at the same speed electrode is melted and deposited into the weld. A small motor with adjustable speed will be used to remove the wire from the reel and fed into the arc.

Generally CO_2 or Argon are used as shielding gases. Mainly the process was developed for welding Aluminium and Titanium. But nowadays it has wide application since it can be used for welding in all the positions; and less skilled operators are required to operate this set up.

II (c) Thermit Welding. This welding process is used in the repair of heavy parts such as spokes of driving wheels, broken connecting rods, other motor parts etc.

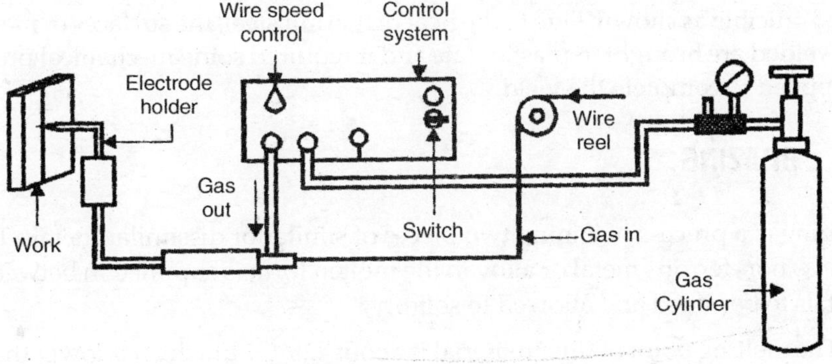

Fig. 8.11. MIG welding set up.

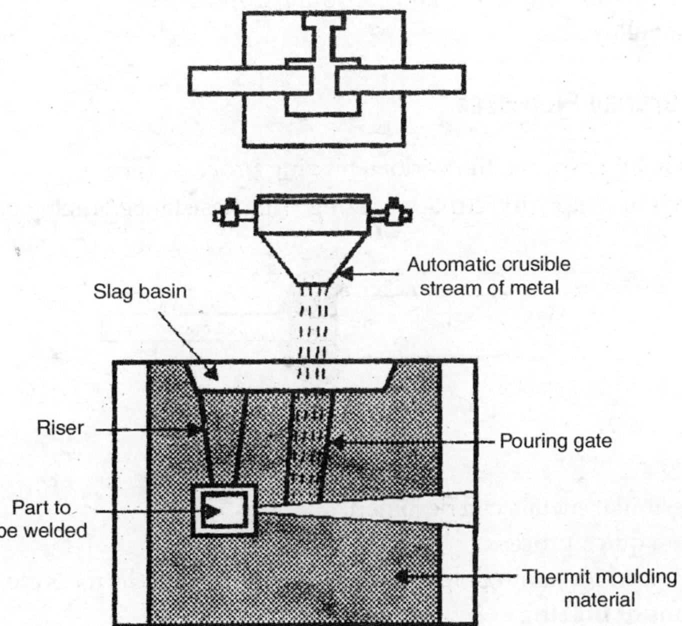

Fig. 8.12. Thermit welding.

Thermit welding is based on the following exothermic chemical reaction.

$$8\ Al + 3\ Fe_3\ O_4 \longrightarrow 9\ Fe + 4Al_2O_3$$

So, oxygen from iron oxide unites with aluminium and forms aluminum oxide or slag and a superheated thermit steel will be formed. The temperature of thermit will be around 2700°C which is almost double when compared to the melting point of steel.

In thermit welding process, the parts which are to be welded are placed in the mould as shown. Then the superheated thermit steel is poured from the refractory lined crucible as shown. Due to the heat of thermit steel, the surfaces of metals to be welded are brought to plastic state and if required some mechanical pressure is applied to complete the weld.

8.5 BRAZING

Brazing is a process of joining two pieces of similar or dissimilar metals. In this case as non-ferrous metal or alloy in the molten form is supplied in between the metals to be joined and allowed to solidify.

The melting point of filler material is about 425°C but which is lower than the melting point of parent metals. During brazing no forging action is present and also the parent metal parts do not melt.

The various brazing metals and alloys are, Copper, Brass, Bronze, Silver alloys, Aluminum alloys etc.

Types of Brazing Processes

Based on heating source the various brazing processes are—

(i) Torch brazing; (ii) Furnace brazing; (iii) Resistance brazing etc.

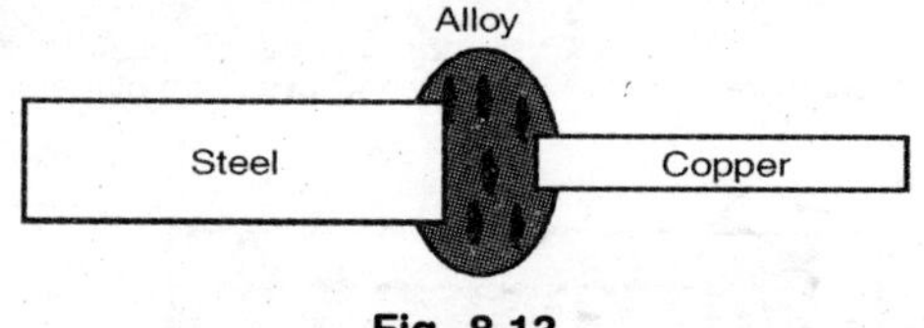

Fig. 8.13

Advantages

(i) Dissimilar metals can be joined.
(ii) It is a quick process.
(iii) It is used for joining pipe fittings, carbide tips on the tools etc.

Limitations of Brazing

1. Brazing cannot be performed on hardened steel.
2. Aluminum brazing needs a special technique and is very expensive.

8.6 SOLDERING

Soldering is a process of joining two pieces of metals. In this case a molten fusible metal called solder is supplied and allowed to solidify.

The melting point of solder is below 420° C. Generally Lead and Tin alloys are used as solder materials. Some of the solders are:

(i) **Soft Solder.** Lead 37% and tin 63%.
(ii) **Medium Solder.** Lead 50% and tin 50%.
(iii) **Electricians Solder.** Lead 58 % and tin 42 %.
(iv) **Plumbers Solder.** Lead 70% tin 30%.

As one might have seen with the radio repairers that, when the current is supplied to the solder gun, the front end of it becomes hot. This hot end when brought in contact with solder material, it melts. This molten material is used for joining wires in radios, taperecorders etc. The molten metal, when cools, produces solder joints.

Types of Soldering Processes

(i) **Soft Soldering.** It is used in sheet metal work and it has less strength.
(ii) **Hard Soldering.** It is employed when stronger joint is required.

Based on the method of heating source, the soldering processes may be, resistance soldering, torch soldering, furnace soldering etc.

Application. It is mainly used in joining wires in radios, taperecorders, printed circuit boards etc.

Advantages

1. It produces liquid and gas-tight joints quickly.
2. The cost of the joint is quite low.
3. The temperature of soldering is low.
4. The equipment used for soldering is simple, cheap and easy to handle.
5. It is the most economical method.

Disadvantages

The joints are not strong enough to withstand the constant jerks and pulls, which is the main drawback of the process.

8.7 GAS CUTTING PROCESS AND EQUIPMENTS OR (OXYACETYLENE GAS CUTTING PROCESS)

As we have studied in gas welding, heat energy required to heat the surfaces is obtained by the combustion of oxygen and acetylene. These gases are mixed in proper proportions in the torch which is provided with regulators. When the gases are burnt a flame is produced.

The flame so produced is used to heat the steel to red hot colour. Then a jet of pure oxygen is blown on the red hot surface. The steel is burnt i.e. iron oxide (slag) is produced, which under pressure falls down and the steel is cut.

Following are the equipments used in Gas welding and gas cutting—

1. **Gas torch or welding torch.** In this gas torch proper mixing of oxygen and acetylene takes place. The torch is provided with regulators to regulate the mixture of gases. When the combustion takes place flame will be produced at the tip.

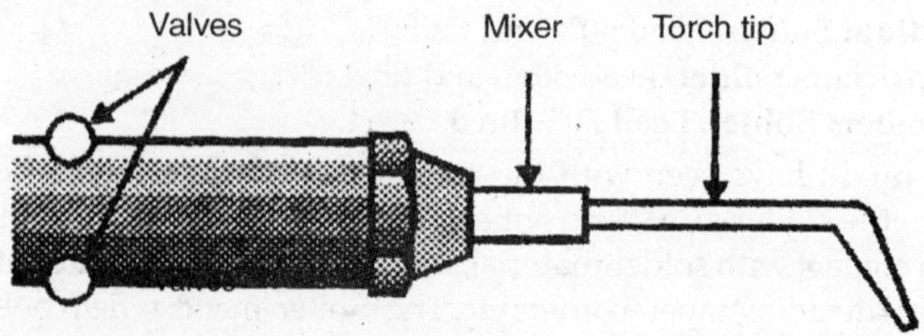

Fig. 8.14

2. **Torch tip.** It is that part of the torch through which gases pass just before their ignition and burning.
3. **Pressure regulator.** The function of a pressure regulator is to reduce the cylinder pressure to the required working pressure and also to produce steady flow of gas.
4. **Hose and hose fittings.** Hose pipes are used to convey gases from the cylinder to the torch They should be strong and durable.
5. **Goggles.** Goggles fitted with coloured lenses are provided to protect the eyes of the operator from harmful heat.
6. **Gloves.** These are used to protect the hands of the operator.
7. **Spark lighters.** These are used for lighting the torch.
8. **Oxygen and acetylene cylinders.** These are commercially available in the market or acetylene gas can be prepared as discussed earlier.

THEORY QUESTIONS

1. What are the kinds of joints, that are normally employed for welding purposes? Give their sketches.
2. How the welding is classified? Explain them briefly.
3. Discuss the method of electric resistance welding. What are its advantages and disadvantages?
4. What do you understand by gas welding?
5. Explain with the help of a neat sketch, the principles of carbon arc and metal arc weldings.
6. Describe the following:
   (1) TIG Welding (2) MIG Welding
7. State the advantages of soldering and brazing.
8. Describe the various processes used in cutting of metals.
9. Write a short note on Gas cutting tools.

❑❑❑

Unit 5

Sheet Metal Working and Design Considerations

Syllabus

Introduction to Sheet Metal Working : Sheet metal forming processes drawing and bending.

Sheet metal cutting processes, Blanking, Piercing and Notching, Tolerance, Limits, Fits.

Design Consideration : Need of design, stress, strain, modes of failure, factor of safety, materials aesthetic and ergonomic considerations.

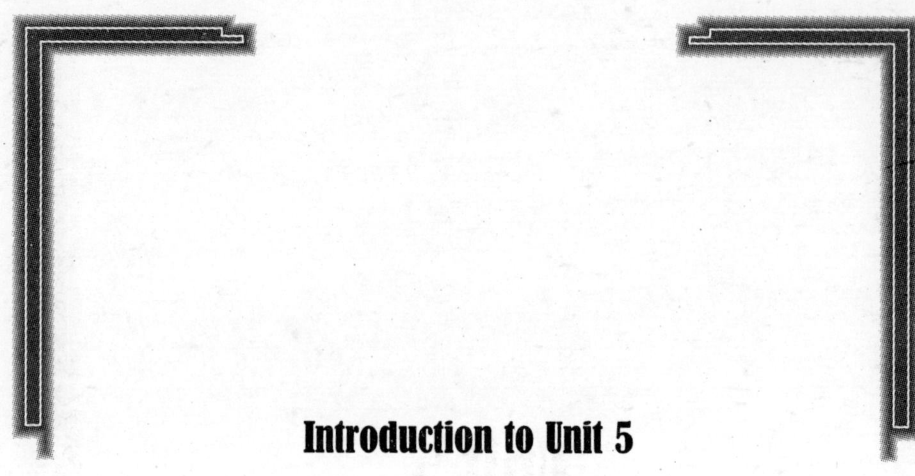

Introduction to Unit 5

In the first part of this unit specific manufacturing methods used in sheet metal working have been covered.

Lot many items like household utensils, packages, buckets, funnels, car bodies, aeroplanes are made out of sheet metals. Various manufacturing aspects used in sheet metal have been covered.

In addition, introduction to limits, tolerance and fits which are so important is mass manufacture have also been covered. For mass manufacture the parts must be interchangeable where concept of tolerances and limits is essential.

The design is the starting point of any engineering concept. Understanding basic concepts and need for design are therefore essential. Knowledge of stress, strain and failure is equally important for proper design. These days the concept and aesthetics, ergonomics is essential. All there aspects have been covered in the later part of this section.

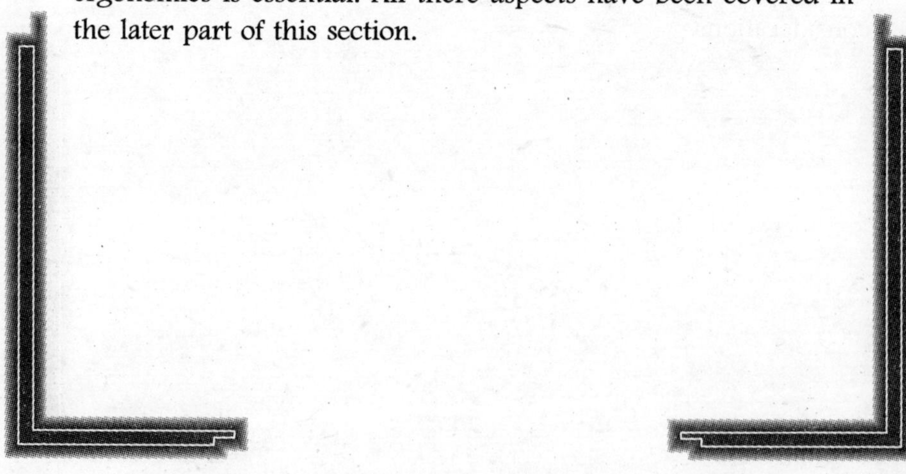

CHAPTER 9

Introduction to Sheet Metal Working

9.1 SHEET METAL

Generally sheet metals have thickness less than 5 mm and are produced by cold or hot rolling processes. The sheet metals can be in various materials like—

(i) **Base metals** – black iron, mild steel, Aluminum, copper, lead

(ii) **Coated metals** – like Galvanized iron, (GI), tin sheets

(iii) **Alloys** – Stainless steel, brass etc.

Sheet metals are available in rectangular sizes varying from 1 to 1.5 m in width and 2 to 3 m in length. The sheets are also available in rolls. The thickness of the sheets is given in gauge numbers. These are given in the table at the end of the chapter.

Most of the sheet metal in use is in mild steel and comes in the form of sheets or strips. The steel may be annealed, soft, half hard, extra spring hard etc. The soft steels are used for deep drawing while harder varieties can be used for blanking.

9.2 USES OF SHEET METALS

Sheet metal parts are used extensively for items such as—

(a) Household utensils, covers, pipes, funnels, boxes, bends etc. for day-to-day requirements.

(b) Automobiles-bicycle, cars, scooters, motorcycles and trucks parts.

(c) Aircraft.

(d) Railway locomotives and coaches.

(e) Farm and construction equipment.

(f) Household appliances.

(g) Office furniture.
(h) Computers and office equipment.
(i) Packaging like containers, boxes.
(j) Corrugated sheets for roofs.

9.3 ADVANTAGES OF SHEET METAL PARTS

The components made out of sheet metals have the following advantages.

(a) High strength
(b) Good dimensional accuracy
(c) Good surface finish
(d) Relatively low cost of manufacture
(e) Economical mass production operations

9.4 METAL WORKING

The required items or components can be made from the sheet metals by hand tools or by press work. But most of the sheet metal work is done on the presses.

9.5 PRESS WORK

It is method of forming sheet metals and plates by cold working into required shapes by applying large force through press tools. The presses may be manually or power operated.

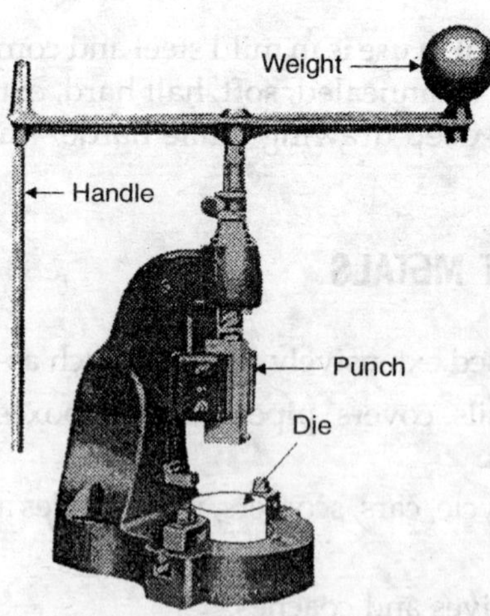

Fig. 9.1. Hand operated fly press.

(a) Hand Operated Press. It is small press which is operated by hand. As the arm of the press is rotated, the ram moves up or down and the stored energy is transferred in doing work on the sheet metal. Hand presses are used for small jobs. (Fig. 9.1)

(b) Power Press. For large jobs and where high production is required hand operated presses are not suitable and power presses are used. There can be many types of power presses on the basis of power used–electric motor driven, pneumatic or hydraulic press etc. or on the type of jobs to be done like—shearing press, coining or punching press etc.

(c) Die and Punch. The sheet metal work on the press would require die and punch depending on the type of job. Die and punch for making a circular disc is shown in the Fig. 9.2. The die and punch are made of materials harder than that of the sheet metals.

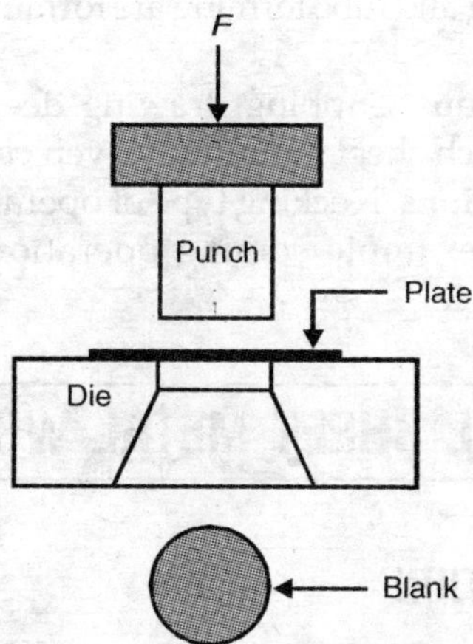

Fig. 9.2. Die and punch for disc manufacture.

(d) Hand Tools. For sheet metal working following hand tools are used—

(i) *Measuring tools*—Like steel rule, folding rules, flexible, push-pull rule, vernier caliper, micrometer, thickness gauge etc. are used to measure the parts during manufacture.

(ii) *Straight Edge*—It is used to scribe long straight lines.

(iii) *Scriber*—It is a long thin rod with pointed end and handle at the other side. It is used to mark lines on the sheet metal.

(iv) *Divider*—Used to draw small circles.

(v) *Trammel Compass*—It is used to draw large circles on sheets.

(vi) *Punch*—To locate centre or to make mar.

(vii) *Hammer*—Used for forming.

(viii) *Mallets*—These are soft hammers with rubber, raw hide, soft metals etc.

(ix) Miscellaneous tools like chisels, scissors etc.

9.6 SHEET METAL WORKING

Most of the sheet metal work is done on presses where a die and punch or other formed tools are required. In press work large force is applied on thin sheet metals to give the required shape or to cut it in to the desired shape. Press work is highly economical method of manufacturing. Various manufacturing methods on sheet metals can be classified as follows.

(a) **Cutting and Shearing.** Here the sheets are cut by using various tools. Blanking, piercing, perforation, notching etc. are various types of shearing operations.

(b) **Bending Operation.** The sheet metals are straining around a straight axis.

(c) **Forming.** Flanging and tube forming are forming operations to give shape to the sheets.

(d) **Drawing Operations.** Cupping, drawing, deep drawing are all drawing operations in which sheet metals are given cup or shell type shapes.

(e) **Reducing Operations.** Necking type of operations.

(f) **Squeezing.** For example coining operation is done in closed dies applying large force.

TYPICAL SHEET METAL WORKING

9.7 DRAWING OPERATION

Drawing operations are described as follows.

(a) It is a forming process by pushing a punch against a flat sheet and forcing it into a die to take the required shape. In other words drawing is the operation of producing thin walled, hollow shaped parts from sheet metal. Both the die and the punch have shape of the part to be manufactured as shown in Fig. 9.3. Classifications of drawing operations could be cupping, redrawing and deep drawing. Components like cups, shells, household utensils etc. can be made by drawing.

(b) In the drawing operation soft material are used since the material is required to undergo large permanent deformation to acquire the intended shapes in the presses. Drawing operations could be for cup making or deep drawing.

(c) For making cup shapes on the press first a circular blank of proper size is taken. The die is usually open from the bottom side and has an ejector system placed below a pad for the removal of the component after pressing it into shape. The pressure pad ensures smooth flow of the blank over the die / punch.

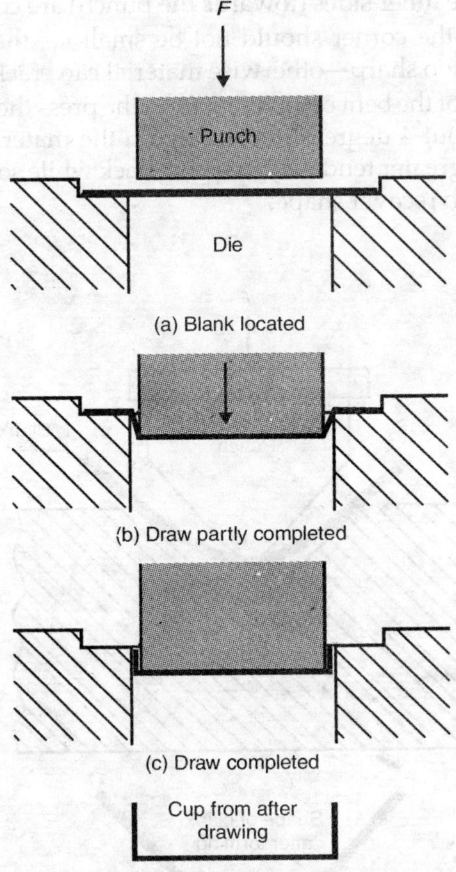

Fig. 9.3. Drawing operation.

(d) The ratio of the diameter of the blank (D) and that of the cup (d) is called draw ratio (D/d). If draw ratio is greater than about 1.8 then the cup cannot be drawn in one draw. Such items may require redrawing in several steps.

(e) In drawing operations lubricants are extensively used.

9.8 BENDING

It is a forming operation performed on a press to bend a sheet metal or a strip through required angle. Figure 9.4 showing bending operation. Salient features of bending are,

(a) It requires a press and tooling to do the bending operation.

(b) The sheet metal strip is permanently deformed to the required shape.

(c) When the sheet metal is pressed for bending in the die the outer portion is stretched while inner sides (towards the punch) are compressed.

(d) The radius of the corner should not be small i.e. the angle at the bent should not be too sharp—otherwise material can crack at the bend.

(e) After removal of the bent component from the press there is usually spring back up to about 4 degrees depending on the material hardness. Hard material have greater tendency to spring back while softer materials have less tendency to recover shape.

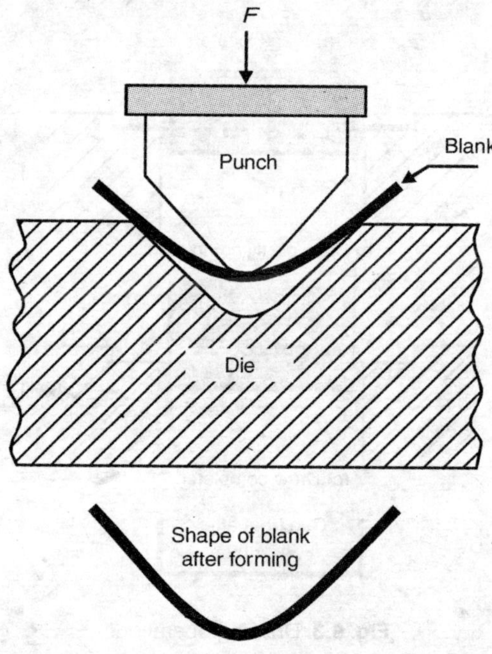

Fig. 9.4. Bending operation.

9.9 BLANKING

Operation of cutting an object is called blanking. Blanking is discussed as following.

(a) Blanking is an operation of cutting an object of given shape from sheet metal strip. It may or may not be necessary to perform further operations on the blank. Piece detached from the metal strip is called blank. Examples are discs, washers.

(b) Blanking operations are extensively used in sheet metal working to make components like washers, discs etc. If needed other operations may be combined with the blanking operation. Blanking is usually the first operation in sheet metal.

(c) A blanking operation to make disc from a steel strip would require—

  (i) A press

  (ii) A punch and die with clearance at the lower side so that the disc can fall down freely. (Ref. Fig. 9.2)

  (iii) The material for blanking operation is usually hard.

9.10 PIERCING

It is a distinct process of making a hole in sheet metals. A pierced hole is shown in Fig. 9.3. It is characterised by

(a) The punch is sharp and pointed tool which is able to penetrate or pierce through the sheet metal.

(b) There is no scrap from the hole.

(c) The hole has rough flanges around the hole.

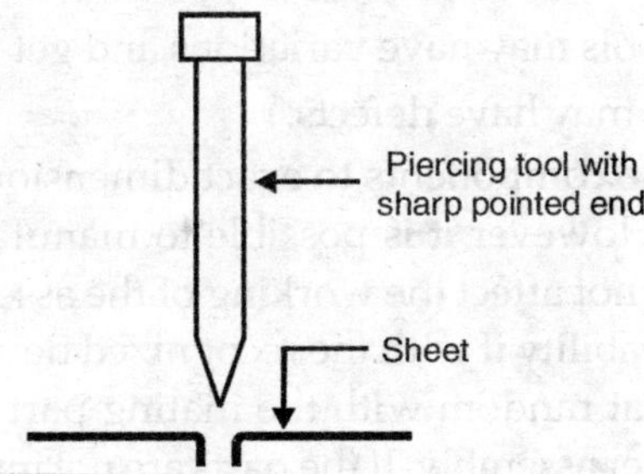

Fig. 9.5. Pierced hole.

9.11 NOTCHING

Notching is removing small quantity of material from the edges of a sheet metal part. Notching may be done to avoid overlapping of material after bending at the seams. Proper notching would give the sheet metal joints a better fit.Various types of notching is shown in Fig. 9.6.

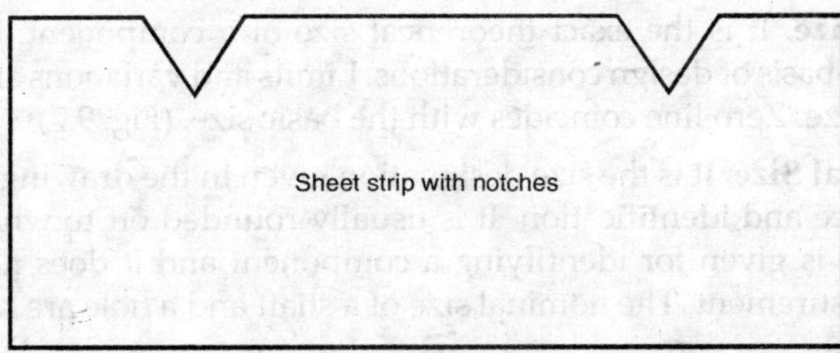

Fig. 9.6. Notches.

LIMITS, TOLERANCE AND FITS

9.12 INTERCHANGEABILITY

Interchangeability is described as following.

Modern industry has been developed on the basis of interchangeable manufacturing. By interchangeability it is meant that the parts manufactured anywhere can be assembled without rework. The production of any component requires interaction between man, machine and material. Defects or variations in any of these would result in variation in the dimensions of component.

The variations may arise due to—

(i) Difference in the skills of the machine operator.

(ii) Machine tool may have inherent inaccuracies.

(iii) The cutting tools may have variations and get worn out.

(iv) The materials may have defects.

Cost of producing the components to exact dimensions would be very high and rather impossible. However it is possible to manufacture them with such variations which would not affect the working of the assembly. A component is said to have interchangeability if with the recognized deviations the component is able to be assembled at random with the mating part and also has required tightness or looseness after assembly. If the parts are not interchangeable selective assembly would be needed which is costly and time consuming. Advantages of interchangeability are as follows:

(i) It is possible to manufacture components on mass scale at various places.

(ii) It is also possible to replace worn out parts.

9.13 ELEMENTS OF INTERCHANGEABILITY

This leads us to the concept of limits, tolerance and fits. In this connection knowledge of following is essential.

(a) Basic Size. It is the exact theoretical size of a component. Basic size is chosen on the basis of design considerations. Limits and variations are in relation to the basic size. Zero line coincides with the basic size. (Fig. 9.7)

(b) Nominal Size. It is the size designation given in the drawing for the sake of convenience and identification. It is usually rounded off to whole number. Nominal size is given for identifying a component and it does not represent accurate measurement. The nominal size of a shaft and a hole are same.

(c) Actual Size. It is the dimension of a part as it actually measures is actual size. Actual size should be within the specified limits.

(d) Zero Line. The limits and fits are represented graphically in relation to the zero line which represents the basic size.

(e) Limit of Size. It is the two extreme limits of sizes between which the actual size may lie.

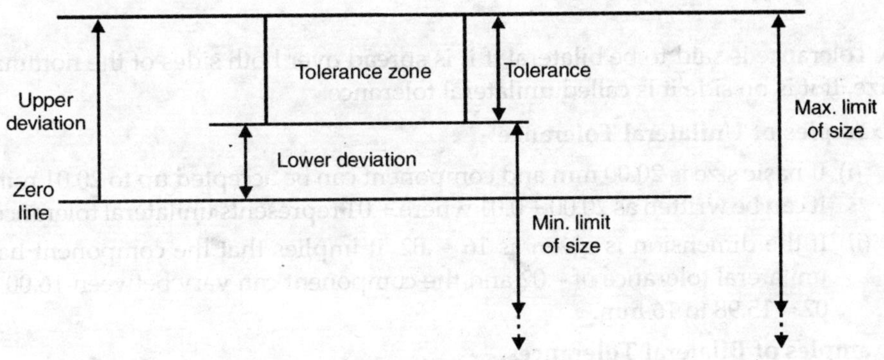

Fig. 9.7. Basic and other sizes.

9.14 LIMITS

Limits are mentioned as below.

(a) As mentioned earlier it is impossible to manufacture large number of pieces to an exact dimension and there will always be some difference in size. Then limits of acceptable dimensions are found so that the component would perform satisfactorily within these limits. The component can differ from the proper size by the small amount and still be able to be used.

(b) For example if a component has basic size of 25.00 mm and it is found that it would perform satisfactorily if the size is between 25.02 mm on the higher side and 24.97 mm on the lower side.

Then upper deviation is: The maximum size minus the basic size (in this case 25.02 – 25.00 = 0.002).

Lower Deviation – Minimum deviation = Basic Size – Lowest size = (in this case 25.00 – 24.97 = 0.003)

(c) Maximum Limit or Upper Limit = Max. dimension which can be allowed = Basic size + Upper deviation = 25.02.

(d) Minimum Limit or Lower Limit Minimum dimension of component which can be allowed = Basic size – Lower deviation = 24.97.

9.15 TOLERANCE

It is the total amount by which a dimension is allowed to vary. The Tolerance is the total amount by which the size of the component can differ from the Nominal

Size. The difference between the maximum and the minimum limits of sizes is tolerance. It is equal to the algebraic difference between the upper and lower deviation.

9.15.1 Types of Tolerances

A Tolerance is said to be bilateral if it is spread over both sides of the nominal size. If it is on side it is called unilateral tolerance.

Examples of Unilateral Tolerance

(i) If basic size is 20.00 mm and component can be accepted up to 20.01 min. It can be written as 20.00 + 0.01 where +.01 represents unilateral tolerance.

(ii) If the dimension is given as 16 – .02, it implies that the component has unilateral tolerance of –.02 and the component can vary between 16.00 – 02 = 15.98 to 16 mm.

Examples of Bilateral Tolerances.

Here the tolerance are spread on both sides of the zero line. For example 20.01 would mean that component can lie between 19.99 to 20.01 mm.

Tolerance Zone. The zone between the two limits of sizes as represented graphically in Fig. 9.7.

9.16 ALLOWANCE

Allowance is basically the gap between components that work together. Allowance between parts that are assembled is very important. For example, the axle of a car has to be supported in a bearing otherwise it will fall to the ground. If there was no gap between the axle and the bearing then there would be a lot of friction and it would be difficult to get the car to move. If there was too much of a gap then the axle would be jumping around in the bearing. It is important to get the Allowance between the axle and the bearing, correct so that the axle rotates smoothly and easily. Allowance is intentional difference between a shaft and hole dimensions for any type of fit.

The allowance is the intended difference in the sizes of mating parts. This allowance may be: positive which means there is intended clearance between parts; negative for intentional interference: or "zero allowance" if the two parts are intended to be the "same size". This last case is common to selective assembly.

9.17 FITS

Tolerance is applied to a single component. When two such components are assembled and are required to work together as a unit—the assembly characteristics is known as fit between mating parts. The relationship between parts when one is inserted in to another with degree of tightness or looseness is called fit. Generally speaking following are most common fits:

(i) **Clearance Fit.** When there is relative motion between the shaft and the hole it is called clearance fit. In clearance fit there is positive clearance between the maximum shaft size and the minimum whole size.

(ii) **Interference Fit.** When there is no relative movement between the shaft and the hole it is called interference fit. Here the shaft size is bigger than the hole.

(iii) **Transitional Fit.** When there is neither interference nor clearance fit—then it is called transitional fit.

9.18 HOLE AND SHAFT BASED FIT SYSTEM

Fits can be based on the basic hole size or basic shaft size.

Basic Hole System	Basic Shaft
1. Design size of the hole is taken as the basic size for both the hole and the shaft.	1. The design size of the shaft is taken as basic size for both the hole and the shaft.
2. It is most common system for limit dimensions. In this system the design size of the hole is taken to be equivalent to the basic size for the pair. This means that the lower limit of the hole dimension is equal to design size. The basic hole system is more frequently used since most hole generating devices are of fixed size like drills, reamers etc.	2. When designing using bought out items with fixed outer diameters like bearings, bushings, etc. a basic shaft System may be used. Taking the basic size of a common reference in both following relationships exist.
3. The position of the tolerance zone of the shaft may be above or below the zero line. Depending on the tolerance zone it can give various types of fits—interference, clearance	3. The position of the tolerance zone of the shaft may be above or within the tolerance zone of the hole.
4. Capital letters H is used for the hole. Shaft have different letters to indicate the position of tolerance zone for desired fit.	4. Small letter *h* is used for the shaft and the holes have different letters

Gauge Thickness Conversion Table

Sheet Metal Thickness		
Imperial Gauge	**Imperial in mm**	**Metric Sheet mm**
10	3.25	3.0
12	2.64	2.5
14	2.03	2.0
16	1.63	1.5
18	1.22	1.2
20	0.91	0.9
22	0.71	0.7
24	0.56	0.6
26	0.45	0.5

THEORY QUESTIONS

1. What is a sheet metal? Give different metals that can be classified under sheet metal.
2. Enumerate the uses of sheet metals.
3. What advantages are obtained from the parts which are formed from sheet metal.
4. Explain the working with a neat sketch of a hand operated flypress.
5. Give the procedures for sheet metal working like Bending, Blanking etc.
6. Write a note on limit, fits and tolerances.

❑❑❑

CHAPTER

10

PRINCIPLES OF DESIGN

10.1 NEED OF DESIGN

We try to understand the meaning of design before studying the need of design. Design is defined in general as following:

(i) The art of constructing (by A.W.Willington 1887)

(ii) The art of directing the great sources of power in nature for the use and convenience of man (by Thomas Tredgold 1828)

(iii) To utilise the natural energy to best advantage so that there may be least possible waste (by Willard A. Smith 1908)

We often come across terms like building design, bridge design, interior design, ship design, system design, curriculum design, machine design etc. The simplest way of understanding design is as follows.

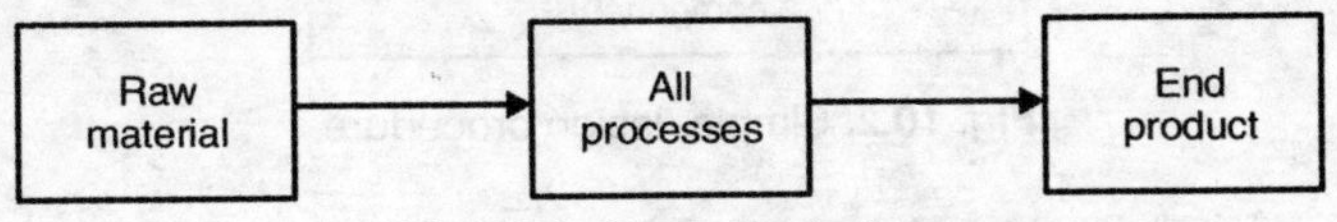

Fig 10.1. Stages in design.

Every machine component or system initially takes birth in the mind of design engineer. The component is brought from the conceptual and visualisation form into reality through following design procedures.

After the design process is over, the drawings are prepared. These are then communicated to production department for manufacturing.

Thus for manufacturing any component or system first step is design .Without proper design and drawing no engineering component or any system is manufactured. Similarly no building or bridge is constructed without design. Hence design is very important.

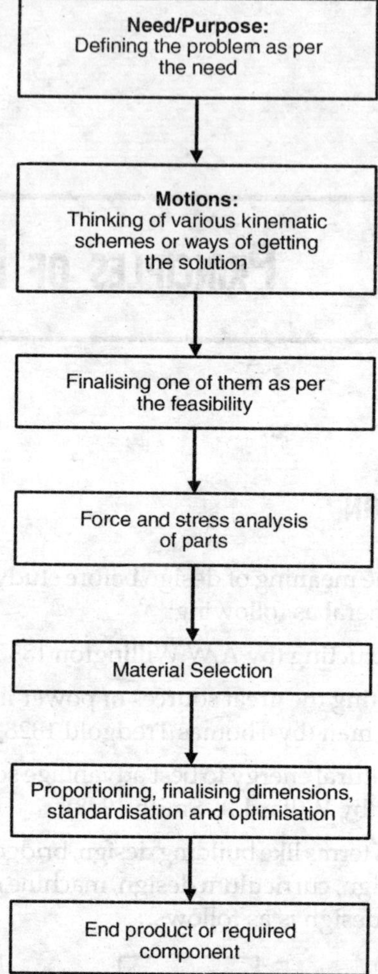

Fig. 10.2. Simple design procedure.

10.2 STRESS AND STRAIN

Along with the study of terms stress strain it is also required to understand the correct meaning of various other related terms.

(a) Stress. If an elastic material is subjected to the action of external forces it undergoes some deformation. Materials like rubber elongate substantially under small loads. On the other hand in materials like steel the deformation could be very small under much heavier loads. Whenever any load is applied to an object-internal forces are created in the material to resist the load. *The resisting force developed in the body per unit area is known as stress. The area is at right angle to the direction of the applied force.*

Stress is denoted by sigma (σ)

$$\sigma = \frac{\text{Load}}{\text{Area}} = \frac{P}{A}$$

In SI units stress is expressed in N/m^2 or pascal

and 1 MPa = $10^6 N/m^2$ = $1N/mm^2$

(b) Types of stresses. Depending on the type of loading the stresses are—

(i) Tensile Stress. When load is such that it tends to exert pull on the body—without moving it. It is subjected to tensile stress.

(ii) Compressive Stress. When the load has tendency to compress the object—the compressive stress arise in the body.

(iii) Shear Stress. Consider a riveted joint as shown; because of forces applied rivet body shears off. Stress developed is known as shear stress (τ) tau.

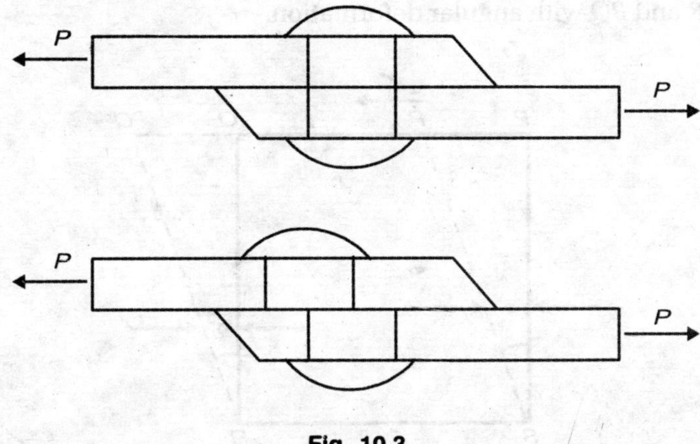

Fig. 10.3

(b) Strain

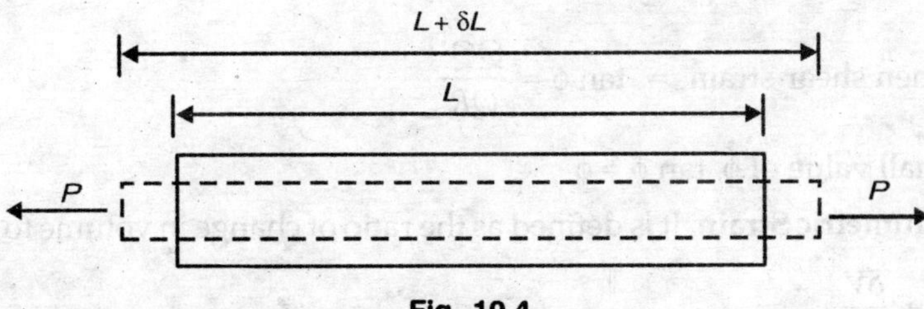

Fig. 10.4

If a member is subjected to axial forces as shown, then it elongates i.e. its length changes and the ratio of change in length to original length is known as strain and is denoted by epsilon (∈).

$$\in \; = \; \frac{\delta L}{L}$$

Strain being a ratio has no units.

The stresses and strains may be tensile or compressive depending upon the forces to which they are subjected.

(i) Tensile or Longitudinal Strain. If a bar of length L gets elongated to $L + \delta L$, the extension is δL then longitudinal strain is given by the ratio of extension to the original length.

$$\varepsilon = \delta L/L$$

(ii) Compressive Strain. In this case if the length gets reduced or shortened under compressive load by δL as shown then

$$\varepsilon = \delta L/L$$

(iii) Shear Strain. Consider above element *PQRS* subjected to shear stress on the faces *SR* and *PQ* with angular deformation.

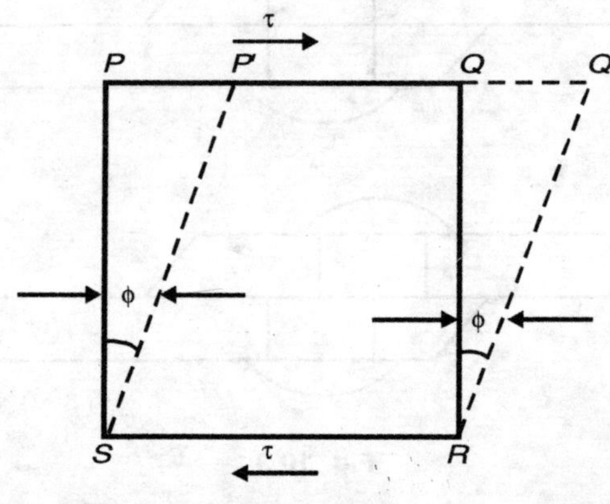

Fig. 10.5

$$\text{Then shear strain} = \tan\phi = \frac{QQ'}{QR}$$

For small value of ϕ, $\tan\phi = \phi$

(iv) Volumetric Strain. It is defined as the ratio of change in volume to original volume i.e. $\dfrac{\delta V}{V}$.

(c) Poisson's Ratio. When a bar is subjected to an axial force as shown above, When the forces are pulling the member axially as shown in Fig. 10.6, then they are called as tensile forces and the stresses developed are called tensile stresses. If the forces are opposite to the tensile forces, then they are called compressive forces. The stresses developed are called compressive stresses. The length of the bar would increases under tensile forces, whereas its cross-sectional area decreases. If the external force is axial and tends to pull out as shown in the figure above, the strain in axial direction is called *longitudinal strain* and the strain across the cross-section is called *lateral strain.* And the ratio of lateral to longitudinal strain is called Poisson's ratio—

$$\frac{\text{Lateral strain}}{\text{Longitudinal strain}} = \mu = \text{Poission's ratio}$$

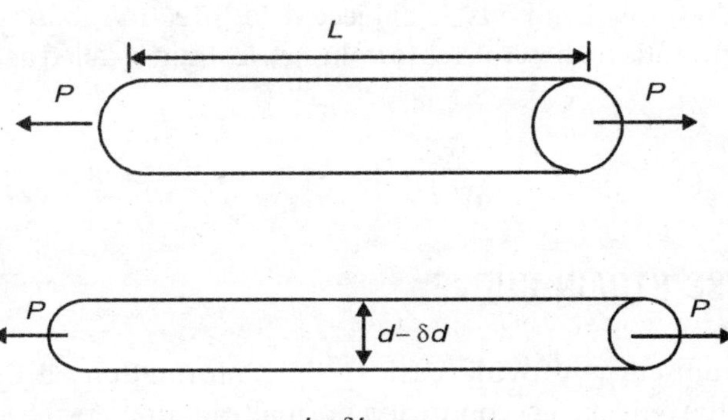

Fig. 10.6

(d) Hooke's Law. For an elastic material stress is directly proportional to strain within the elastic limit. The ratio of stress to strain remains constant and is called as Modulus of elasticity or Young's Modulus. It is denoted by *E*.

$$E = \frac{\text{Stress}}{\text{Strain}} = \frac{\sigma}{\varepsilon}\,\frac{N}{\text{mm}^2}$$

Equation for change in length

We know that Strain $\varepsilon = \dfrac{\delta L}{L}$

$$\delta L = \varepsilon . L \qquad \ldots(1)$$

From Hooke's law $E = \dfrac{\sigma}{\varepsilon}$

or $\varepsilon = \dfrac{\sigma}{E}$

Substituing in (1) we get

$$\delta L = \frac{\sigma}{E} . L$$

$$\delta L = \frac{P/A}{E}.L$$

$$\therefore \quad \delta L = \frac{PL}{AE}$$

(e) Modulus of Rigidity. Shear stress is proportional to shear strain upto elastic limit. The ratio is called Modulus of rigidity and is denoted by *G*.

$$G = \frac{\text{Shear Stress}}{\text{Shear Strain}} = \frac{\tau}{\phi}$$

(f) Bulk Modulus. If a body is subjected to three mutually perpendicular stresses, then the ratio of direct stress to volumetric strain is called as Bulk modulus *K*.

$$K = \frac{\sigma}{\delta V / V}$$

10.3 STRESS-STRAIN CURVE

The stress strain curve provide basic design information on the strength of materials . In the tension test conducted normally on universal testing machine. Specimen of the test piece is subjected to a continually increasing uni-axial tensile force while simultaneous observations are made of the elongation of the specimen. The stress-strain curve is constructed from the load elongation measurements (Fig. 10.7). It is obtained by dividing the load by the original area of the cross section of the specimen. The stress strain curve can be broadly divided into elastic and plastic regions.

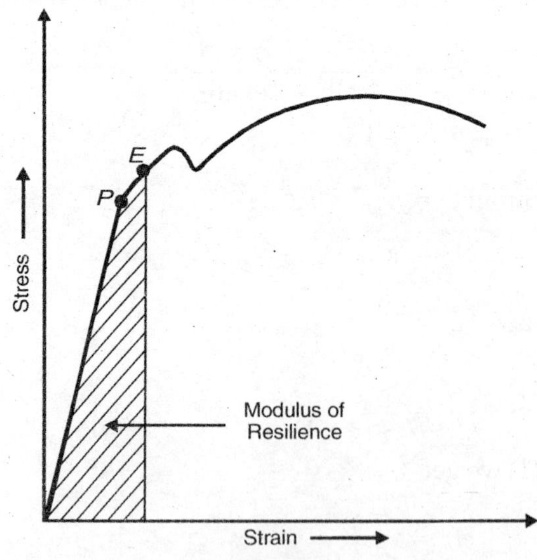

Fig. 10.7

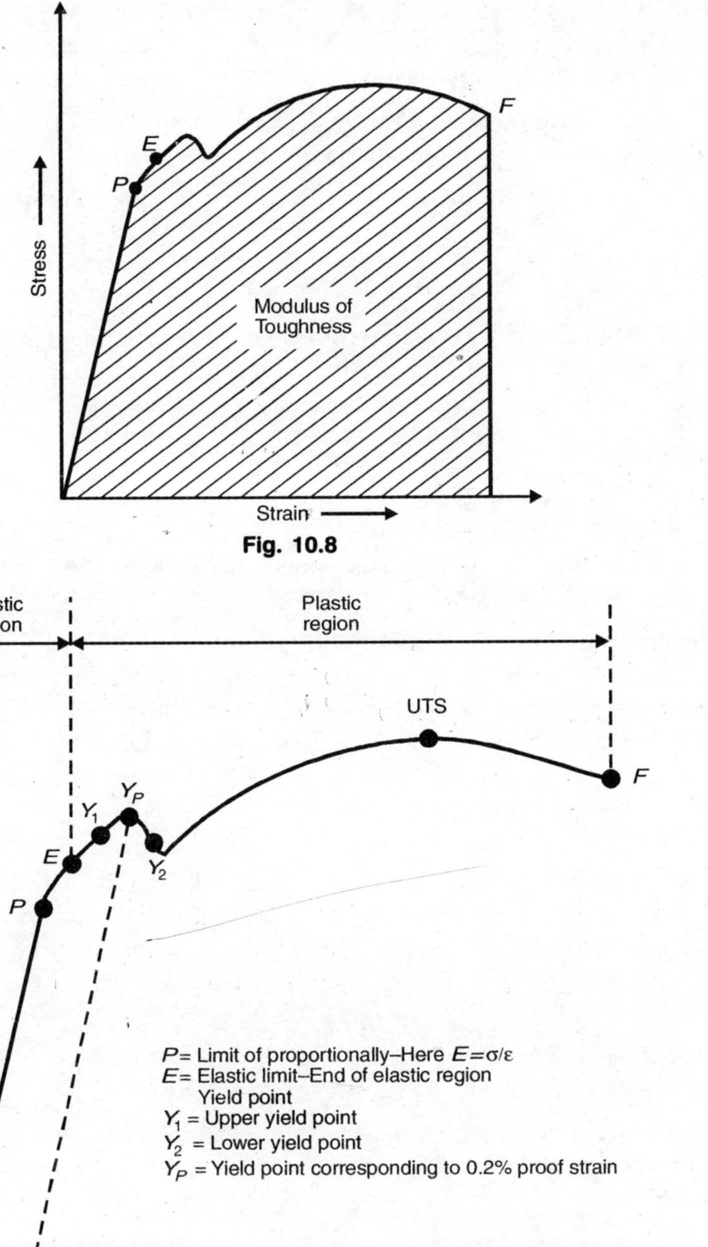

Fig. 10.8

Fig. 10.9

(a) Elastic Region. In the elastic region which is up to the point elastic limit (E). In this portion the specimen will return to its original shape without any deformation when the load is removed.

(b) Plastic Region. If loading is applied beyond yield point—the material undergoes permanent deformation till failure.

(c) Important Points on Stress Strain Curve and their Significance.

10.3.1 Elastic Region

Limit of Proportionality (P). The stress-strain curve is straight line up to the point P which is called limit of proportionality. In this portion the stress is directly proportional to strain which is Hooke's law is followed.

or $$\sigma \propto \varepsilon$$

or $$E = \sigma/\varepsilon$$

where E is called **Young's modulus.** For steel E is about 200 kPa. The value of E for common engineering materials along with other properties are given in table at the end of this chapter.

Elastic Limit (EL). Beyond point P the stress-strain relationship does not remain linear but gets curved. However till the point EL the region is elastic. In case the load is removed the material will regain its original shape i.e. strain will disappear. Elastic limit is the maximum stress which will cause no permanent deformation.

10.3.2 Plastic Region

Yield Point. If the load is increased beyond the elastic limit—the material will start yielding and substantial deformation will occur. This will results in permanent deformation and the object will not regain its original shape from here onward. *Yield point is where there is marked deformation without increase in load.*

Upper Yield Point. At point Y1 there is marked increase in the strain with slight increase of the load. Y1 is located beyond EL.

Lower Yield Point Y2. In materials like steels it is observed that more strain occurs accompanied by reduction of load. These two yield points are very close to each other.

Ultimate Tensile Strength. With further loading there appreciable increase in strain and the stress also increases due to strain hardening. The point UTS with maximum stress is called Ultimate Tensile Strength (UTS). With further loading failure occurs at point F.

(d) Importance of Yield Strength. *The yield strength* is the stress required to produce a small-specified amount of plastic deformation corresponding to the yield points as above. However in some materials it is difficult to locate the yield points. Then *the yield strength* is determined by offset method. It would be the stress corresponding to the intersection of the stress-strain curve and a line parallel to the elastic part of the curve offset by a specified strain of usually 0.2%. It is obtained by drawing a line parallel to the straight portion of the stress-strain curve at 0.2% strain. This point has been shown point *Yp*.

(e) Information from Stress Strain Curve

(i) Tensile testing of the material sample gives information about the limit of proportionality, yield point, ultimate strength, percentage elongation, resilience, toughness etc.

(ii) The strain used for the engineering stress-strain curve is the average linear strain, which is obtained by dividing the elongation of the gauge length of the specimen by its original length.

(iii) Since both the stress and the strain are obtained by dividing the load and elongation by constant factors, the load-elongation curve will have the same shape as the engineering stress-strain curve. The two curves are frequently used interchangeably.

(iv) The stress-strain curve of a metal will depend on its composition, heat treatment, temperature, and state of stress imposed during the testing. The parameters, which are used to describe the stress-strain curve of a metal, are the **tensile strength, yield strength or yield point, percentage elongation, and reduction of area.**

10.4 MODES OF FAILURES

The failure of a component can generally take place under three basic modes of failures:

(a) Failure by Elastic Deflection. In case of a shaft supporting either a pulley or a gear, because of elastic deflection gear teeth do not mesh properly. Where transmission is affected shaft does not perform its basic function. It is called failure by elastic deflection.

(b) Failure by Yielding. Many components made of ductile material like steel etc fail because of plastic deflection in dynamic condition. This reduces engineering usefulness of the component. It is called "Failure by Yielding".

(c) Failure by Fracture. Many components made of brittle materials like cast iron etc stop functioning because of sudden fracture without plastic deformation. They fail because of fatigue cracks. Failures of ball bearings, valves, springs, gears etc are examples of this type of failure.

10.5 FACTOR OF SAFETY

If a material has a tensile yield strength of say 400 N/mm≤ allowable stress or working stresses should not exceed 400 N/mm≤. This means that working stresses must be less than 400 N/mm≤. But question arises like how much less?

At the same time, force analysis may not be exact. There may be material or manufacturing defects and the strength may not be uniform. In order to make allowance for the above mentioned factors and more and also to ensure better safety to the operator, a factor known as factor of safety (FS) is used.

$$\text{Factor of Safety} = \frac{\text{Material strength}}{\text{Allowable or working stress or design stress}}$$

For ductile materials

$$\text{Factor of Safety} = \frac{\text{Yield strength}}{\text{Allowable or working stress}}$$

For brittle materials

$$\text{Factor of safety} = \frac{\text{Ultimate strength}}{\text{Allowable or working stress}}$$

For fatigure loading

$$\text{Factor of safety} = \frac{\text{Endure limit}}{\text{Allowable or working stress}}$$

10.6 SELECTION OF FACTOR OF SAFETY

It is chosen after considering the following factors:

(a) Accuracy or exactness of force analysis.
(b) Component life.
(c) Level of safety to human life.
(d) Allowance for misuse by operator.
(e) Loading conditions.
(f) Modes of failures.
(g) Degree of economy desired.
(h) Manufacturing defects.
(i) Material reliability.
(j) Unskilled operator.
(k) Unexpected disturbances and variations.

10.7 MATERIAL PROPERTIES AND SELECTION

The choice of material to be used in the design of a machine component depends on the material properties under the given working conditions especially its mechanical properties, availability, cost etc. Mechanical properties of the materials of interest in design are—

(a) **Strength.** The strength of the material is essential to determine the size of the part. The strength is indicated from stress values corresponding to the yield point (Syp) or the ultimate tensile strength Sut.

(b) **Fatigue Strength.** For components subjected to variable loading for long time the fatigue strength is also an important consideration.

(c) **Elasticity.** It is the property of materials whereby they regain original shape without any deformation after removal of load. Generally the machine parts are designed to remain well within the elastic limits.

(d) **Stiffness.** It is the property to resist deformation under load. Modulus of elasticity E is measure of the stiffness or rigidity. Steel with E value of about 200 kPa is much more rigid as compared to aluminum alloy having E= 70 kPa.

(e) **Resilience.** The energy stored within the elastic limit is called resilience. If the load is removed the object regains its original shape. (Fig.10.6)
Modulus of Rigidity is Resilience per unit volume

(f) **Toughness.** It is the total energy stored in the material before fracture. Toughness is important to withstand impact loading. It is given by the area up to F in the stress-strain curve. (Fig. 10.7).

(g) **Malleability.** It is the property of material to deform to large extent without cracking or fracture under *compressive loading*. This is essential for rolling, extrusions, wire drawing etc.

(h) **Ductility.** It is the ability of material to deform to large extent without failure or cracking under *tensile loading*.

	Malleability	Ductility
Property to deform under	Compression	Tensile loads
Effect of temperature	Increases with temp	Decreases with temp
Uses of property	Forging, rolling, extrusion	For forming or drawing
	All ductile materials are malleable	All malleable materials may not be ductile

(j) **Hardness.** It is the property which resists deformation or scratching, penetration or abrasion. It is measured in Brinell Hardness Number (BHN).

10.8 MATERIAL SELECTION

Engineering materials include metals like steel, brass, aluminum, plastics, composite etc.

The materials which meets the design requirement at minimum cost is most suited for the purpose. Following factors determine the choice of the material.

(a) Strength and usage requirements. It should be possible to manufacture the object to meet the design requirements using the selected material. It should have required mechanical properties.

(b) Availability. The material should be easily available in the required quantities when needed. As such standard materials should be used as far as possible.

(c) Cost factor. (i) Material cost form substantial part of the component cost. As such the chosen material should not be costly as far as possible. (ii) The type of material also some time is important in the method of manufacturing. Both the material and manufacturing costs should be as less as possible.

10.9 COMMON ENGINEERING MATERIALS

Following are the materials mostly used in Common Engineering.

(a) Cast Iron. (i) It is an iron carbon alloy containing more than 2% carbon. It is easily available and can be easily cast. Cast iron is strong in compression but weak to take tensile loads. It is easy to machine and has good resistance to wear. Cast iron is extensively used in casting of machine parts and intricate shapes. Cast irons provide good damping of vibrations. Cast iron is available in different varieties like grey cast iron, malleable cast iron, nodular cast iron etc. Grey cast iron is most common.

(ii) *Designation.* Grey cast iron are designated as FG 150, FG 200 , FG 260, FG 300, FG 350 and FG 400. In FG 200- FG denotes grey iron and 200 indicates its UTS in N/mm^2.

(b) Alloy Cast Steels. For higher strength applications alloy cast steels with alloying elements like nickel, chromium etc. are used. This improves the tensile strength of the cast steel.

(c) Plain Carbon Steel. (i) Steels with less than 1.7% carbon are plain carbon steels. The plain carbon steels have high tensile strength, good ductility and high resilience/toughness. However they can not be cast and have poor wear resistance.

(ii) The plain carbon steels are designated for example as 45 C 8. Here first two digits 45 indicates 100 times % of carbon(0.45%) and the figure following C is for 10 times the Mn content. Thus the above steel has 0.45% carbon and 0.8% Mn. Carbon steels are the most extensively used engineering material. These are used for variety of engineering purposes depending on the strength.

(d) Alloy Steels. (i) These steels contain usually more than 1% Mn and other alloying elements like Ni, Cr, Mo, V etc. to give improve strength or give specific properties. Alloy steels are relatively costly but have much higher strength than plain carbon steels. All alloy steels are required to be heat treated increase the strength, hardness etc.

(ii) The alloy steels are designated by specifying their chemical composition. For example 40 Nil Cr 60 denotes — Alloy steel with average of 0.40% carbon, 1% Ni and 0.60% Cr.

(e) Aluminum. (i) Aluminum and its alloys find extensive use in engineering and aeronautical industry. Aluminum has low specific weight, good thermal and electrical conductivity, have good mach inability and can be cast.

(ii) Aluminum alloys are designated by a system of 4 digits for castings and 5 digits for wrought Al alloys. For example in 4250 for casting material—the first digit 4 stands for major alloying element –(here Si), second digit gives half of the rounded off value of the major alloying element in this case it would mean 4% Si —while the third, fourth and fifth digits indicate the symbols for minor alloying elements in descending order. In wrought Al the second digit stands for % of major alloying element (not half as in cast alloys)

(f) Copper, Brass and Other Metals. These metals can be used where specifically required. Generally these materials much more costly than steels and aluminum.

(g) Plastics. Plastic is also being used quite often in engineering application. They are light and plastic components are easy to make.

10.10 AESTHETICS

Apart from proper design any item should look beautiful to impress the buyer. Aesthetics is set of principles concerning beauty of product during designing.

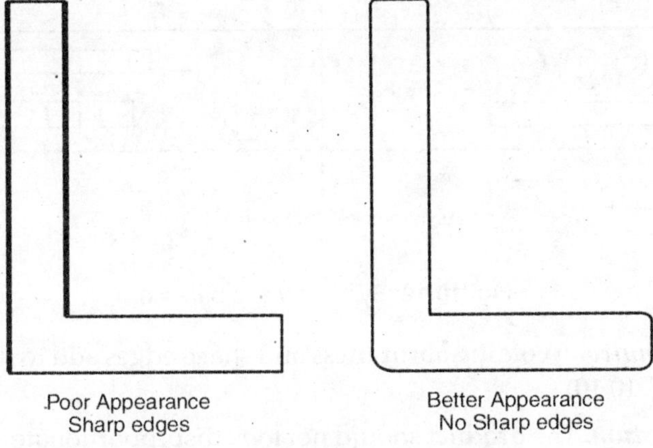

Fig. 10.10

(a) Appearance. The appearance of a product can add to the functioning of a product. For example plating, painting can enhance the beauty of a product apart from providing protection from corrosion. Similarly aerodynamically shaped car body has aesthetic appeal and it also reduces air resistance.

(b) Factors affecting Aesthetics. Following factors can improve the aesthetic appeal of a product.

(i) Shapes. Primarily the shapes could be step, taper, shear, streamlined etc. as given in Fig. 10.11.

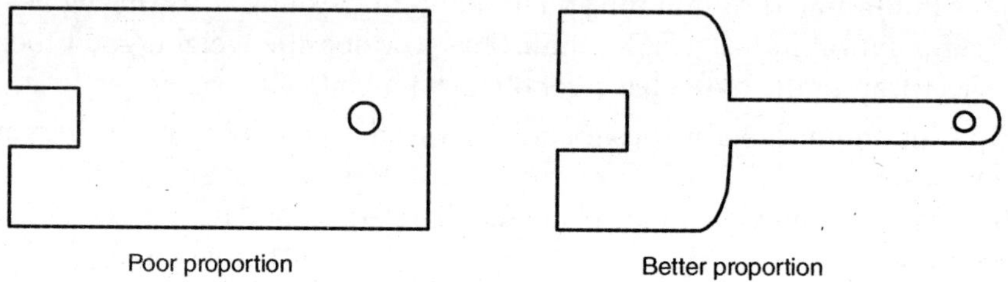

Fig. 10.11. Various Shapes for aesthetic design.

(ii) Symmetry. Symmetry about an axis always adds to aesthetic appeal. Where symmetry is not possible balancing by proper placement of objects can be attempted. This is illustrated in Fig. 10.12.

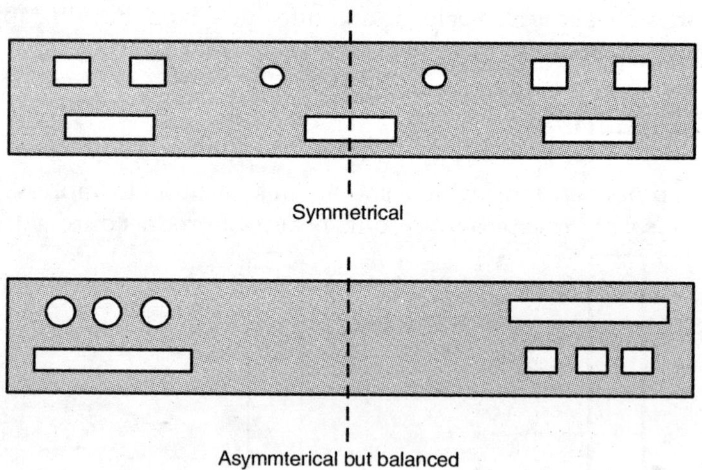

Fig. 10.12. Symmetry and balance.

(iii) Continuity. Avoiding abruptness and sharp edges add to the beauty of a product. (Fig. 10.10)

(iv) Proportion. The product should not look disproportionate. (Fig. 10.11)

(v) Style. The style should be contemporary and impressive.

(vi) Surface Finish. Surface finish is very important. The surface finishes could be painting, plating, anodizing etc.

(a) Colour. Colour is quite important. Following colours reflect the associated meanings.

Colour	Red	Orange	Yellow	Green	Blue	Grey
Indicates	Danger Hot	Possible danger	Caution	Safe	Cold	Dull

10.11 ERGONOMICS

The term "ergonomics" is derived from two Greek words: "ergon", meaning work and "nomoi", meaning natural laws. Ergonomists study human capabilities in relationship to work demands. Any man operated machine must be so designed that elements of human limitations and environment are taken into consideration. Ergonomics is application of human anatomical, physiological and psychological principles for best use of man-machine-working environment relationship. Ergonomically designed machine is easy to operate and results in less fatigue. Following aspects are covered under ergonomics.

(i) Communication between operator and machine

(ii) Working environment

(iii) Human anatomy and posture during operation

(iv) Energy expenditure by hand or foot operation

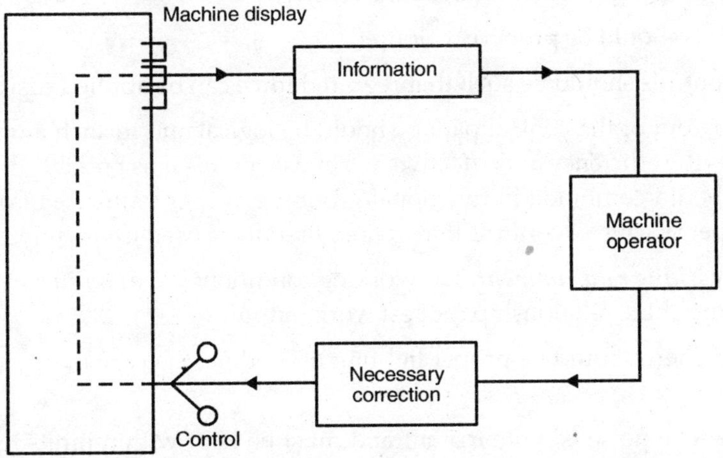

Fig. 10.13. Man and machine make close loop system.

(b) Communication with the Machine. While operating a machine—there is close interaction between the operators who together with the machine form a close loop system. The operator monitors various gauges and instruments to take corrective actions if needed. The communication between the man and machine consists of displays and control.

(i) Displays. These are instruments through which the operator gets information about the working of machine. Displays can be qualitative or quantitative. Examples of Qualitative displays are temperature indicators in cars, traffic signals etc. The Quantitative Displays—such displays give the value of the parameter being measured. Examples are—speedometer, pressure gauges etc. The displays can be with moving pointers, moving scales and fixed pointers or digital displays. The displays should be easy to read and avoid fatigue to the operator. Following should be taken into account. These should have clear and easily readable scales.

(ii) Controls. Controls are devices by which operator convey instructions to the machine. Design of controls depends on the requirements. There can be various types of controls such as—cranks, levers, foot pedals, hand wheels, star wheels, knobs, push buttons, toggle switches etc. The controls must be designed properly taking into account the following.

- Controls should be suitably positioned
- Be easily accessible
- Should require minimum and smooth movement
- Use minimum energy

The controls should be easy to handle the portion which comes in contact with hand or fingers should have matching shape.

Controls should be properly coloured.

The controls should be such that required force can be applied easily.

The lay out of the control panels should be logical and in such a way that it improves the efficiency and effectiveness of usage. As far as possible the control panels should combined in functional grouping, as per sequence of operation and proper placing of controls that are required to be used frequently.

(iii) Working Environment. The working conditions have significant effect on the man machine relationship for best work output.

Light. There should be proper lighting needed for the type of operation the machine.

Noise. The noise is quite irritant and must be kept within limits by proper maintenance, use of silencers etc.

Temperature/Humidity. These are important for proper working conditions.

Air-Conditioning. Some machines and computers require air conditioning for proper working.

(iv) Posture. The postures which minimize unnecessary static work and reduce the forces acting on the body are most suited. The risk of injury can be reduced significantly reduced by following the principles given below.

All work activities should permit the worker to adopt several different, but equally healthy and safe postures.

Where muscular force has to be exerted it should be done by the largest appropriate muscle groups available.

Work activities should be performed with the joints at about mid-point of their range of movement. This applies particularly to the head, trunk, and upper limbs.

THEORY QUESTIONS

1. Explain need of design?
2. Give definitions of stress, strain etc.
3. Explain stress–strain curve.
4. What are the modes of failure ?
5. What do you mean by factor of safety and how it is selected?
6. Write a note as material properties and selection.
7. Write notes on aesthetics and ergonomics.

Unit 6

Mechanical Devices & Machine Elements

Syllabus

Mechanical Devices : Drives — Individual and group drives, belt drive, rope drive, chain drive, gear drive and friction clutches and brakes (Types and applications only).

Machine Elements — Power transmission shafts, axles, keys coupling, bush and ball bearings flywheel and governor (Types and applications only).

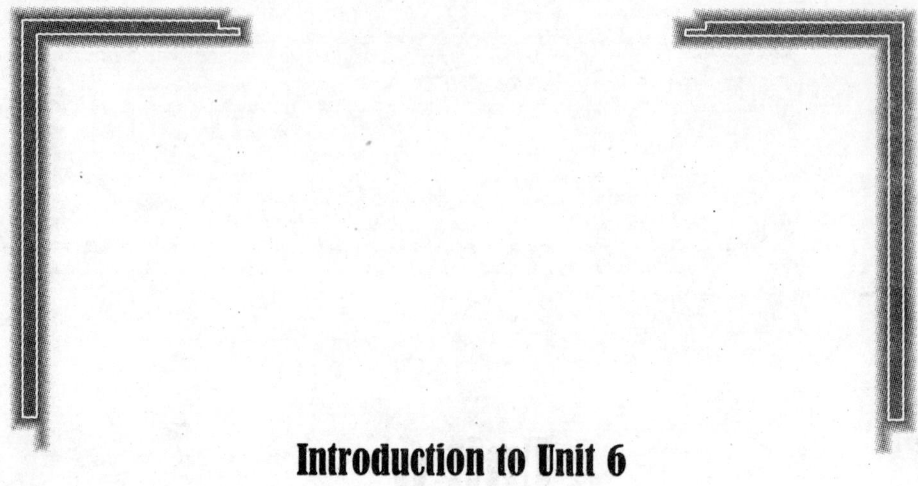

Introduction to Unit 6

To any machine, energy supplied by a prime mover. The major task now is, to transfer this energy to different parts of machine as per their requirements in terms of type of motion (rotating/ reciprocating/oscillating), sense of direction, speed of rotation etc. For this purpose various devices are employed. These are known as mechanical drives

This last unit is about Mechanical devices. In every machine there are elements like belt drives used say in floor mills, rope drives for example in lifts, chain drives as in motor cycles, and gear drives used extensively in lathes, motor vehicles etc. In addition clutches and brakes are used in automobiles for better control over power transmission. Similarly study of basic machine elements like shafts, keys, coupling and bearing is quite important and these have been covered in detail.

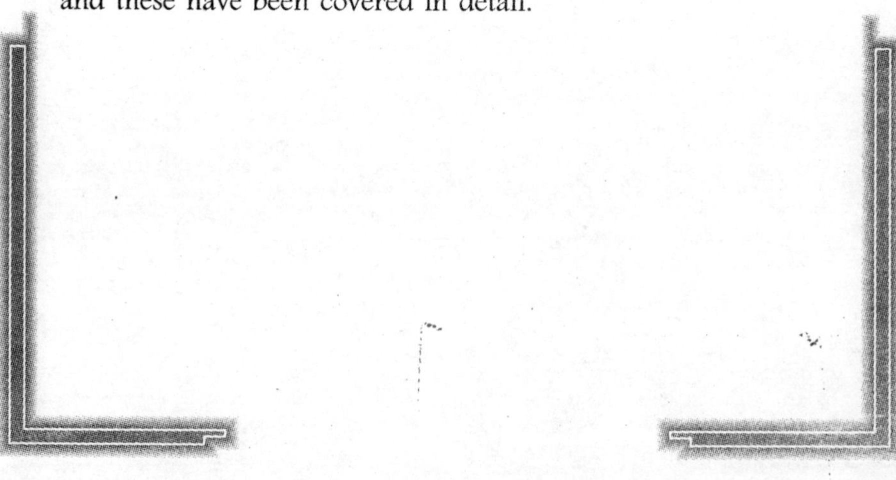

CHAPTER

MACHINE DEVICES

DRIVES

11.1 INDIVIDUAL AND GROUP DRIVE SYSTEM

The arrangement used to transfer the power from a prime mover to the machine is known as *'driven system'*. Thus this system connects the output of the prime mover to the input of machine.

11.1.1 Individual Drive

When output from one prime mover is used to drive one and only one machine, the arrangement is known as *individual drive,* e.g. electric motor driving a compressor or a pump.

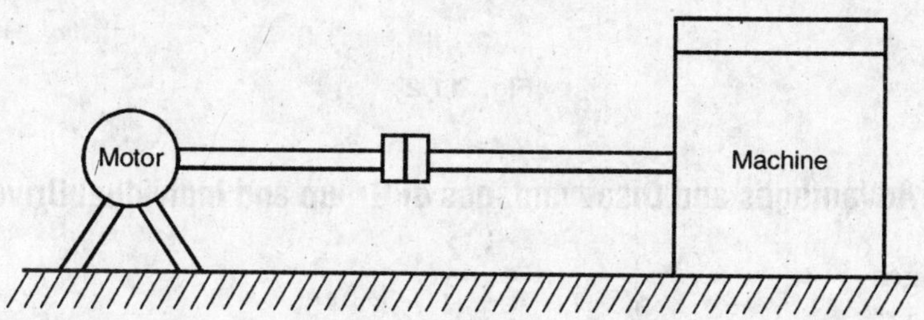

Fig. 11.1

11.1.2 Group Drive

When the output from one prime mover is used to drive number of machines, the arrangement is known as *group drive*. e.g. a car engine is used to drive the car through a gear box. Simultaneously it also operates the cooling water pump of radiator, cooling fan, compressor of air conditioner etc.

This type of arrangement is generally used in machine shops as shown in Fig. 11.2. The prime mover is first connected to the line shaft with belt and pulley arrangement. The line shaft is supported by bearings and runs over all the machines which are to be driven. On the line shaft, for each machine two pulleys are mounted. Out of those, one is known as loose pulley. It can run freely over the shaft while the other is known as *fast pulley* which is keyed to the line shaft and rotates along with it. Thus using this arrangement the machine can be started or stopped just by shifting the belt on 'fast' or 'loose' pulley respectively.

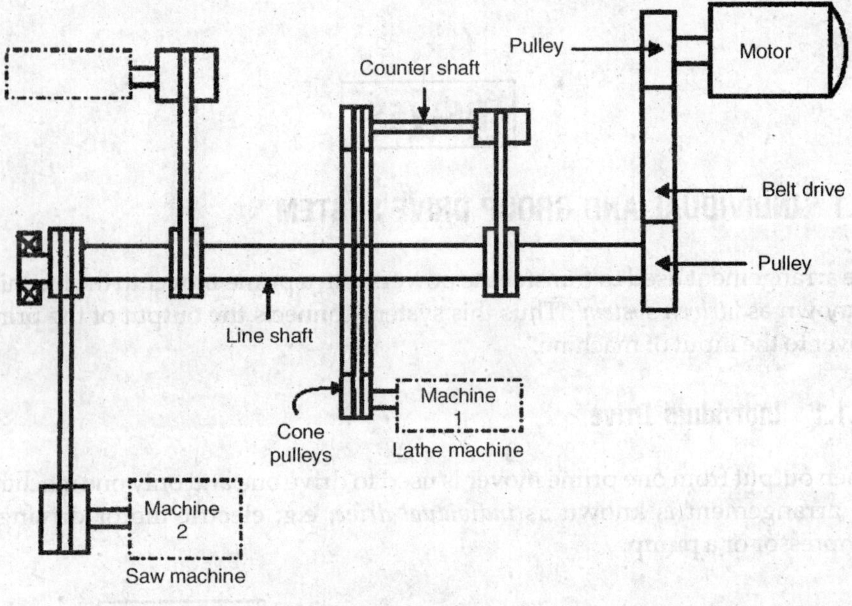

Fig. 11.2

11.1.3 Advantages and Disadvantages of Group and Individual Drives

Advantages

1. Only one prime mover generating sufficient power to drive all machines. Hence the initial cost is less.
2. Number of machines, types of machines can be increased without arranging for additional prime mover.
3. Prime mover can be located out of the machine shop, giving pollution free, noiseless working place.

4. Power losses in prime movers, friction are less giving high overall efficiency.
5. Suitable where number of machines are required to operate simultaneously and the total load remaining more or less constant.
6. Less floor space is required for same number of machines.

Disadvantages

1. The prime mover must be kept working even when one of the machines is to be operated.
2. Failure of prime mover entirely paralyses all the operations.
3. Not suitable when number of machines is very small or where the total load is much fluctuating.

11.2 BELT DRIVE

This arrangement is used to transmit the power from one shaft to another (Fig. 11.3) when distance between them is large.

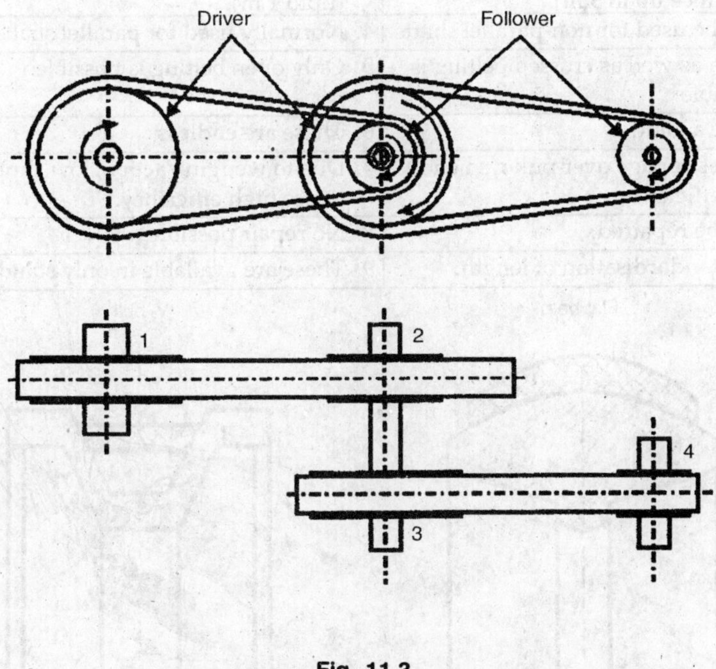

Fig. 11.3

The main components of this drive are discussed as follows.

1. Belt. These are made of material like canvas, leather, rubber etc. These materials give high friction coefficient between them and pulleys. Sometimes belts are reinforced with steel wires to improve the strength. They are manufactured with different cross sections and accordingly are known as *Flat*

Belt (with rectangular c/s) and *V* belt (trapezoidal c/s). The low initial cost, no maintenance are major *advantages*, while non-positive transmission and low transmission efficiency (due to slipping of belt) are major disadvantages of belt drives.

2. Pulley. It is a circular disc with a flat surface or a groove. It is mounted on shaft with help of a key. Thus it rotates with the shaft. The belt is made to pass over it. Due to the friction between belt and pulley surface, the power is transmitted.

11.2.1 Comparison of Flat and V-belt

The comparison between flat and V belt is tabulated below.

Flat Belt	V-Belt
1. Cross section is rectangular.	1. Cross section is trapezoidal.
2. Made of canvas or leather.	2. Made of molded rubber, reinforced with steel wires, covered with canvas sheath.
3. Can be used for distant shafts (distance up to 5 m).	3. Are used for relatively closer shaft (upto 1 m).
4. Can be used for non-parallel shafts.	4. Normally used for parallel shaft.
5. Open as well as crossed belting is possible.	5. Only open belting is possible.
6. Have a joint.	6. These are endless.
7. High slippage over pulleys hence low efficiency.	7. Due to wedging action, low slippage, hence high efficiency.
8. Can be repaired.	8. No repair possible.
9. No standardisation of length.	9. These are available in only standard

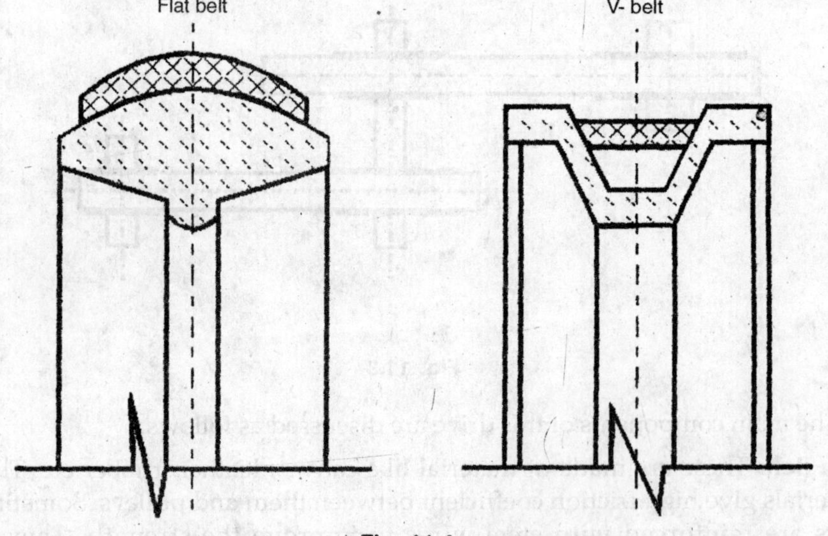

Fig. 11.4

11.2.2 Advantages of V-belt over Flat Belt

Following are the few advantages of V-belt over flat belts.

1. Transmission efficiency of V-belt is more than flat belts as it is also in contact from sides.
2. V-belts are comparatively silent during operation.
3. In case of V-belts the contact area is more when compared with flat belts as they are at contact from sides.
4. V-belts are endless and they generally used in the power transmission in the industry to run pumps, compressors.
5. The life of V-belts is more when compared with flat belts.

11.2.3 Application of Flat Belt and V-belts

1. Flat belts are generally used in floor mills and Lathe machines to transmit power.

2. V-belts are used to transmit power in industries i.e. to run compressors, pumps, blowers etc.

3. In automobiles V-belt is used to drive the fan to blow air over the radiator etc.

11.3 ROPES

Rope is a bundle of steel wires twisted in a special way. These are used to transmit power when the distance between shafts is very large (8–10 m). In this drive also, the rope is made to pass over pulleys. Pulleys used in rope drive have a circumferential groove in order to prevent *running out* of rope. Ropes are used in cranes, elevators, etc.

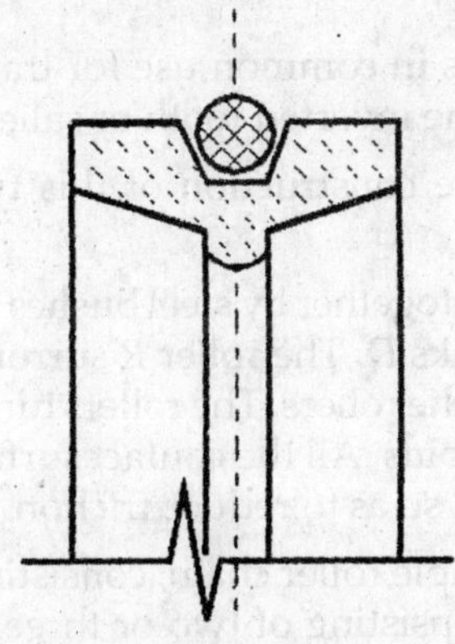

Fig. 11.5

11.4 CHAIN DRIVE

A chain is built up of rigid links, which are hinged together in order to provide the necessary flexibility for wrapping action around the driving and driven wheels. These wheels have projecting teeth, which fit into suitable recesses in the links of the chain and thus enable a positive drive to be obtained. They are known as chain sprockets and bear a superficial resemblance to spure gears.

The pitch of the chain is the distance between a hinge centre of one link and the corresponding hinge centre of adjacent link.

The pitch circle diameter of the chain sprocket is the diameter of the circle on which the hinge centres lie, when the chain is wrapped round the sprocket.

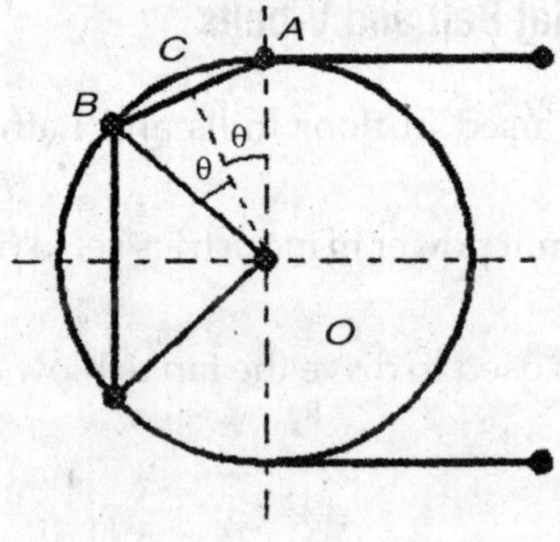

Fig. 11.6

Referring to Fig. 11.6 it will be seen that, since the chain links are rigid, the pitch becomes a chord, not an arc of the pitch circle.

11.1.4 Types of Chain

There are two types of chains in common use for transmitting power, namely (a) The roller chain and (b) The inverted tooth or salient chain.

(a) The Roller Chain. The construction of this type of chain is shown in Fig. 11.7.

The inner plates *A* are held together by steel bushes *B*, through which pass the pins *C* riveted to the outer links *D*. The roller *R* surrounds each bush *B* and the teeth of the sprockets bear on the rollers. The rollers turn freely on the bushes and the bushes turn freely on the pins. All the contact surfaces are hardened so as to resist wear and are lubricated so as to reduce friction.

Figure 11.7 (a) shows a simple roller chain, consisting of one strand only, but duplex and triplex chains, consisting of two or three strands, may be build as shown in Fig. 11.7(b), each pin passing right through the bushes in the two or three strands.

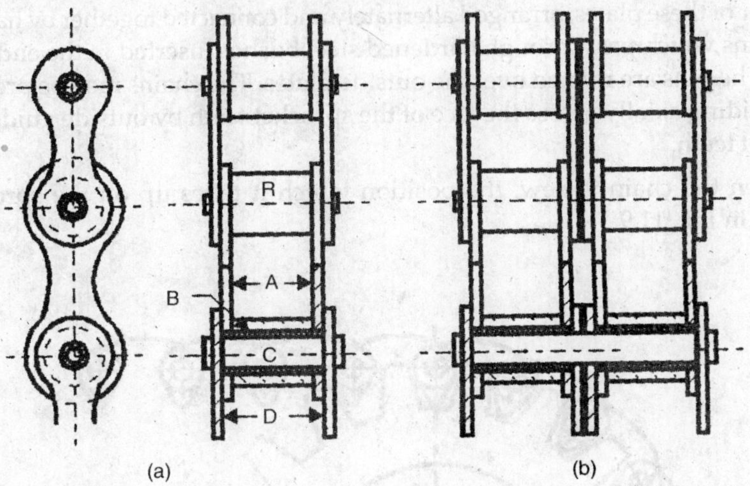

Fig. 11.7

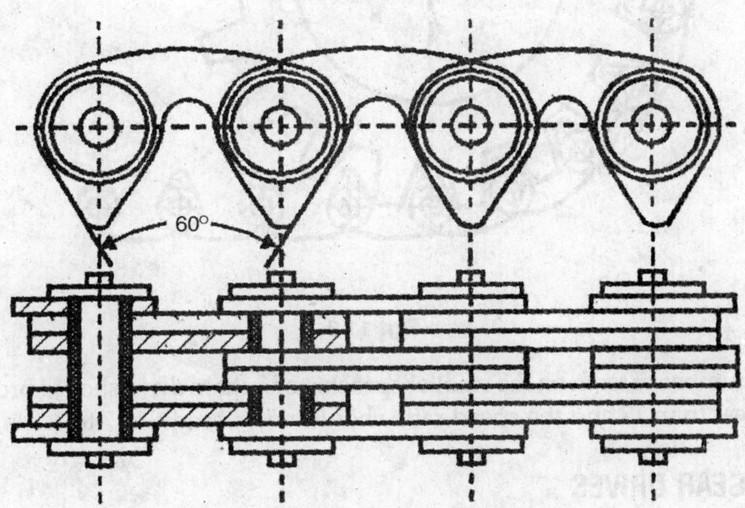

Fig. 11.8

(b) The Inverted Tooth or Salient Chain. The construction of this type of chain is shown in Fig. 11.9. It is built up from a series of flat plates, each of which has two projections or teeth. The outer faces of the teeth are ground to give an included angle of 60° and they bear against the working faces of the sprocket teeth. The inner faces of the link teeth take no part in the drive and are so shaped as to clear the sprocket teeth. The required width of chain is built up from a number of these plates, arranged alternately and connected together by hardened steel pins which pass through hardened steel bushes inserted in the ends of the links. The pins are riveted over the outside plates. The chains may be prevented from sliding axially across the face of the sprocket teeth by outside guide plates without teeth.

When the chain is new, the position which it takes up on the sprocket is shown in Fig. 11.9.

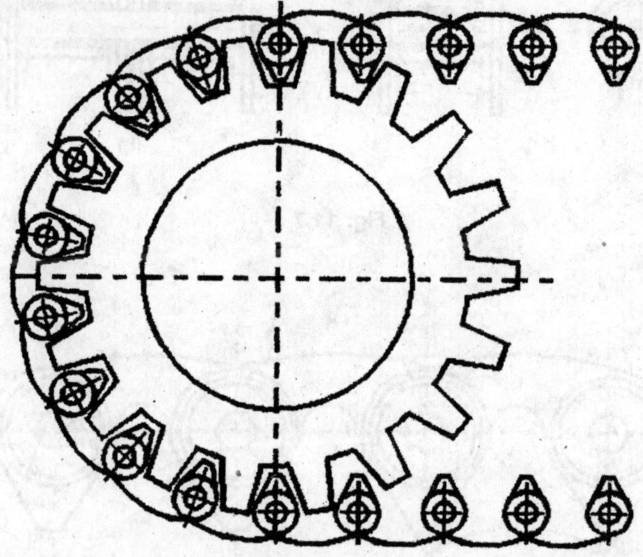

Fig. 11.9

The number of teeth on the smaller sprocket of a chain drive should preferably be not less than 17 and the speed ratio should not exceed 6 or 7 to 1.

11.5 GEAR DRIVES

Following types of gears are widely used in practice.

1. Spur Gear. In this type, the teeth are cut on cylindrical surface of hub. The teeth in this gear are parallel to the axis of the shaft. These are used to transmit power between two parallel shafts. (Refer Fig. 11.10)

The manufacturing of these gears is relatively simple. These are used where the gear ratio (reduction in speed) is upto 5.

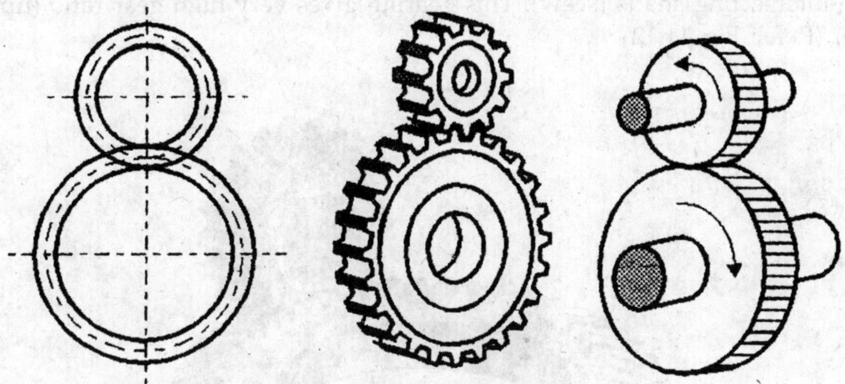

Fig. 11.10. Spur gear.

Application. In small gear boxes.

2. Helical Gear. The only difference in spur gear and helical gear is in helical gear the teeth are cut making certain angle with axis of shaft. Gradual loading of tooth is characteristic of the gear, which results in smooth, noiseless operation. These gears have high power transmitting capacity and are used upto gear ratio of 10. (Refer Fig. 11.11).

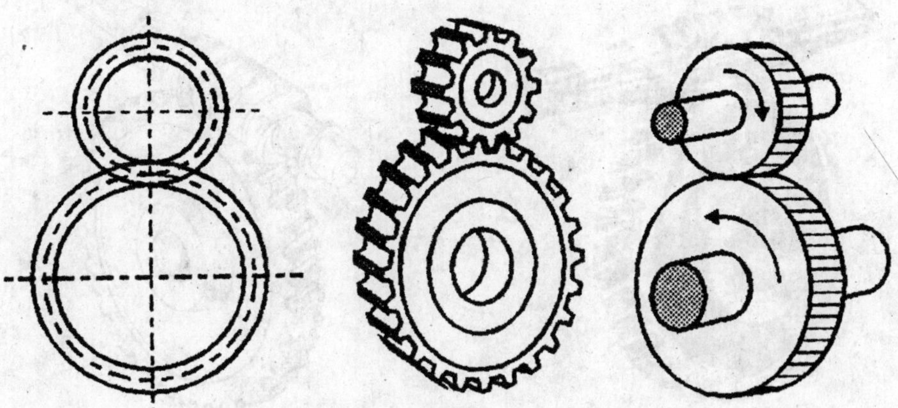

Fig. 11.11. Helical gear.

Application. In special duty gear box as marine gear box.

3. Bevel Gear. In this type, the teeth are cut on conical surface. These are used to transmit power between two shafts having intersecting axes. (Refer Fig. 11.12).

Application. Differential of 4 wheeler automobile.

4. Worm and Worm Wheel. These are used to transmit the power between to non-intersecting shafts (scew). This gearing gives very high gear ratio (up to 100). (Refer. Fig. 11.12)

Fig. 11.12. Bevel gear.

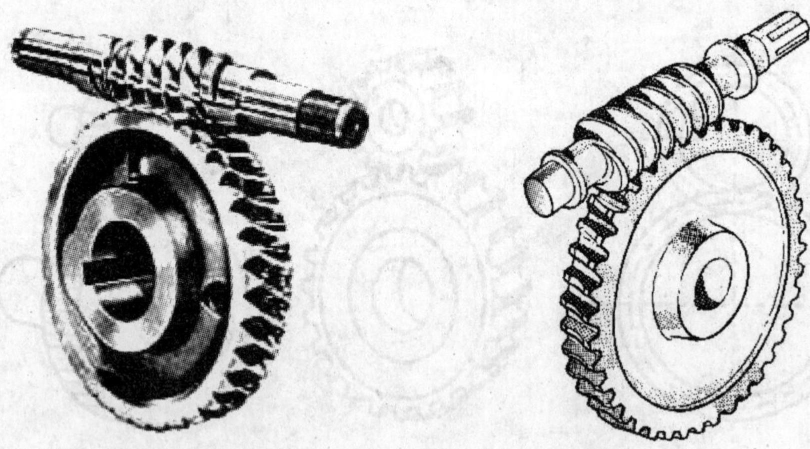

Fig. 11.13. Worm and worm wheel.

Application. These are used in Lathe machine, feed shaft.

5. Rack and Pinion. The toothed flat strip is known as *rack* and small circular toothed wheel is known as *pinion*. This arrangement is used to convert rotary motion into linear motion.

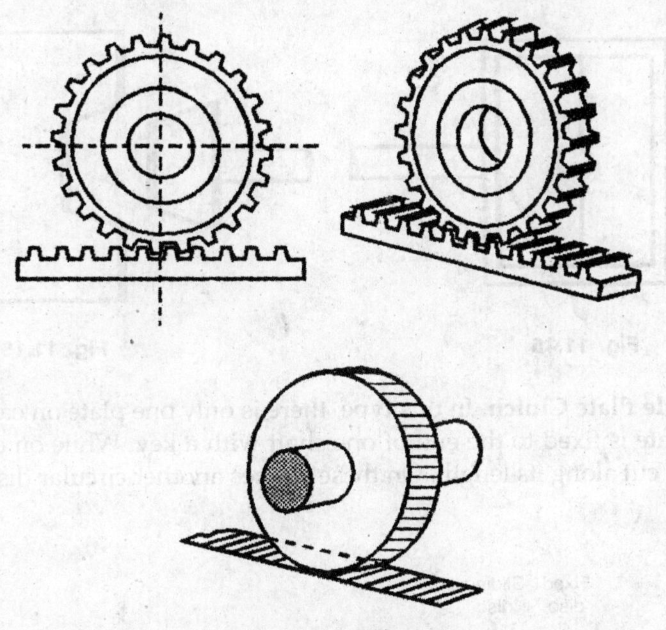

Fig. 11.14. Rack and pinion.

Application. Plastic hand molding machine. (Fig. 11.14)

11.6 CLUTCHES

Clutch is a device used to connect or disconnect two coaxial shafts. Clutches have very wide application in automobiles.

With help of clutch we can bring the vehicle to rest, without stopping, the engine, just by disconnecting the input and output shafts. It also enables change of gear without creating a shock load on engine.

11.6.1 Types of Clutches

Clutches are classified in two types as described below.

1. Positive Clutch. This type of clutch when engaged, connects two shafts rigidly.

2. Non-Positive Clutch. These normally use frictional force to transmit power. Hence these are also known as *friction clutch.* There are three types of friction clutch as:

(a) Single plate clutch

(b) Cone clutch

(c) Centrifugal clutch.

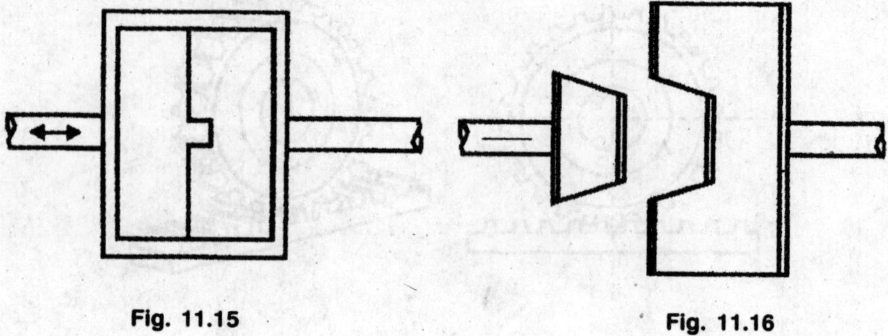

Fig. 11.15 Fig. 11.16

(a) Single Plate Clutch. In this type, there is only one plate on each shaft. A circular plate is fixed to the end of one shaft with a key. While on other shaft, splines are cut along its length. On these splines another circular disk can slide freely.

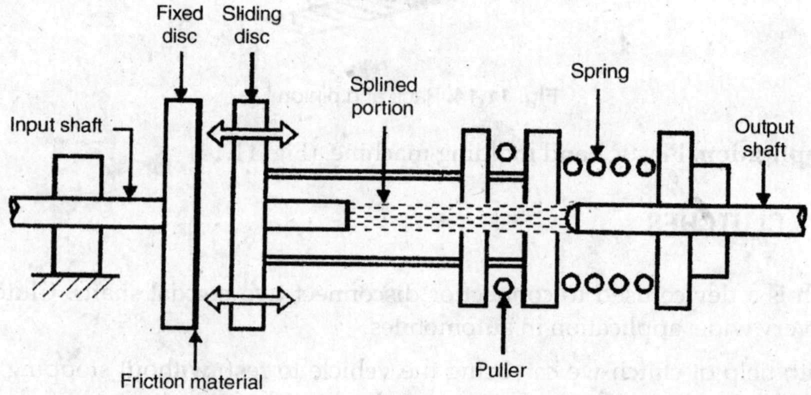

Fig. 11.17

The surfaces of one of the plates facing other is lined with friction material. The sliding plate is forced to remain in contact with the fixed plate with spring force. While this plate can be pulled with help of *puller* which is fixed to *neck*. Thus when the sliding plate is pulled against spring force, it result in disconnecting the two shafts.

Due to low power transmitting capacity it has not much practical importance.

(b) Cone Clutch. If we want to increase the power transmitting capacity of a clutch, we have to increase the pressure between the contact surfaces or the area of contact surface.

But there is limit for the higher value of pressure which the friction linings can withstand, hence one has to increase the contact area.

This is achieved by making the friction linings on inner or outer conical surfaces of the plates mounted on each shaft, as shown in Fig. 11.18.

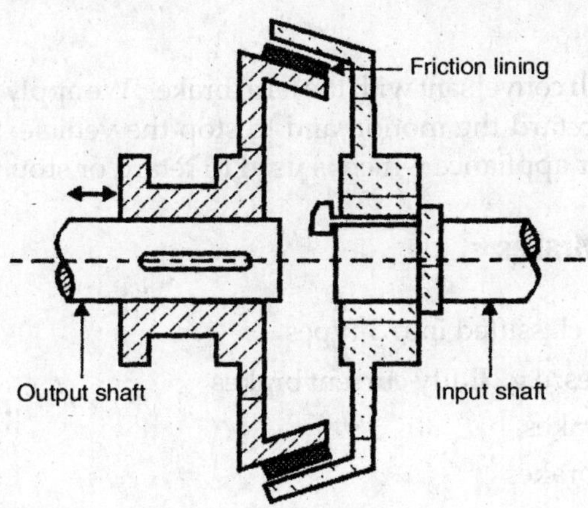

Fig. 11.18

Similar to the plate clutch, one cone is fixed on one shaft while the other can slide. The operation in same as the plate clutch.

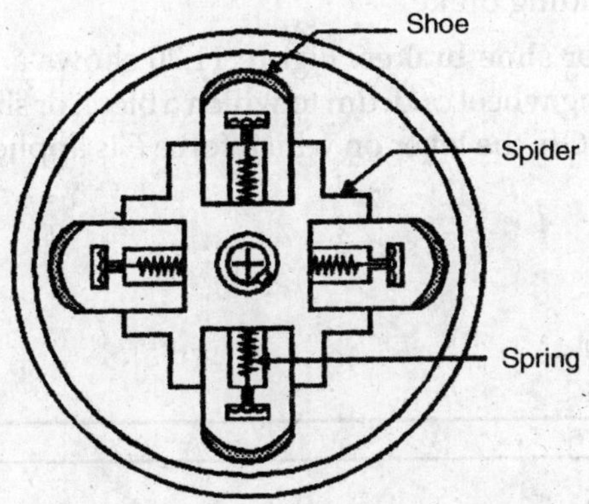

Fig. 11.19

(c) Centrifugal Clutch. This gets engaged due to the centrifugal force. Thus it operates automatically.

In this the shoes are held in the position as shown in Fig. 11.19 with springs, which are fixed to the shaft. When the shaft starts rotating due to centrifugal force the shoes start moving in radially outward direction and at a specific speed they make contact with the dish from inside. This dish is fixed to the output shaft. This type of clutch is widely used in small automobiles (below 50ºC).

11.7 BRAKES

We all are very much conversant with the term brakes. We apply brakes in bicycles, motor vehicles to retard the motion and to stop the vehicle. So we can define brake as a device or appliance which is used to retard or stop the vehicle.

11.7.1 Types of Brakes

Brakes are broadly classified into 3-types.

1. Electric brakes: e.g., Eddy current brakes
2. Hydraulic brakes
3. Mechanical brakes

Mechanical brakes are further classified as:

(a) Single block or shoe brakes
(b) Double block or Double shoe brake
(c) Band brake
(d) Band and Block brake
(e) Internal Expanding brake.

(a) Single block or shoe brakes. Figure 11.20 shows a single block or shoe brake. *A* is the rotating wheel or drum to which a block or shoe. *B* is pressed so as to retard its speed. *OC* is the lever on which force *F* is applied to retard and stop the wheel.

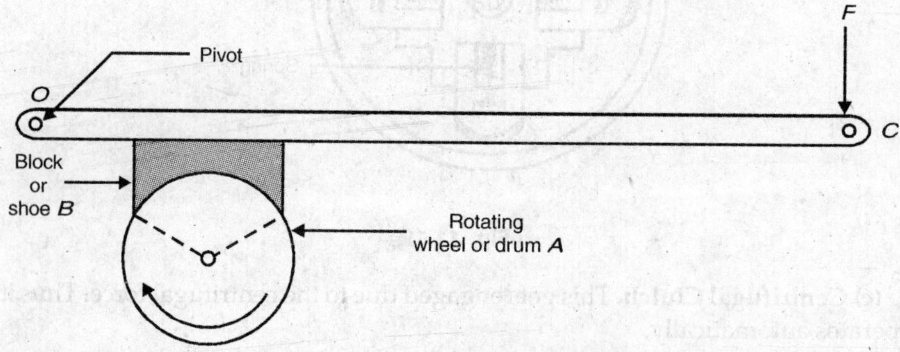

Fig. 11.20

(b) Double Block or Double Shoe Brake. Figure 11.21 shows a double block or double shoe brake. A is the rotating wheel or drum to which two blocks B are pressed so as to retard its speed. OC and $O'C'$ are the two levers on which forces F and F' are applied to retard and stop the wheel.

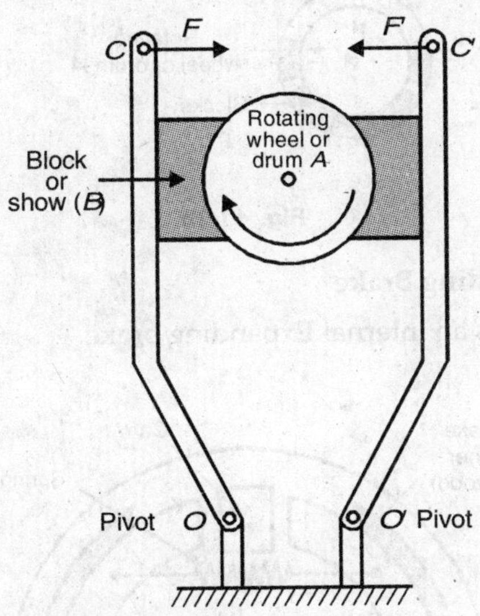

Fig. 11.21

(c) Band Brake. Figure 11.22 shows a Band brake. A is the rotating wheel or drum to which a band generally belt / rope / flexible band (B) is pressed so as to retard its speed. OC is the lever on which force F is applied. T_1 and T_2 are the tensions in rope or belt. Similarly we can have the drum rotation in anti-clockwise direction also.

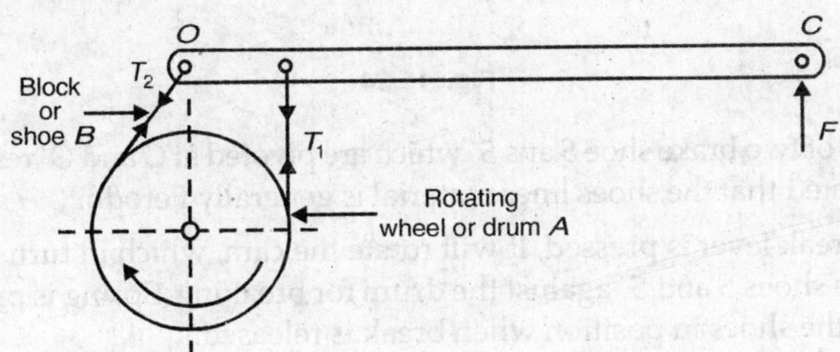

Fig. 11.22

(d) Band and Block Brake. Figure 11.23 shows Band and Block brake. As the name implies it consists both belt or flexible band and the wooden blocks as shown. Force F is applied for braking.

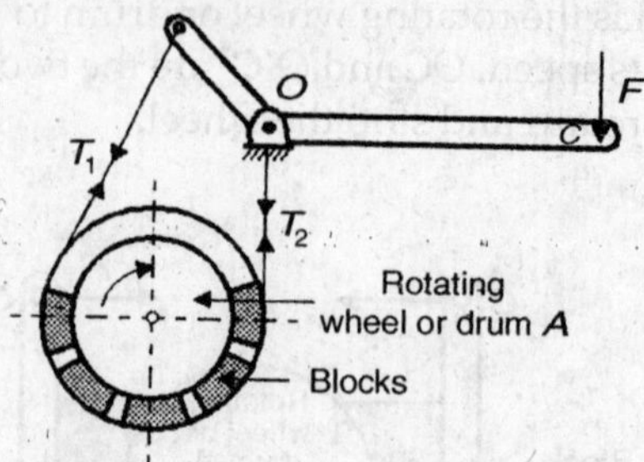

Fig. 11.23

(e) Internal Expanding Brake

Figure 11.24 shows an internal Expanding brake.

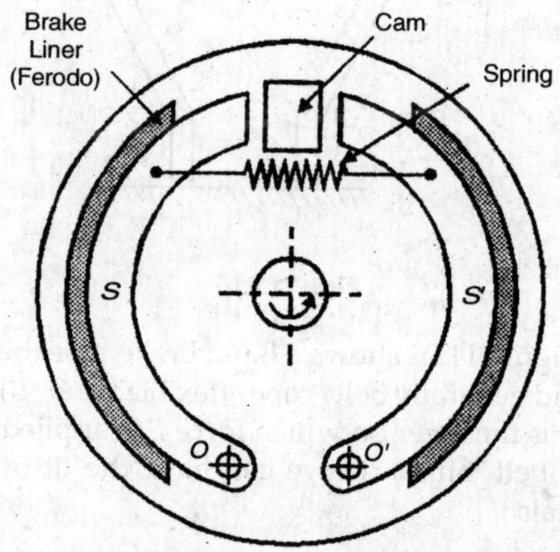

Fig. 11.24

It consist of two brake shoe S ans S' which are pivoted at O and O' respectively. It is to be noted that the shoes liner material is generally Ferodo.

When break lever is pressed, it will rotate the cam, which in turn presses or expands the shoes S and S' against the drum for breaking. Spring is provided to bring back the shoes in position when break is released.

Application. Breaks are used in bicycles, motor vehicles, tram cars, railways etc.

THEORY QUESTIONS

1. Explain individual and group drive system with sketch.
2. Compare the advantages and disadvantages of group drives over individual drives.
3. Compare Group drive and Individual drive.
4. Write a short note on 'Belt drives'.
5. Draw with neat sketches different types of belts commonly used in workshops.
6. State the function of the following mechanical element : V-belt
7. State the advantages of V-belt over Flat Belt.
8. Enumerate various types of belts with their applications.
9. Write a short note on rope drive.
10. What are different types of gears? Mention their applications.
11. Draw the sketches of the following types of gears and state their applications.
    1. Spur gear
    2. Helical gear
    3. Bevel gear
    4. Worm and worm wheel
12. Explain with diagrams which types of gears are used in the following situations:
    1. Parallel shafts
    2. Intersecting shafts
    3. Non-parallel and non-intersecting shafts.
13. Compare the belt drive, chain drive and gear drive on the basis of following: layout / design, applications, advantages and limitations. You may present your answer in tabular form.
14. What is a clutch, what are its different types?
15. Explain the difference between a coupling and a clutch.
16. Write a short note on single plate clutch, cone clutch and centrifugal clutch.
17. Draw a neat sketch of single plate clutch and describe its working.
18. Draw the sketches of the Helical gear and state their applications.
19. Explain chain drive in detail.
20. Explain what is brake? What are its types? Explain in detail.

CHAPTER 12

MACHINE ELEMENTS

12.1 POWER TRANSMISSION SHAFTS

Shaft is a machine element, which is used to transmit power from one member to another member. When a tangential force is acting on the shaft, a turning moment is set up. This makes the shaft to rotate for transmitting power from one shaft to another. Pulleys, gears etc. are mounted on the shafts through keys.

Generally the cross-section of the shafts will be circular (i.e. cylindrical in length) also they are solid throughout their length. However square or rectangular shafts are also available depending upon specific applications.

12.2 TYPES OF SHAFTS

They are broadly classified into two types:

(i) Power transmission shafts. As the name implies, they transmit power from one member to another member.

(ii) Machine shafts. As the name suggests, they are the part of machine only. For example crank shaft.

12.3 SHAFT MATERIAL

Generally for regular shafts carbon steels are used. However for higher strengths Ni, Ni–Cr steels are used.

12.4 APPLICATION OF SHAFTS

As discussed earlier whenever power is to be transmitted from one member to another member, shafts are used.

Figure 12.1 shows a flywheel or gear mounted shaft. Shaft is supported in the bearing at *A* and *B*.

12.5 AXLE

A non-rotating member similar to that of a shaft is called as an axle. It simply supports a rotating body.

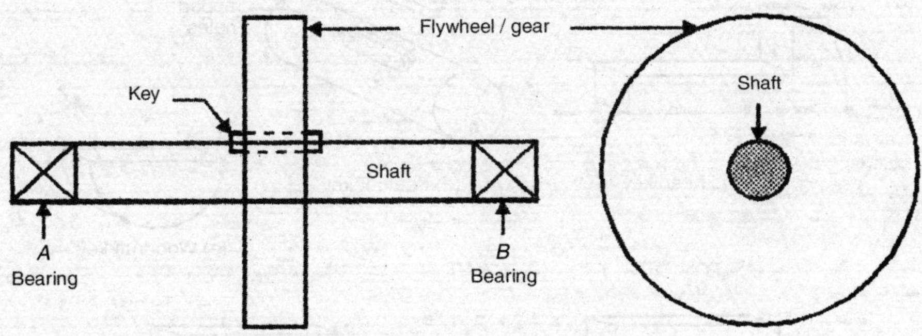

Fig. 12.1. Flywheel mounted on a shaft.

12.6 APPLICATION

One might have seen the pulley mounted on the axle on the wells, where pulley is rotating, when the rope is pulled and the axle is stationary.

Key is a member used to prevent relative rotational motion between shaft and flange or pulley. It is a piece of metal, rectangular in cross-section. The slots on outer surface of shaft and inner surface of the bore are cut along the length. When these slots are matched they give a cross section same as that of the key. The key is then inserted in this hole which prevents the rotational motion of flange or pulley on the shaft.

12.7 KEYS

Various types of keys are used in practice as shown in Fig. 12.2. Key is a metallic piece inserted mating shaft and hole to prevent their relative motion. Pulleys, gears, fly wheels etc. are secured to the shaft by means of keys.

Key way. Keys are inserted in the key ways cut in the mating parts i.e. shaft and hub (or gears, pulleys, fly wheels etc.).

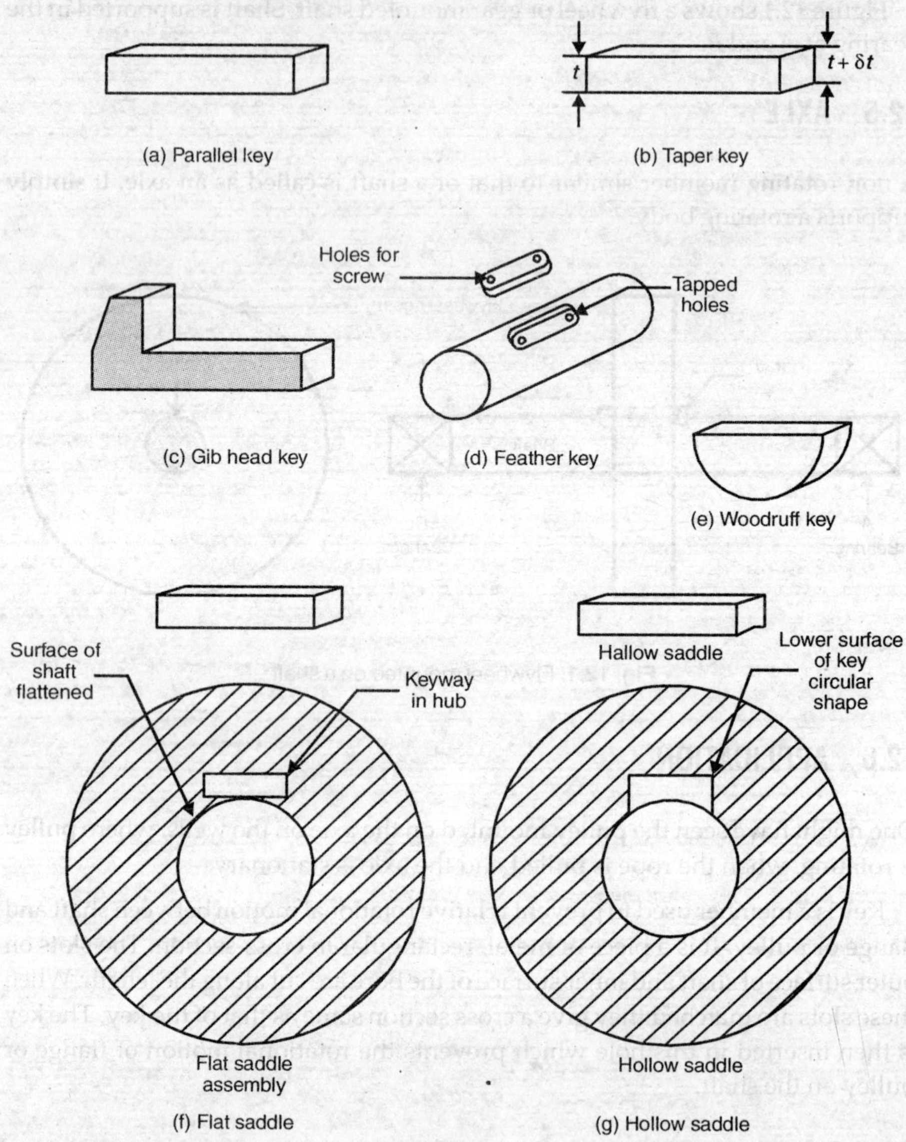

Fig. 12.2

Types of Keys

Main types of keys are describes as below:

(a) Sunk Keys. Here the keys are inserted in the key ways such that they are half in the shaft and remaining half in the hub portion. Various types under the sunk key category are:

(i) **Parallel key.** The key has uniform rectangular or square cross section throughout its entire length. Such keys are used for securing pulleys and gears.

(ii) **Taper key.** This is similar to the parallel sunk key excepting that it has taper in its thickness along the length. The taper may be 1 in 100. The width is uniform and only the thickness has the taper as shown. If the key has thickness, width and length of say 10, 20 and 100 mm at one end its thickness at the other end will be 9 mm while all other dimensions will remain same if it has a taper of 1 in 100. Due to taper insertion and removal is easy and it also prevents relative circular and axial moment.

(iii) **Gib Head key.** It is taper sunk key with head at one end. The head makes it easy to remove the key. Such keys are used where the shaft and hub are required to remove frequently.

(iv) **Feather key.** It is a special type of parallel sunk key and is secured to the shaft by screws as shown. The key allows transmission of turning moment as well as axial force. It is generally used in drilling machines and clutches.

(v) **Woodruff key.** It has semicircular cross section which partly sits in similar key way in the shaft portion. It can be used to assemble any shaft or hub with taper.

(vi) **Round key.** It has a circular cross section and fits into the circular key way in the shaft and hub. There are chances of slip under heavy loads. Therefore such keys are used for light loads.

(b) Saddle keys. Saddle keys are attached to the hub portion and just sit over the shaft. There is key way only in one mating portion generally the hub. These are used for light loads. Basically there are two types:

(i) **Flat saddle.** The key is flat like parallel key which sits on flattened surface of shaft. There is key way in the hub. (Fig. 12.2f)

(ii) **Hollow saddle.** The lower side of the key is hollow i.e. it has the same contour as that of the shaft. The holding force is the friction only. (Fig. 12.2g)

12.8 COUPLING AND THEIR TYPES

The arrangement used to connect two rotating shafts with each other is known as *coupling*.

The simplest type of coupling is muff coupling. This is used for two co-axial shafts. The 'muff', which is a hollow cylindrical piece of metal is keyed to both shafts and thus transmits the motion and power.

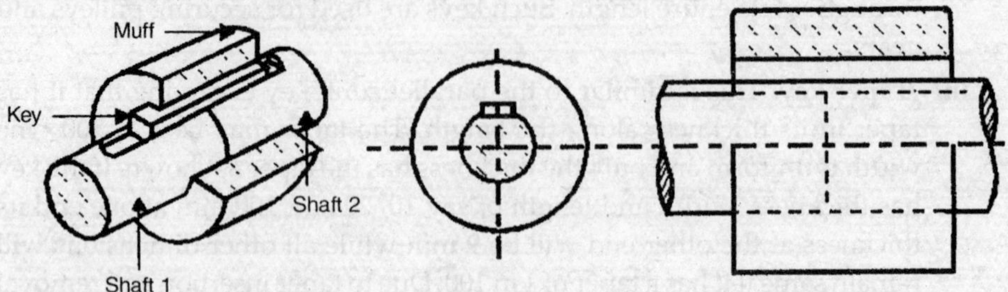

Fig. 12.3. Muff coupling.

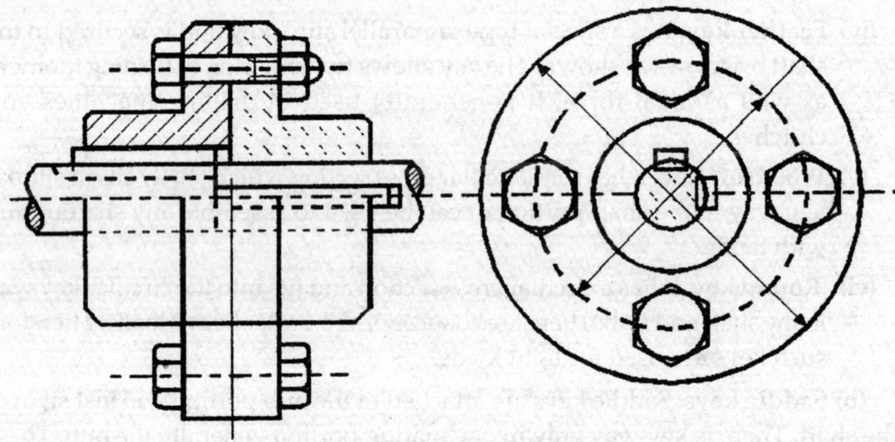

Fig. 12.4. Flanged coupling.

The muff may be manufactured in one piece or in two halves. In the latter case, coupling is known as split muff coupling.

12.9 FLANGED COUPLING

Flange couplings are used to connect two co-axial shafts.

Flange is a circular disc with a bore at centre. A keyway is cut on the bore along its length. Bore diameter is slightly greater than shaft diameter. Number of holes are drilled on the flange. One flange each is keyed to the shaft, while the flanges are connected to each other with bolts.

The types are:

(1) Unprotected flange coupling. (Ref. Fig. 12.4)

(2) Protected type flange coupling. A projection is provided on the outer surface of each flange. This projection prevents any part of nut, bolt when broken, from moving away, because of centrifugal force and hence prevents the hazards.

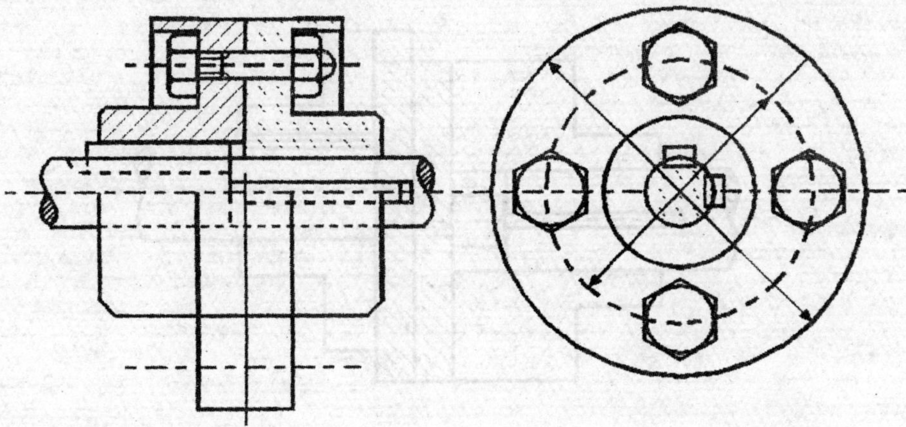

Fig. 12.5. Protected flanged coupling.

(3) Flexible coupling

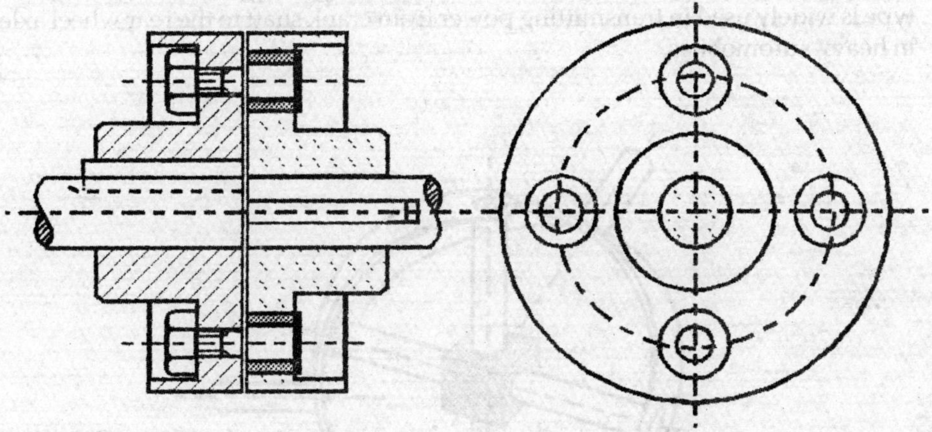

Fig. 12.6. Flexible coupling.

If the load is fluctuating simple flange coupling will transfer these fluctuation to the prime mover. To avoid this, rubber bushes are put in the holes of one of the shaft. These bushes act as shock absorbers.

12.10 OLDHAM'S COUPLING

This is used to connect two parallel shafts with small distance between them. (Ref. Fig. 12.7)

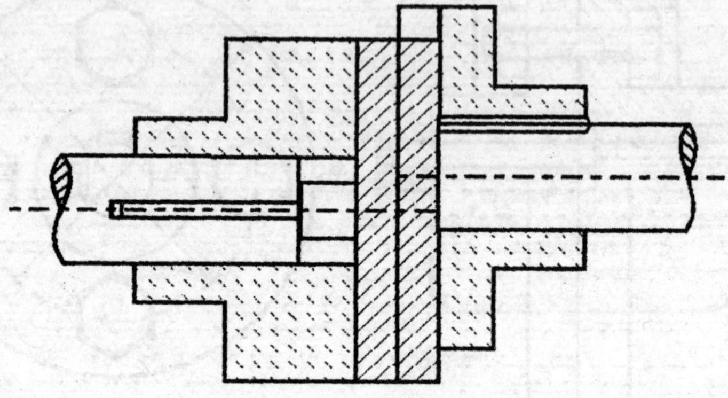

Fig. 12.7. Oldham's coupling.

12.11 UNIVERSAL COUPLING OR HOOKE'S JOINT

This type of coupling is used to connect two shafts having interesting axes. This type is widely used in transmitting power from crank shaft to the rear wheel axle in heavy automobiles.

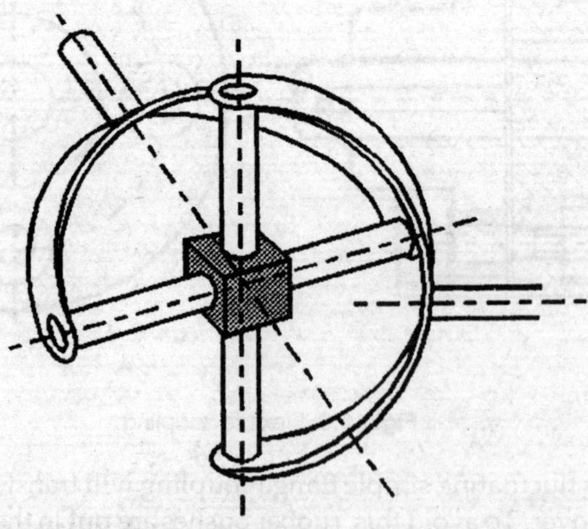

Fig. 12.8. Universal coupling or Hooke's joint.

In this type a fork is fixed to end of each shaft. Two forks are connected together with a cross head and pins.

The disadvantage is that we get non-uniform angular velocity at output. This is overcome by using two couplings in line.

12.12 BEARINGS AND THEIR TYPES

Bearing, as its name suggests, bears the load of shaft without offering any resistance to its rotational motion.

Bearings are used to support the shaft and to transfer the load to the ground. But if the shaft is supported on a rough surface, large amount of power will be lost in overcoming the friction. Hence in *bearing* by means of proper *lubrication*, the direct contact between shaft and supporting surface is prevented. Thus the frictional losses are reduced by very large extent.

Normally two types of bearings are used.

(i) Journal Bearing. In this type the Journal bearing or the *plummer block* is lined with special material to reduce the friction. In this, the shaft theoretically makes line contact. Hence this type has high load bearing capacity.

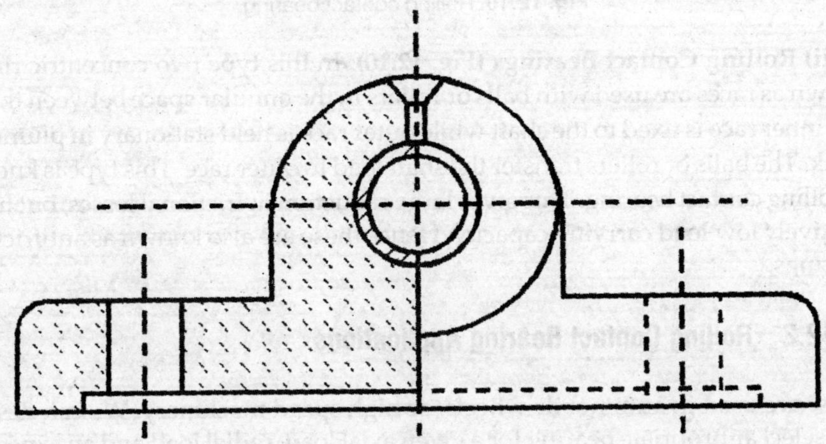

Fig. 12.9. Journal bearing.

12.12.1 Journal Bearings Applications

Used for shafts for general purpose machinery with medium loads and medium speeds. For example, compressor shafts, overhang transmission shafts, shafts used with group drives in floor mills.

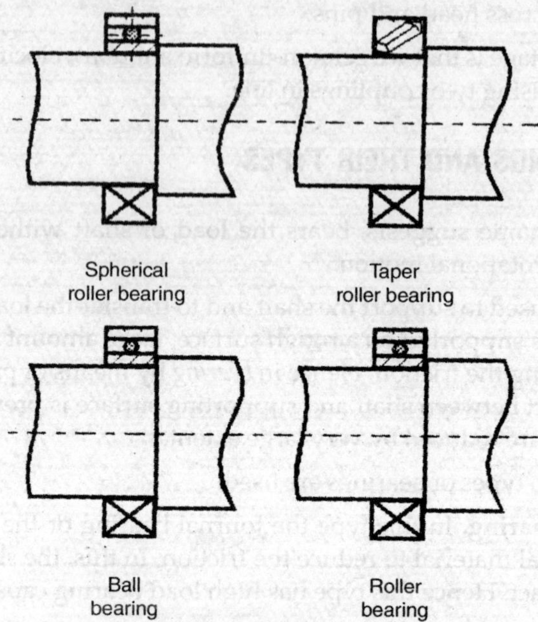

Fig. 12.10. Rolling contact bearing.

(ii) Rolling Contact Bearings (Fig. 12.10). In this type two concentric rings, known as races are used with balls or rollers in the annular space between them. The inner race is fixed to the shaft while outer race is held stationary in plummer block. The balls or rollers transfer the shaft load to outer race. This type is known as rolling contact bearing. This gives large reduction in frictional losses, but have relatively low load carrying capacity. Hence these are also known as antifriction bearings.

12.12.2 Rolling Contact Bearing Applications

These are used in automobile wheels, or high speed machinery. We can design and select appropriate bearing for a given axial load, radial load and an expected life of bearing. These bearings are safe, as they make a squeaky noise before failing.

12.12.3 Thrust Bearings

There are two types of thrust bearings viz. foot step and collar thrust.

Foot step bearing is used to take the axial load (acting along its length) of the shaft.

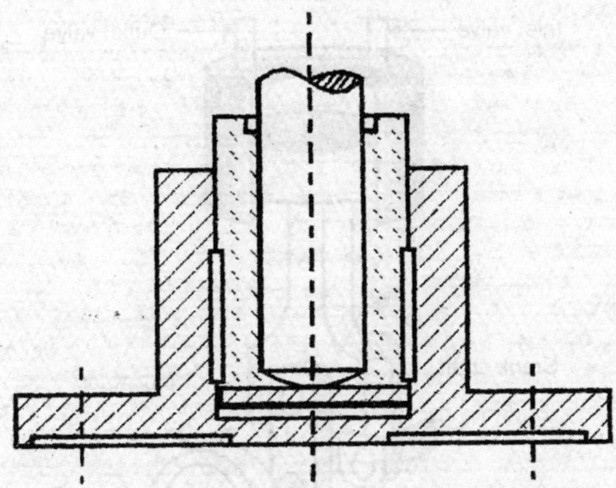

Fig. 12.11

It comprises of three parts. Two parts are identical circular discs having a groove along their diameters on one face. These discs are mounted on the shafts with keys. The third part is central disc. It is also circular disc of same size but it has two projections along its diameter but at right angle to each other on its either side. When these parts are assembled as shown in Fig. 12.11, transfer the rotating motion of one shaft to the other by means of sliding of the projection on central disc in the grooves in the end discs.

12.13 FLYWHEEL

Flywheel stores energy when more energy is being produced and gives back the same energy when less energy is being produced.

Or in the other words, "it is an energy reservoir, it stores energy when more energy is produced and gives back when less energy is produced, so that the crank shaft has uniform speed".

If no fly wheel is used then the working of the engine will be jerky.

As we know in case of 4-stroke cycle engines there are four strokes suction, compression, power generation and exhaust. And power is produced only during the power stroke and no power is produced during suction, compression and exhaust strokes.

The energy produced during the power stroke will be stored in the flywheel in the form of inertia and is given back during other strokes the crack shaft so as to have the uniform speed.

As shown in Fig. 12.12(b) a wheel is mounted on the shaft and which is being rotated by hand. If the hand is removed wheel will continue rotating for some time depending upon energy imparted to it.

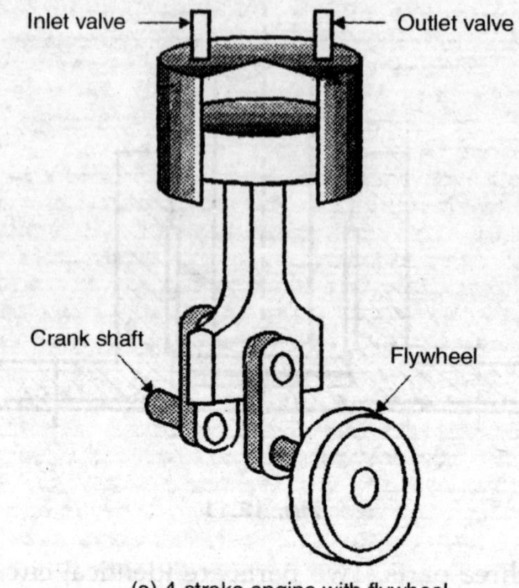

(a) 4-stroke engine with flywheel

(b) Wheel mounted on a shaft and rotated by hand

Fig. 12.12

12.14 FLYWHEEL CONSTRUCTION AND TYPES

Flywheels are made out of castings. Depending upon the diameter and speed they are classed into 3-types.

(i) Smaller flywheel with web and without arms for sizes upto 550–650 mm diameter.

(ii) Medium sized flywheel for sizes upto 2.25 m to 2.50 m dia. and arms provided as shown in Fig. 12.12 (a) and (b).

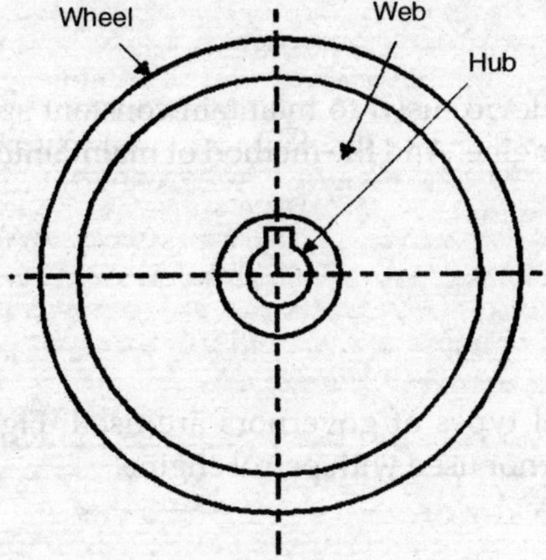

(a) Flywheel with web.

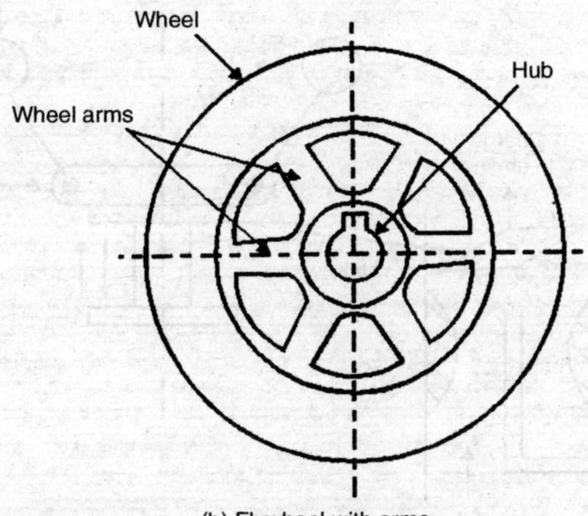

(b) Flywheel with arms.

Fig. 12.13

(iii) For sizes above 2.5 m diameter and for higher speeds split type of CI fly wheels are used as shown in Fig. 12.14.

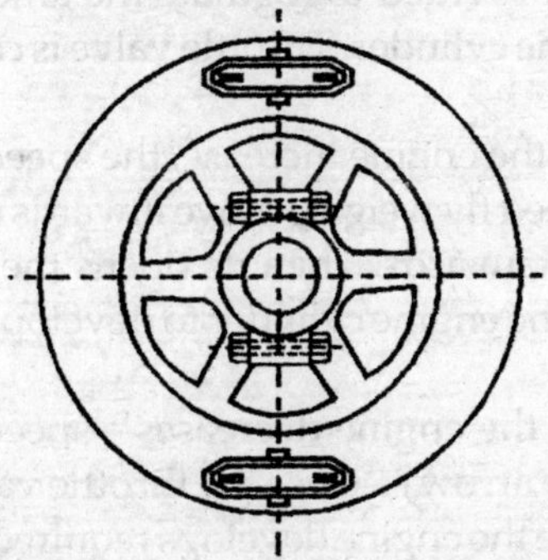

Fig. 12.14. Reproduce mirror image.

12.15 GOVERNOR

It is the mechanism or device used to maintain constant speed irrespective of changes of load on the engine. And the method of maintaining constant speed is called as *Governing*.

Classification

(i) Centrifugal

(ii) Inertia

Generally centrifugal types of governors are used. Figure 12.15 shows a centrifugal type of governor used with petrol engine.

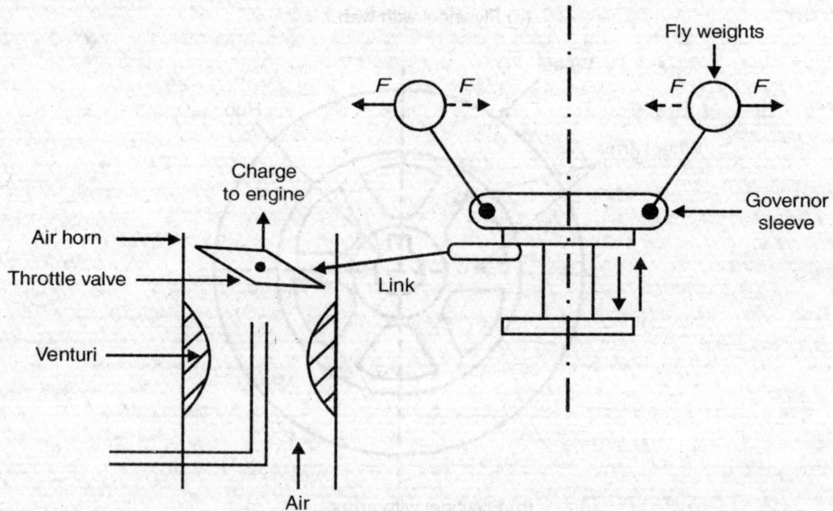

Fig. 12.15. Centrifugal governor.

Cylindrical part is the air horn and the narrowest cross-section of which is the venturi. Throttle valve is provided to regulate the amount of charge (air + fuel mixture) entering the engine cylinder. Throttle valve is connected to the Governor as shown.

Now when the load on the engine increase, the speed of the engine decreases and because of reduced speed fly weights move inwards (shown by dotted arrows) and the sleeve moves downwards. Then its opens the throttle valve more and hence more charge enters the engine cylinder to develop more power and to bring engine speed constant.

And when the load on the engine decreases—speed increases—fly weights move outwards (shown by arrow)—close the throttle valve accordingly—reduce the supply of charge. Hence the engine develops required less power and maintain constant speed.

Centrifugal Governor Classification

Following are the improtant centrifugal governors.

(i) Watt Governer (Simple Conical Governer). It is named after Watt, when the load decreases-speed increases fly weights move outwards as shown by arrows and lift the sleeve. On the other hand when the load increase speed decreases, weight move inwards (shown by dotted arrows) and sleeve moves downwards. (Ref. Fig. 12.16)

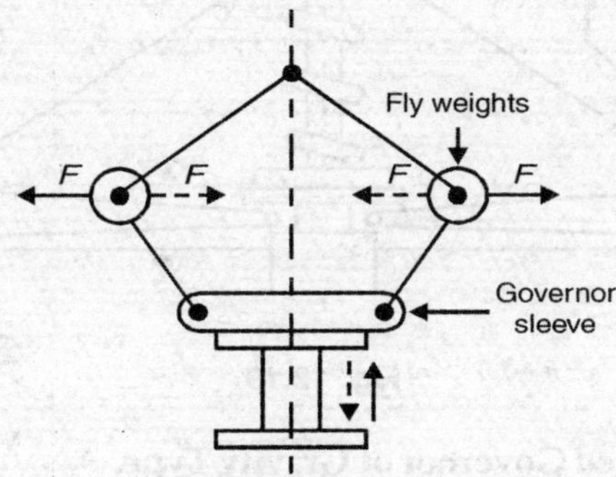

Fig. 12.16

(ii) Porter Governor. If the central load is provided on the sleeve then watts governor is called its Porter Governor. (Ref. Fig. 12.17)

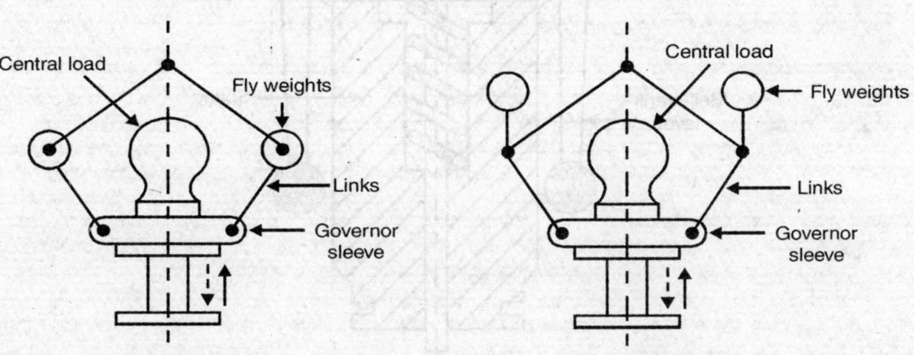

Fig. 12.17 **Fig. 12.18**

(iii) Proell Governor. In this governor fly wights are provided on the extension of lower links as shown and it contains central load. (Ref. Fig. 12.18)

(iv) Hartnell Governor. It mainly consists of 2-bell crank levers and compression spring as shown. To the one end of bell crank levers, fly weight are provided and other end rollers. The levers are pivoted at *O*. When the speed increases fly weight move outwards and the lever which is provided at *O* lifts the rollers and sleeve against the compression of spring.

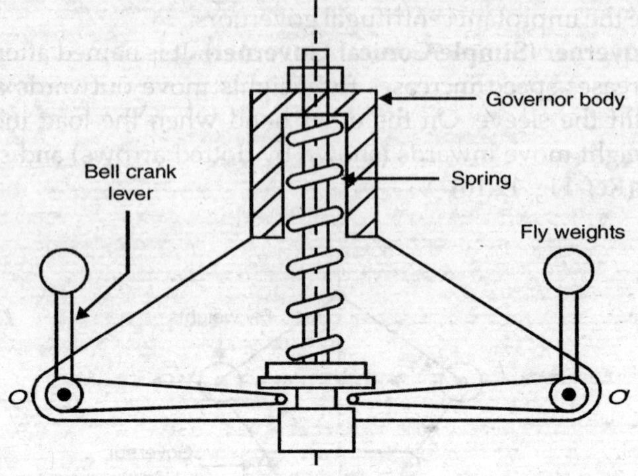

Fig. 12.19

(v) Spring Controlled Governor of Gravity Type.

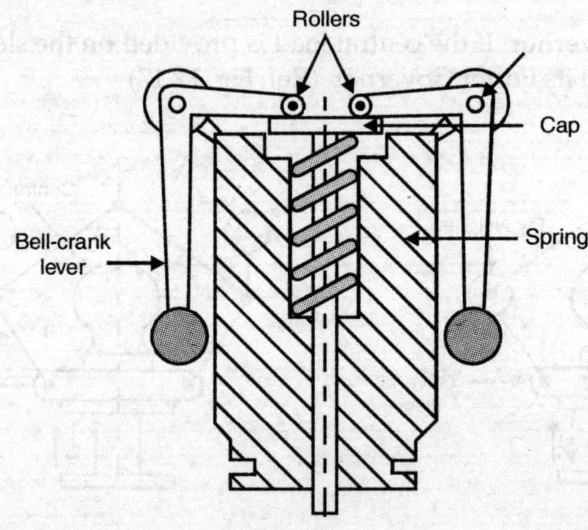

Fig. 12.20

Figure 12.20 shows spring controlled gravity governor. It mainly consists of 2-bell crank levers. They are provided with rollers at one end and fly weights at the other. Rollers will be resting over the cap which forms top end of shaft. Under the cap spring is provided as shown.

As we know the load decreases, speed increases and fly weight move outwards and the other press cap which in turn compresses the spring.

Application of Governors

Used in Petrol engines, Diesel engines, Steam engines and Steam turbines etc.

THEORY QUESTIONS

1. What is a 'key'. Draw sketches of different types of keys.
2. What is a 'coupling'? Draw a sketch of Muff Coupling and Flanged Coupling.
3. Explain the difference between a Clutch and Coupling.
4. Draw a sketch of flanged coupling.
5. Sketch different types of flanged coupling showing their features.
6. Draw the sketches of the Helical gear and state their applications.
7. Explain with the help of a suitable diagram the construction, working and applications of a bush pin type flexible coupling.
8. Compare rigid and flexible coupling.
9. What are the different types of shaft couplings?
10. With a neat sketch explain universal coupling.
11. What is function of bearing? What are its types?
12. Draw a neat sketch of a ball bearing and name its parts. What are its advantages?
13. What is a thrust bearing? Draw sketch of its types.
14. Explain the different types of bearings used to support shafts.
15. Draw neat sketches of:
    (a) Bush bearing
    (b) Thrust bearing
    (c) Sunk and Woodruff key
16. State the advantages of rolling contact bearings over sliding contact bearings.
17. State the function of the following mechanical element: Bush Bearing.
18. Compare bush bearing and antifriction bearings.
19. State where Thrustbearing is used.
20. State the advantages of ball and roller bearings over plain journal bearings.
21. What is a 'key'. Draw sketches of different types of keys.
22. What is a 'coupling'? Draw a sketch of Muff Coupling and Flanged Coupling.
23. Explain the difference between a Clutch and Coupling.
24. Draw a sketch of flanged coupling.

25. Sketch different types of flanged coupling showing their features.
26. Explain with the help of a suitable diagram the construction, working and applications of a bush pin type flexible coupling.
27. Compare rigid and flexible coupling.
28. What are the different types of shaft couplings?
29. With a neat sketch explain universal coupling.
30. What is function of bearing? What are its types?
31. Draw a neat sketch of a ball bearing and name its parts. What are its advantages.
32. What is a thrust bearing? Draw sketch of its types.
33. Explain the different types of bearings used to support shafts.
34. Draw neat sketches of:
    (a) Bush bearing
    (b) Thrust bearing
    (c) Sunk and Woodruff key
35. State the advantages of rolling contact bearings over sliding contact bearings.
36. State the function of the following mechanical element: Bush Bearing.
37. Compare bush bearing and antifriction bearings.
38. State where Thrustbearing is used.
39. State the advantages of ball and roller bearings over plain journal bearings.

Practicals Experiments (TW 25 Marks)

Experiment 1 : Study of steam Generators :

(i) Babcock Wilcox Boiler [Ref. Article 3.5.5]

(ii) Lancashire Boiler [Ref. Article 3.5. 1]

Experiment 2 : I. C. Engines [Ref. Article 3.10]

Experiment 3 : House hold refrigerator and Window air conditioner [Ref. Article 4.12, 4.14]

Experiment 4 : Shell and tube heat exchanger [Ref. Article 6.6]

Experiment 5 : Solar water heater

Aim : To study solar water heater.

In solar water heater, the principle followed is to expose a dark surface to solar radiation so that the radiation is absorbed. This absorbed radiation is then transferred to water.

When no optical concentration is done, a device in which collection is achieved is called a flat plate collector. It is the most common device as it is very simple in design, has no moving parts and requires negligible maintenance. It is nowadays widely used for hot water applications with temperature range of 40°C to 80°C.

Figure P-1 shows a typical arrangement of a liquid solar flat plate collector. It consists of an absorber plate on which solar radiations fall after coming through a transparent cover (glass). The absorbed radiation is partly transferred to water flowing through the tubes that are fixed to the absorber plate. The remaining part of absorbed radiation are lost by convection and re-radiation from the top surface. The transpartent cover helps in reducing convection and re-rediation losses while, the bottom insulation reduces losses due to condction. A solar (water heater) collector is usually held tilted and at Pune tilt angle is 18° facing south. The capacity of the water drum is usually 100 to 200 litres and the collector surface area is 2m². If required, the capacity may be enhanced by arranging several individual collectors in row having a common outlet pipeline.

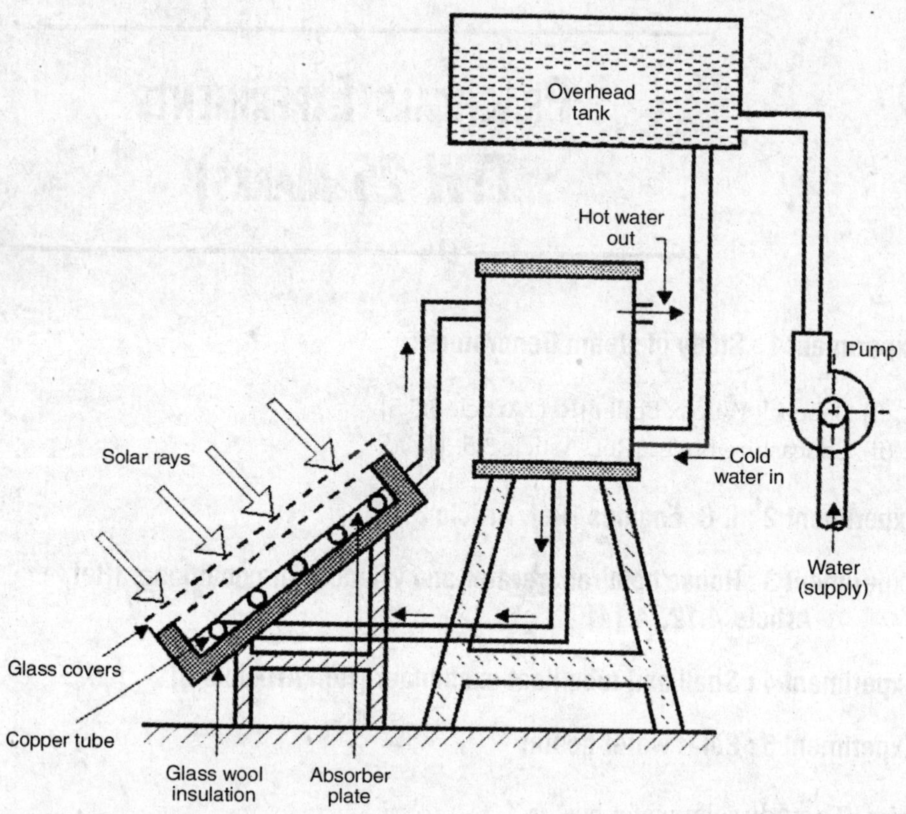

Fig. P-1

Apart from solar water heater, other applications of solar power are

1. Solar dryers
2. Solar cookers
3. Solar photovoltaic converters etc.

Advantages of solar energy

1. This energy is freely available and in abundance.
2. No environmental pollution is caused by solar energy.
3. Equipments require negligible maintenance.
4. Can be stored in dry batteries (as electricity) or in oil pipes as heat energy)

Disadvantages

1. Initial investment cost in solar equipment is high.
2. Preservation of solar energy in the form of either heat or electricity is difficult.

Due to above mentioned advantages, solar operated equipments are nowadays very popular in our country and the future is bright.

Experiment 6 : Lathe [Ref. Article 7.2]

Experiment 7 : Drilling machine [Ref. Article 7.5]

Experiment 8 : NC/CNC machine [Ref. Articles 7.10 to 7.16]

Experiment 9 : Power transmitting elements like shafts, couplings and gears [Ref. Articles 11.5 and 12.1, 12.2, 12.8]

Experiment 10 : Material Handling Equipments

Aim : To study various Material Handling Equipment.

Introduction : Materials and equipment is required to be moved during production within a factory or at other places like power stations, mines, airports etc, Light materials can be handled manually but where there has to be continuous requirement of the supply of material or the components are heavy material handling equipments are needed.

Requirement of Material Handling

(a) The materials should be supplied through the shortest distance.
(b) Use of gravity be made for the movement if possible.
(c) As far as possible materials be handled both ways. For example a fork lifter can bring finished item after supplying components.

Advantages of Material Handling

(a) Better productivity — This means more production from the resources
(b) Increased safety.
(c) Less fatigue.
(d) Reduced handling and product cost.

Considerations for Selection of Material Handling Equipment

(a) Type of material to be handled
(b) Type of material flow — Is it continuous or intermittent
(c) Restrictions if any may be considered

Types of Material Handling Equipment :

There are various types of material handling equipments. We can study following representative such equipments.

1. Belt Conveyor
2. Gantry Crain
3. Fork lifter

1. Belt Conveyor. The belt conveyors are used to transport the materials over a fixed route. For example in thermal power stations the coal is generally supplied by belt conveyor system. The belt conveyor system is quite economical where constant supply of materials is needed. Figure P-2 shows a typical belt conveyor.

Construction of Belt

(a) The belt is made of canvass or nylon plies and vulcanized with rubber. The belt is continuous. It goes around two rollers located at the delivery and supply points. Depending on the load and other considerations there are idlers in between to support the load.

(b) The thickness and the width of the belt will deped on the load.

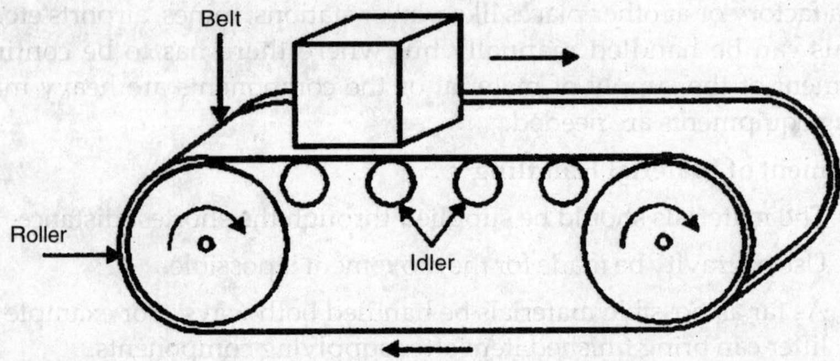

Fig. P-2. Belt Conveyor.

(c) If the belt conveyor has to operate at higher temperature the belt can have steel wire reinforcement.

(d) The roller at one end can be driven by motor and gears at the required speed.

Advantages of Belt Conveyor

(a) Gives continuous supply

(b) Can be used both for bulk and small item (Unlike in roller conveyors here small item can be handled)

(c) Belt conveyours can be used for long distances

(d) Requires less power for operation

2. Cranes. Basically the cranes are used to lift and lower heavy and bulky loads. There are large varieties of cranes for different job. Some of the types of cranes are—

(a) Stationary Cranes. These cranes are generally kept located at one place and are used to lift and lower the loads within a limited area. Jib, Cantilever, Dettick and Pillar cranes are such cranes.

(b) Mobile Crane. Crawler and Portable elevators can be categorised as mobile cranes. Even verstile fork lifter has limited lifting and lowering capability.

(c) Other Cranes. Like Gantry and Bridge cranes.

Jib Pillar Crane. Jib crane consists of a pillar on which a swing arm with chain hoist is attached as shown in Fig. P-3. The load is connected to the hook of the chain hoist which are hand operated. The jib cranes are used to move heavy loads across an arc within a restricted area. Figure P-3 shows the Jib crane.

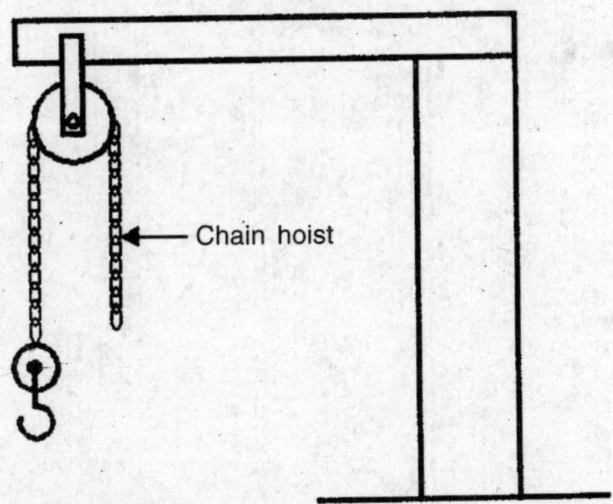

Fig. P-3. Jib pillar crane.

(c) Fork Lift. The fork lift truck is one of the most verstile materials handling equipment used to move material form one place to another and also with ability raise and lower the load as required. These are used within the factory premises only. The fork lifter trucks are generally powered by diesel engines or can be battery operated. Details of a Fork lifter are given in Fig. P-4.

Construction

(a) The fork lifter has two forks in the front side which can be raised or lowered on the guide rail mast. The distance between the two forks also can be adjusted. The control for raising and lowering of the fork is with the operator who drives the fork lift truck.

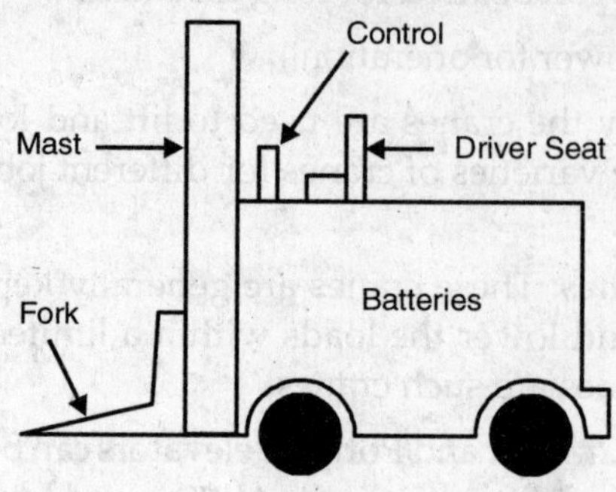

Fig. P-4

(b) This is generally picked up from the ground level and adjused on the two forks evenly for stability. The forks are then raised so that the load clears the grounds. Then the fork lift truck is driven as normal vertical at slow speed to its destination. The load is the lowered to the work place or can be raised to be stacked.

MODEL QUESTION PAPER 1

Solve all the questions

SECTION A

Q.1(a) Explain what is meant by point function and path function. Give two examples of each.

(4 Marks)

(b) Explain microscopic and macroscopic point of view. **(5 Marks)**

(c) A U-tube manometer with one arm open to atmosphere is used to measure the pressure of steam flowing through a pipe. Hg is used in the manometer. The height of the column open to the atm is 9.75 cm greater than the Hg level in the arm connected to the steam pipe. Steam condenses in the manometer on the steam side.

The column of condensate is 3.4 cm. The atmospheric pressure is 76 cm of Hg. Find the absolute pressure of steam. Take,

$$g = 9.806 \text{ m/sec}^2 \text{ and}$$

$$\rho Hg = 13596 \text{ kg/m}^3$$

$$P_{abs} = P \text{ of steam} = ?$$

(8 Marks)

Q.2(a) Prove that internal energy is a property of the system. **(4 Marks)**

(b) State and explain the steady flow energy equation. **(5 Marks)**

(c) Following data refers to a steady flow process

	Enthalpy (kJ/kg)	**Velocity (m/sec.)**	**Height (m)**
At entrance	4000	50	50
At exit	4100	20	10

Mass flow rate $\dot{m} = 1$ kg/sec

Heat transfer rate = 200 kJ/sec to the system.

Heat transfer rate = 200 kJ/sec to the system.

Determine the power capacity of the system and state whether it is a power producing system or otherwise. **(8 Marks)**

Q.3(a) With a neat sketches describe the working of a two stroke petrol engine what are its main advantages and disadvantages. **(9 Marks)**

(b) Discuss the working of a domestic refrigerator with neat sketch. **(8 Marks)**

Q.4(a) Describe the principle and working of centrifugal pump with neat sketch. **(8 Marks)**

(b) Describe the working of water tube boiler with neat sketch. **(9 Marks)**

Q.5(a) Explain the construction and working of a flat plate solar collector. **(4 Marks)**

(b) What is fin ? Give its types and application. **(4 Marks)**

(c) A temperature of the inner sum of a furnace wall is 450°C and that of the outer surfaced 200°C. The ambient temperatures is 50°C. It is required to reduce the heat loss from the outer surface by doubling the thickness of the brickwork. Assuming constant values convective heat transfer coefficient of air and thermal conductivity of wall material calculate percent decrease in heat loss. **(8 Marks)**

Q.6(a) What are the advantages of wind energy ? Explain wind mill. **(4 Marks)**

(b) Explain conduction connection and radiation in brief. **(6 Marks)**

(c) Explain Stefan Boltzmann's of Radiation. **(6 Marks)**

SECTION B

Q.7(a) Draw a neat sketch of a lathe machine, and explain its main parts. **(8 Marks)**

(b) Explain soldering and brazing. **(9Marks)**

Q.8(a) Explain NC/CNC machine in detail. **(8 Marks)**

(b) Explain TIG welding. **(9 Marks)**

Q.9(a) Write notes on pillar drilling machine. **(9 Marks)**

(b) Define stress, strain, factor of safety, yield point, endurance limit. **(8 Marks)**

Q.10(a) What are the different types of sheet metal cutting process explain. **(8 Marks)**

(b) Write notes on limits, fits and tolerance. **(9 Marks)**

Q.11(a) Name the drive and justify its use for the following applications. **(6 Marks)**

(i) Driving of grinding and floor mill.

(ii) Drive for hour and minute hands of a mechanical watch.

(iii) Driving a centrifugal blower by a turbine at the same speed. **(6 Marks)**

(b) Differentiate between flywheel and governor. **(4 Marks)**

(c) Write notes on bush and ball bearings. **(6 Marks)**

Q.12(a) Draw neat sketches of the following and label the parts and give there applications.

(i) Two types of keys

(ii) Rolling contact bearing

(iii) Friction clutch **(10 Marks)**

(b) Explain the different types of gears with neat sketches. **(6 Marks)**

MODEL QUESTION PAPER 2

Solve all the questions

SECTION A

Q.1(a) Explain thermodynamic properties, processes, and cycles. **(6 Marks)**

(b) What is intensive and extensive properties? **(4 Marks)**

(c) A vertical composite liquid column with its upper end exposed to atmospheric pressure, comprise of 45 cm of Hg (sp. gr. 13.6); 65 cm of water and 80 cm of oil (sp. gr. 0.8). Calculate the pressure in bar,

(i) At the bottom of the column.

(ii) At the inter surface of oil and water.

(iii) At the inter surface of water and Hg.

Assume atmospheric pressure as

100 kPa and $g = 9.81\ \text{m/sec}^2$ **(7 Marks)**

or

Q.2(a) Explain Joules experiments. **(6 Marks)**

(b) Explain continuity equation. **(4 Marks)**

(c) Derive steady flow energy equation on time basis. **(7 Marks)**

Q.3(a) Explain Babcock wilcox boiler with a neat sketch. **(8 Marks)**

(b) Explain 4-stroke petrol engine. **(9 Marks)**

Q.4(a) Explain reciprocating pump. **(6 Marks)**

(b) Explain any rotary compressor. **(4 Marks)**

(c) Explain refrigerator with a neat sketch. **(7 Marks)**

Q.5(a) Explain Thermal Power Plant. **(6 Marks)**

(b) Explain bio-gas plant. **(6 Marks)**

(c) Explain fuel cells. **(5 Marks)**

Q.6(a) (1) Explain Newton's law of cooling (2) Stefan Boltzmann's law of radiation. **(8 Marks)**

(b) A brick wall of a building is 30 cm thick and has an inside surface temperature of 24°C and an outside surface temperature of –6°C. The wall is 2.75 m high and 6.1 m long. Estimate (a) the heat transfer by conduction through the wall per hour and (b) the conductance and resistance of the wall. The coefficient of conductivity of the brick material is 2.6 kJ/hr.m°C. **(8 Marks)**

SECTION B

Q.7(a) With a neat sketch of a center lathe. Explain all the parts. **(8 Marks)**

(b) With neat sketches explain NC/CNC machines. **(9 Marks)**

Q.9(a) Explain TIG welding with sketch **(8 Marks)**

(b) Explain soldering and brazing. **(9 Marks)**

Q.9(a) Explain drawing and bending. **(8 Marks)**

(b) Explain limits, fits and tolerance. **(9 Marks)**

Q.10(a) What is need of design explain. **(4 Marks)**

(b) Explain stress strain, stress strain curve. **(8 Marks)**

(c) Explain factor of safety. **(5 Marks)**

Q.11(a) Explain cone clutch. **(8 Marks)**

(b) Explain 3 types of brakes. **(8 Marks)**

Q.12(a) Explain bush coupling. **(6 Marks)**

(b) Explain flywheel and governor. **(10 Marks)**

Index

A

Absolute Zero Temperature, 30
Adiabatic and Diabatic Systems, 8
  Index, 74
Advantages of NC Machines, 271
  of Sheet Metal Parts, 290
  of Superheating the Steam, 146
Aesthetics, 313
Air Acetylene Welding, 279
  Compressors, 168
  Conditioning, 178
Allowance, 298
Alloy Cast Steels, 312
  Steels, 313
Application, 339
  of Shafts, 338
  of Energy Equations, 83
Artificial Draught, 136
Axial Flow Compressor, 169, 174
Axle, 339

B

Babcock Wilcox Boiler, 142
Barometer, 24
Base Metals, 289
Basic Types of Heat-Exchangers, 214
Bearings and Their Types, 345
Belt Drive, 323
Bending, 293
Bevel Gear, 329
Bio-Gas Plant, 193
Biomass, 191
  Resource, 192
Blanking, 294
Bottom Dead Centre, 170
Boundary of the System, 4
Bourdon Pressure Gauge, 22, 25
Brakes, 334
Brazing, 284
Bulk Modulus, 306

C

Carburising Flame, 279
Celsius Scale, 29
Centre Lathe, 253
Centrifugal Compressor, 173
  Pump, 165
Chain Drive, 326
Change of State, 9
Classification of Boilers, 134
  of Sources of Energy , 185
Closed System, 6

Clutches, 331
CNC Machines, 272
Cochran Boiler, 141
Common Engineering Materials, 312
Comparison between Water Tube and Fire Tube Boilers, 147
Components of NC Machine, 270
Composite Walls, 210
Compressed Air Motors, 160
Compressive Strain, 304
Stress, 303
Conditions of Steady Flow System, 77
Conduction, 203
Cone Clutch, 332
Continuity Equation or Law of Conservation of Mass, 88
Convection, 204
Conventional and Non-conventional Energy Resources, 183 – 202
Cornish Boiler, 140
Counter Boring, 263
Coupling and Their Types, 341

Die and Punch, 291
Different Drilling Operations, 263
Forms of Stored Energies, 70
Disadvantages of NC Machines, 271
Drawing Operation, 292
Drilling Machines, 258-59
Dynamic Compressors, 169

Elastic Limit (El), 308
Region, 308
Electric Hand Drills, 262
Electrolyte, 200
Elements of Interchangeability, 296
Energy — a Property of the System, 69
Energy in Transit, 33
Enthalpy (H), 72
Equations for Work Done in Various Processes, 17
Ergonomics, 315
Essentials of a Good Boiler, 146
Extended Surfaces (Fins), 218
Externally Fired, 135

Factor of Safety, 309
Fahrenheit Scale, 29
Film Coefficient of Heat Transfer, 208
Fire Tube Boilers, 134
Fits, 298
Fixed Boundary, 5
Flanged Coupling, 342
Flat Plate Heaters, 185
Plate Solar Heaters, 186
Flexible Coupling, 343
Flow Process; Control Volume; Control Surface, 75
Work or Flow Energy, 76
Flywheel, 347
Construction and Types, 348
Forms of Matter, 131
Fourier's Law of Conduction, 204
Friction Clutch, 331
Fuel Cell, 199
Electrochemistry, 200
Power Plants, 201
Fusion Welding, 274

Gas Cutting Process and Equipments or (Oxyacetylene Gas Cutting Process), 285

Metal Arc Welding, 282
Tungsten Arc Welding, 281
Turbine Power Plant, 159
Welding, 277
Gear Drives, 328
Geothermal Energy, 189
Governor, 350
Grinding Machine, 264
Group Drive, 321

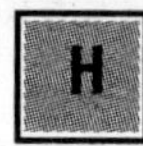

Hand Operated Press, 291
Pump, 163
Tools, 291
Heat, 21
Exchanger, 212
Helical Gear, 329
High Pressure boilers, 136
Hole and Shaft Based Fit System, 299
Hooke's Law, 305
Household Refrigerator, 175
Hydraulic Pumps, 163
Turbines, 159
Hydro Electric Power Plant, 198

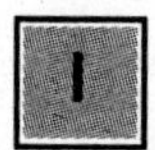

Illustrative Example of NC Programming, 271
Imaginary Boundary, 5
Impeller, 165
Importance of Electric Energy, 183
Insulation, 216
Interchangeability, 296
Internal Combustion Engines, 151
Energy (U), 71
Internally Fired, 135
Inverted Tooth or Salient Chain, 327
Irreversible process, 32
Isolated System, 7

Joule's Experiment, 67
Journal Bearing, 345

Lancashire Boiler, 137
Lathe Operations, 257
Limitations of First Law of Thermodynamics, 89
Limits, 297
Locomotive Boiler, 140
Low Pressure boilers, 136

Machine shafts, 338
Tool, 253
Manometers, 26
Material Properties and Selection, 311
Selection, 312
Mechanical and Thermodynamic Work, 13
Metal Working, 290
Microscopic and Macroscopic Point of View, 31
Modes of Failures, 309
of Heat Transfer, 203
Modulus of Rigidity, 306
Moving Boundary, 5

Natural draught, 136
Need of Design, 301
Neutral Flame, 279
Newton's Law of Cooling, 206
Non-Positive Clutch, 331
Notching, 295

Nuclear Power Plant, 195

Ocean Energy, 188
  Thermal Energy, 188
Oldham's Coupling, 344
Oxidising Flame, 279
Oxyacetylene Welding, 278

Parallel key, 341
Path functions, 9
Pelton Turbine, 159
Piercing, 295
Plain Carbon Steel, 312
Plastic Region, 308
  Welding, 274
Poisson's Ratio, 304
Positive Clutch, 331
  Displacement Type Compressor, 169
Power Press, 291
  Saw, 268
  Transmission Shafts, 338
Precision or Fine Grinding, 266
Press Work, 290
Pressure, 22
  Exerted due to a Column of Fluid, 23
  Measurement, 22
Prime–movers, 131
Principle of Temperature Measurement, 28
Projection Welding, 277
Protected type flange coupling, 343
Pumps, 163

Quasi-Static Process, 32

Rack and Pinion, 330
Radiation, 204
Real Boundary, 4
Reciprocating Air Compressor, 169
  Pump, 164
Recuperative Heat Exchangers, 213
Regenerative Heat Exchanger, 213
Requirements of Good Welded Joints, 273
Roller Chain, 326
Rolling Contact Bearings, 346
Room Air Conditioner (Window Air Conditioner), 178
Roots Blower, 171
Ropes, 325
Rotary Positive Displacement Compressors, 171

Scale of Temperature, 29
Screw Type Compressors, 172
Seam Welding, 277
Selection of Factor of Safety, 310
Shaft Material, 338
Shear Strain, 304
  Stress, 303
Sheet Metal, 289
  Working, 292
Simple Steam Power Plant (or Thermal Power Plant), 194
Solar Energy, 185
  Power Plant, 186
Soldering, 284
Sources of Energy, 183
Spur Gear, 328
Square butt weld, 274
Stand Grinder, 265

Steady Flow and Unsteady Flow
Systems, 7
Energy Equation (S.F.E.E.), 77
Steam Boilers, 134
Power Plant, 195
Turbines, 147
Stress and Strain, 302
Strain Curve, 306
Sunk Keys, 340
Surrounding, 4
Swing Frame Grinders, 265

Tensile Stress, 303
Thermodynamic Cycle, 10
Systems, 3
Thermometric Property and
Thermometers, 28
Tidal and Wave Energy, 189
Tolerance, 297
Top dead centre, 156
Twist Drill, 259
Two Stroke SI Engine, 155
Types of Boilers, 137
of Insulations, 217
of Lathes, 257
of Shafts, 338

Units and Dimensions, 11
Universal Coupling or Hooke's
Joint, 344
Unprotected flange coupling, 342
Uses of Compressed Air, 168
of NC Machines, 271
of Sheet Metals, 289

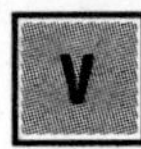

Vane type Rotary Compressor, 171
Volute casing, 167
Vortex casing, 167

Water Tube Boilers, 135
Welding, 273
Applications, 273
Classification, 274
Wind Energy, 187
Working Substance, 11
Worm and Worm Wheel, 330